AF349268

Control Engineering

For further volumes:
http://www.springer.com/series/4988

Ashish Tewari

Aeroservoelasticity

Modeling and Control

 Birkhäuser

Ashish Tewari
Department of Aerospace Engineering
Indian Institute of Technology, Kanpur
Kanpur
Uttar Pradesh
India

ISSN 2373-7719   ISSN 2373-7727 (electronic)
Control Engineering
ISBN 978-1-4939-2367-0   ISBN 978-1-4939-2368-7 (eBook)
DOI 10.1007/978-1-4939-2368-7

Mathematics Subject Classification (2010): 70Q05, 74H10, 74H15, 74F10, 74S05, 74S25, 76G25, 76H05, 76J20, 76M22, 76M23, 76M25, 76M40, 93B15, 93B30, 93B36, 93B52, 93B55, 93C05, 93C10, 93C15, 93C35, 93C40, 93C80, 93D05, 93D20, 32D99

Library of Congress Control Number: 2015933384

Springer New York Heidelberg Dordrecht London

Printed on acid-free paper

Springer is part of Springer Science+Business Media (www.springer.com)

To the order and beauty of nature,
and its finely crafted control laws.

Preface

This book presents aeroservoelasticity (ASE) as a well-formed discipline with a systematic framework. While many research articles have appeared on the special applications of ASE–such as active flutter suppression and gust load alleviation– there are no textbooks and monographs on the general and systematic procedure to be followed in the modeling and analysis of aeroservoelastic systems. This book is a first step in trying to fill this important gap in the aerospace engineering literature. This book introduces the basic math modeling concepts and highlights important developments involved in structural dynamics, unsteady aerodynamics, and control systems. It also attempts to evolve a generic procedure to be applied for ASE system synthesis. The treatment includes finite-element structural modeling and detailed unsteady aerodynamic modeling at various speeds for deriving the necessary aeroelastic plants, with sample control applications to active flutter suppression, load alleviation, and adverse ASE coupling.

A general aeroelastic plant is derived via the finite-element structural dynamic model, unsteady aerodynamic models for various regimes in the frequency domain, and the associated state-space model by rational function approximations. For more advanced models, the full-potential, Euler, and Navier–Stokes methods for treating transonic and separated flows are also briefly addressed. Essential ASE controller design and analysis techniques are introduced to the reader. Introduction to robust control-law design methods of LQG/LTR and H_2/H_∞ synthesis is followed by a brief coverage of nonlinear control techniques of describing functions and Lyapunov functions.

The fundamental concepts are presented in such a way that the most important features can be easily deduced. The breadth of coverage is sufficient for a thorough understanding of ASE.

The focus of this book is on aeroservoelastic modeling, including a brief presentation on robust and optimal control methods that can be applied to important aeroservoelastic design problems. It is not possible to give a more comprehensive ASE treatment in a single book, and it is envisaged that a future book can be devoted to more advanced topics such as adaptive and nonlinear control design techniques.

This book is aimed at graduate students and advanced researchers in aerospace engineering, as well as professional engineers, technicians, and test pilots in the aircraft industry and laboratories. The reader is assumed to have taken basic undergraduate courses in mathematics and physics—particularly calculus, complex variables, linear algebra, and fundamental dynamics—and is encouraged to review these concepts at several places in the text.

A book on ASE is difficult to write due to the breadth of topics it must necessarily address. While this book covers the essentials of modeling aspects of ASE with some control applications, it gives sufficient motivation to a reader with specific research interests to further explore the relevant topics. Furthermore, the treatment of topics is such that a novice can quickly build up his/her understanding of ASE without much difficulty. References are selected keeping both the types of readers in mind.

This book has been long in writing, with the intention first having occurred to the author about 15 years ago. Not having access to the industrial codes for finite-element and unsteady aerodynamics necessary for building such an exposition, and not finding the time to write one's own codes, the project continued to be delayed until about 2 years ago, when courage was finally gathered for this purpose. Testing and validating the codes for the many examples in the book was itself a formidable task, which required many hours of patient programming.

I would like to thank Walt Eversman for his course on aeroelasticity, and for advising me in my graduate studies. The editorial and production staff of Birkhäuser have offered many constructive inputs during the preparation of the manuscript, for which I am indebted to them. I am also grateful to The MathWorks, Inc. for providing the latest MATLAB/Simulink version utilized for the examples throughout the book.

November 2014   Ashish Tewari

Contents

Contents

Chapter 1
Aeroservoelasticity

1.1 Introduction

Aeroservoelasticity (ASE) is the interface of unsteady aerodynamics, structural dynamics, and control systems, and is an important interdisciplinary topic in aerospace engineering. It is a study of dynamic interactions among air loads, structural deformations, and automatic flight control systems commonly experienced by the modern aircraft. The relevance of ASE to modern airplane design has increased considerably with the advent of flexible, lightweight structures, increased airspeeds, and closed-loop automatic flight control. Since aeroservoelasticity lies at the interface of aerodynamics, structures, and control, its impact on aircraft design and operation requires a thorough understanding of these core areas as far as they contribute to building an accurate mathematical model (Fig. 1.1).

While aeroelastic interactions have been studied for nearly a century, the impact of an active control system on dynamic aeroelasticity is a relatively new topic, and has come into focus with the advent of modern fly-by-wire designs. In such aircraft, the controller bandwidth can encroach the upon aeroelastic modal spectrum, thereby leading to resonance-like behavior in certain flight conditions. The most common example of such an interaction between the control and aeroelastic systems is the closed-loop flutter—a catastrophic dynamic coupling between the elastic motion, the unsteady aerodynamic loading, and a controller-actuated surface. Many airplane accidents (such as Taiwan IDF fighter and Lockheed YF-22 prototypes) have been blamed on unforeseen and unstable ASE couplings. In order to understand how such a phenomenon can remain unforeseen in the modern technical era, let us consider a well-designed car with the best engine, chassis, and electronics, and thoroughly tested for the most adverse road conditions that can be expected. However, when put into production, the same car could experience poor performance and even engine stalling due to a minor feature such as cable routing. In such a case, engine vibration at a certain speed can interact with the natural frequency of one of the spark plug cables, thereby leading to its coming loose and causing an even greater engine vibration. The engine controller would detect the poor combustion as a lean or cold mixture condition, and try to correct it by increasing the fuel volume injection. The

© Springer Science+Business Media, LLC 2015
A. Tewari, *Aeroservoelasticity,* Control Engineering,
DOI 10.1007/978-1-4939-2368-7_1

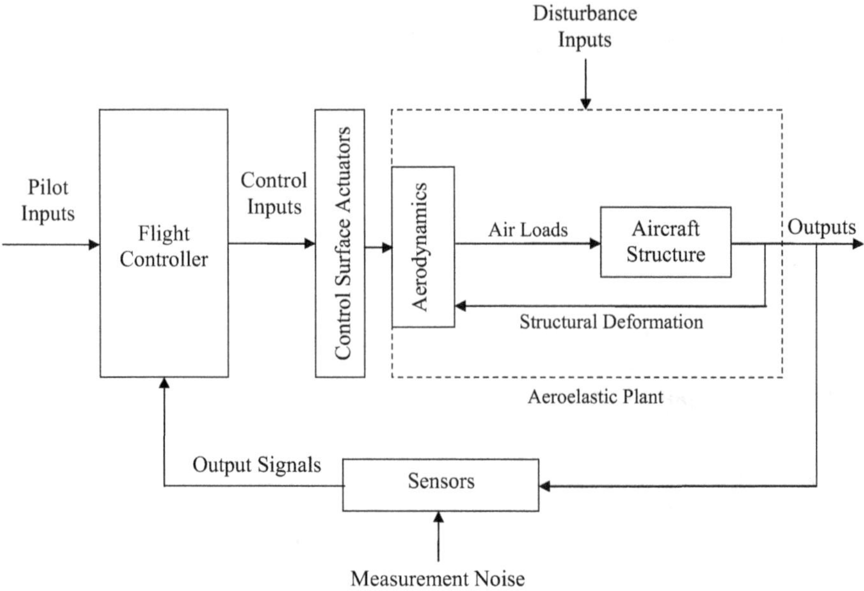

Fig. 1.1 Block diagram of a typical aeroservoelastic system

result would be an even rougher idle with black smoke, fouled spark plugs and injectors, and possibly an engine failure. Troubleshooting such a condition would be a nightmare, and the fix is either an expensive redesign of the engine, a reprogramming of the fuel controller, or an identification and change of the culprit natural frequency by merely redesigning the cable routings and clamps. If a mathematical model is constructed of such a dynamic interaction between electromechanical connectors, fuel control system, and engine dynamics, such a model is likely to be a formidable interdisciplinary exercise.

ASE is a fledgling discipline when compared to the other traditional aerospace areas of aerodynamics, structures, propulsion, and flight mechanics. Its formal beginning can be traced to the early 1970s, when ways of addressing the problem of flutter were being investigated in earnest due to the several new aircraft designs evolving in that era. Highly maneuverable fighters such as the Lockheed F-16 and the McDonnell Douglas F/A-18, as well as efficient passenger transports such as the Boeing-767 and the Airbus A-320 that were being developed, had inbuilt automatic flight control systems, which could be programmed relatively easily to achieve secondary tasks, such as active flutter suppression and maneuver/gust load alleviation. Prior to that era, a passive redesign of the structural components was the only way to avoid flutter, whose analysis often required thousands of hours of painstaking and dangerous flight flutter testing, and wind-tunnel tests of aeroelastically scaled models, thereby increasing the already high costs of prototype development. Consequently, flutter analysis and prevention was a stumbling block in developing novel aircraft configurations.

In order to overcome the inadequacy of passive techniques, and to fly at a velocity greater than the open-loop flutter velocity for greater speed and efficiency, the concept of active flutter suppression was developed in the 1970s [2, 3, 126, 143, 152]. Herein, an automatic control system actuates a control surface on the wing, in response to the structural motion sensed by an accelerometer, which increases the flutter speed in a closed loop by changing the stability characteristics of the open-loop system.

Active flutter suppression requires accurate knowledge of the aeroelastic modes that cause flutter, which are then actively changed in such a way that closed-loop flutter occurs at a higher flight velocity. Although the classical flutter of a high aspect-ratio wing—such as that of a Boeing 747 or an Airbus A-380—is caused by an interaction between the primary bending and torsion aeroelastic modes, the flutter mechanism of a low aspect-ratio wing, such as that of an F/A-18 (or F-22) fighter airplane is rather more complicated, comprising a coupling of several higher aeroelastic modes. In order to actively suppress flutter, it is necessary that an accurate aeroelastic model based on modeling of the unsteady aerodynamic forces as a transfer matrix be derived. The most common method of obtaining the unsteady aerodynamic transfer matrix is the use of optimized rational function approximations [43, 50, 85, 175] for its terms, fitted to the frequency-domain aerodynamic data in the harmonic limit. After the transfer matrix is derived, a linear, time-invariant, state-space model for the aeroelastic system, including the control surface actuators, can be obtained. The multivariable feedback controller for active flutter suppression can then be designed by standard closed-loop techniques, such as eigenstructure assignment and linear optimal control [168]. Since it is crucial that the derived control laws must be robust with respect to modeling uncertainties and measurement noise in a wide range of operating conditions, robust and adaptive controllers are specially sought for a practical application.

The first practical demonstration of active flutter suppression was carried out by the US Air Force in 1973 in their Load Alleviation and Mode Stabilization (LAMS) program, which resulted in a Boeing B-52 bomber flying 10 knots faster than its open-loop flutter velocity. This was accompanied by flight flutter testing of aeroelastic drones under NASA-Langley's Drones for Aeroelastic Testing (DAST) program. These pioneering developments in active flutter suppression received an impetus at NASA-Langley and Ames laboratories [4, 5, 118, 122] with novel control laws developed at Ames being tested and further developed in Langley's transonic dynamics wind tunnel. These developments in the 1970s were greatly enabled by the optimal control theory advancements [131, 163] of that era. ASE design and analysis efforts continued in the 1980s and 1990s [116, 117, 191], which were given a further boost by the newly developed robust multivariable control theory [62, 108]. Buoyed by their early achievements in active flutter suppression, the US Air Force and Rockwell, initiated the ambitious Active Flexible Wing (AFW) program [113, 127], wherein the objective was to utilize favorable aeroservoelastic interactions to produce performance, stability, and control improvements on a highly flexible and overly instrumented wind-tunnel wing model, employing multiple control surfaces and gain scheduled control laws. The survey paper of Mukhopadhyay [119] gives an excellent overview of how the seemingly independent technical developments over the preceding half century in the otherwise disparate areas of structural dynamics, unsteady aerodynamics, and control systems, converged and came into a sharp focus in the field of aeroservoelasticity. An offshoot of aeroservoelastic design is the evolving area of multidisciplinary optimization for synthesizing lightweight wing structures [102], an example of which was the first forward-swept wing experimental prototype, the Grumman X-29A [65].

The main challenge in ASE mathematical analysis and design is in deriving a suitable unsteady aerodynamic model of aircraft wings and tails (or canards). The aeroelastic plant for flutter suppression of a thin wing-like surface is derived at subsonic and supersonic speeds using small-disturbance, potential aerodynamic models [7, 21, 61, 194, 180, 181] with a harmonic (frequency response) theory. Such a model is linear, and can be directly employed in developing an aerodynamic transfer matrix, and finally a linear, time-invariant state-space model through analytic continuation in the Laplace domain. However, there are important flow regimes where such a linearized model is inapplicable. The ASE applications which involve unsteady separated flows and transonic shock-induced flows, are inherently nonlinear in nature and require advanced computational fluid dynamics (CFD) modeling techniques [45, 166]. An example of nonlinear aeroelasticity is the post-stall buffet arising due to a sudden and large increase in the angle of attack, either by an abrupt maneuver, or a vertical gust. The ASE plant for such a case is further complicated by the separated wake and/or leading-edge vortex from the wing interacting with the tail, resulting in an irregular and sometimes catastrophic deformation of the tail— either on its own or driven by rapid and large deflections of the elevator. Such a wing-tail-elevator coupling of a post-stall buffet, or a shock-vortex interaction requires a fully viscous flow modeling that is only possible by a Navier–Stokes method. Another example of nonlinear ASE is the control of unsteady control surface buzz and shock-induced flow separation encountered by an aircraft maneuvering in the transonic regime, leading to nonlinear flutter or limit-cycle oscillation (LCO) [18, 99]. An appropriate CFD model in such a case would require a full-potential code [73, 162], coupled inviscid/boundary-layer method [46], or a Navier–Stokes method [123, 195], depending upon the geometry, structural stiffness, Mach number, and Reynolds number. Sometimes, semiempirical models are devised from wind-tunnel test results for separated and shock-induced flows [47, 156], because they do not require unsteady CFD computations to be performed in loop with structural dynamic and control-law calculations. However, the veracity of such a correlation must be checked carefully before being deployed in ASE design and analysis. The use of CFD models (Euler/Navier–Stokes) to derive linearized aerodynamic transfer functions/matrices has also been suggested in the literature [12, 136, 197]. This is evidently aimed at using the same linearized model for a range of flight conditions (Mach number, angle of attack, etc.) rather than having to repeat a CFD computation for every such condition. The coefficients of the approximation can be adjusted by an auto-regressive moving average (ARMA) model [137, 197], which brings such a method quite close to an adaptive control application (albeit in the open loop). However, approximating a nonlinear system by a linear transfer function can be inaccurate, even in a narrow range of operating conditions. An alternative method is to employ flight-test data for deriving an ASE model, such as the neural-network identification method proposed by Boely and Botez [26].

Transonic flutter/buffet analysis is fraught with shock-induced oscillations, and is thus inherently nonlinear. It is also crucial in the aeroelastic stability analysis of most of the modern transport type aircraft, because their cruising speeds are just under the sonic speed. Fortunately, the nonlinear effects of transonic flutter are accurately captured by a transonic small-disturbance (TSD) model, even for a thick supercritical wing [176, 199] of a modern airliner. This fact offers the promise of coupling the aeroelastic stability analysis with a TSD code [14, 37, 76]. Alternative methods proposed for the time-linearized (low-frequency) transonic case involve a field-panel TSD method [105], or a full-potential method [130] applied in a doublet-lattice [61] type calculation of aerodynamic influence coefficients relating the pressure and normal velocity (upwash) in the harmonic limit. However, while such an approach can be applied to relatively lower natural frequencies of the primary bending and torsion modes of a high aspect-ratio wing (like that of a modern subsonic airliner), the concept of time linearization would fail when applied to the much higher frequencies of a low aspect-ratio, fighter-type wing, which is much stiffer in both bending and torsion.

Despite the early successes in demonstrating active flutter suppression/load alleviation, ASE has remained largely an experimental area [151] and has still not reached operational status on any aircraft. This remarkable failure is mainly due to the difficulty of designing a multivariable control system, which is sufficiently robust to the parametric uncertainties in the underlying unsteady aerodynamic model. Clearly, the aircraft designers and operators are reluctant to take risks until (what might be considered) suitably reliable ASE modeling and analysis methods become prevalent.

Due to the inherent uncertainty of an unsteady aerodynamic model, a closed-loop controller for ASE application must be quite robust to modeling errors. Furthermore, such a controller must also adapt to changing flight conditions. Hence, an ASE control law must not only be robust, but also self-adaptive, which renders it mathematically nonlinear even for a linear aeroelastic plant of the subsonic and supersonic regimes. Furthermore, designing a control law based upon nonlinear aeroelastic iterative models can be a very cumbersome and computationally intensive process. Instead, adaptive control techniques can be used for extending the subsonic and supersonic linear feedback designs to predict and suppress the transonic flutter. Adaptive control has been an area of active research in the past few decades [10, 89], and many useful design techniques have emerged that can be applied to ASE. However, these remain "application specific" (rather than general), if not completely ad hoc in many cases. Thus, ASE control-law derivation for a particular case is as challenging as the problem of aeroelastic modeling. For this reason, ASE has remained a formidable technological problem. For a linear aeroelastic plant, a high-gain linear feedback generally gives robustness with respect to modeling uncertainties in the control bandwidth, but degrades the response to the high-frequency measurement noise. Several linear feedback strategies are in vogue for striking a compromise between robustness to plant uncertainty and noise rejection. These include the linear quadratic Gaussian (LQG) compensation with loop-transfer recovery (LTR) [108], H_2/H_∞ control [64], and structured singular value synthesis [30, 42].

Separated and shock-dominated flows nonlinearly interact with the aircraft structure, resulting in unstable oscillations, hysteresis, or limit cycles. While there are some papers and monographs on nonlinear aeroelastic modeling, such as Dowell and I'lgamov [41], their emphasis is on structural nonlinearities rather than on aerodynamic ones. Furthermore, the control aspects of ASE are seldom covered in such articles. Thus it is important to include nonlinear aerodynamic effects in a discussion on ASE, which is one of the tasks of this book. Control of nonlinear aeroelastic plants requires either describing function approximations [158, 182], Lyapunov-based controllers [60, 132], feedback linearization [78], or a sliding-mode (variable structure) control [53]. Furthermore, the issue of robustness is important for nonlinear plants [77]. Adaptation of controller parameters for both linear and nonlinear plants introduces a further complexity in the ASE design and analysis, and can be handled by gain scheduling, self-tuning regulation, model reference adaptive laws [10], or recursive backstepping [89], depending upon the specific application. There is little mathematical treatment of nonlinear ASE effects in the literature, and a future book is planned to cover this important gap.

The present book is a monograph on the basic concepts relevant to ASE modeling and analysis. Chapter 2 addresses the problem of basic structural modeling, whereas Chap. 3 is largely devoted to the techniques employed in deriving frequency-domain (or harmonic) aerodynamics of low-speed, subsonic, transonic, and supersonic flight regimes. It also covers the discussion of unsteady vortex-lattice model as an example of a simple potential flow model that can be applied to large amplitude movements (flapping) of an airfoil with thickness and camber. Such a treatment is somewhat a departure from what is considered the traditional aerodynamics of an oscillating flat-plate wing, which is commonly used in most linearized ASE models. The objective of such a model is its possible application in flapping-wing flight. Chapter 4 is concerned with the derivation of transient aerodynamic models by rational function approximations (RFA) that can be readily converted to state-space, linear time invariant (LTI) aeroelastic models. Chapter 5 is a presentation of linear control law derivation and analysis techniques, with some typical aeroseroelastic application examples. Finally, Chap. 6 is a discussion of nonlinear ASE topics. The chapters are organized in such a way that a reader can directly go to a specific topic without having read a previous one. However, for comprehensive understanding, it would be ideal to read the chapters sequentially. A basic knowledge of aerodynamics and control theory is assumed of the reader. Furthermore, familiarity with numerical methods [111, 154, 165] will also be helpful. The reader will find the bibliography arranged alphabatically, and thus easy to refer to. While it is impossible to list all the relevant references on this active and productive research topic, some thought has been applied to select useful articles for the benefit of the reader.

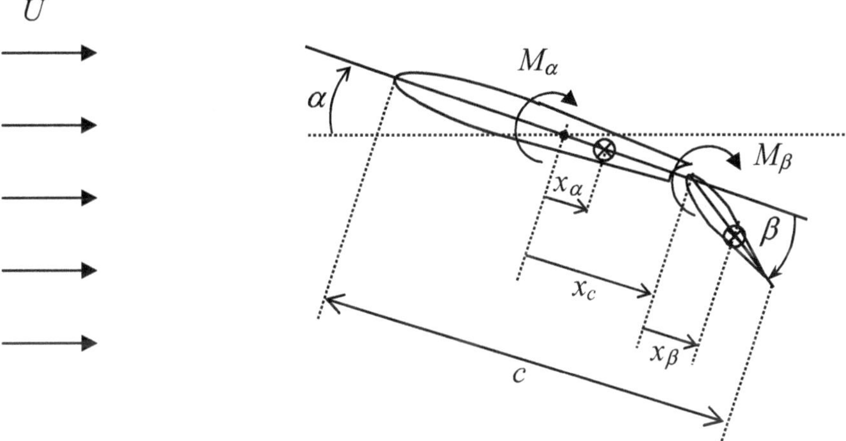

Fig. 1.2 A pitching airfoil with trailing-edge control surface

1.2   An Illustrative Example

While the general aeroelastic motion of an aircraft wing involves many degrees of freedom (Chap. 2), in order to illustrate the concept of aeroservoelasticity, let us consider a two-dimensional wing section (airfoil) that can only rotate about a fixed axis by an angle α (called the angle of attack). The airfoil is equipped with a trailing-edge control surface that can rotate by an angle β relative to the airfoil's chord plane, as shown in Fig. 1.2. The entire setup is mounted on a frictionless torsional spring of stiffness k_α, whereas the control surface hinge is also frictionless and has a rotational spring of stiffness k_β. The hinge line of the control surface is located at a distance of x_c behind the pitch axis. The mass of the control surface is m_c, and the distance of the control surface's center of mass behind its own hinge line is x_β (Fig. 1.2). The moment of inertia of the setup about the pitch axis is I_α, whereas that of control surface about its hinge is I_β. The setup is placed in a uniform, incompressible flow of speed U and density ρ. The control surface is equipped with a direct current (DC) motor that can apply a torque u about the control surface hinge line relative to the airfoil. The linearized equations of motion of the structure can be expressed as follows, by either Newton's second law, or the energy approach:

$$\mathsf{M}\ddot{\mathbf{q}} + \mathsf{K}\mathbf{q} = \mathbf{Q} + (0,\ 1)^T u, \tag{1.1}$$

where $\mathbf{q} = (\alpha, \beta)^T$ is the generalized coordinates vector, $\mathbf{Q} = (M_\alpha, M_\beta)^T$ is the generalized air loads vector, and M, K are the following generalized mass and stiffness matrices, respectively, of the structure:

$$\mathsf{M} = \begin{pmatrix} I_\alpha & m_c x_c x_\beta \\ m_c x_c x_\beta & I_\beta \end{pmatrix}, \tag{1.2}$$

$$\mathsf{K} = \begin{pmatrix} k_\alpha & 0 \\ 0 & k_\beta \end{pmatrix}. \tag{1.3}$$

The generalized aerodynamic loads vector consists of the pitching moment M_α, and the control surface hinge moment M_β. For simplicity, it is assumed that the aerodynamic moments can be modeled primarily as first-order lag (or circulatory) effects of the unsteady wake shed by the airfoil, as well as the noncirculatory contributions of the aerodynamics to inertia (called the *apparent mass* effect), damping, and stiffness. Such a model of unsteady aerodynamics is given by the following relationship in the Laplace domain:

$$\mathbf{Q}(s) = \frac{1}{2}\rho U \left(a_1 + a_2 s + \frac{a_3 s}{s+b} \right) \begin{Bmatrix} A_1 \\ A_2 \end{Bmatrix} w(s), \tag{1.4}$$

where s represents the Laplace variable, and $w(s)$ is the following upwash (flow velocity component normal to the wing) at a specific location on the airfoil:

$$w(s) = (C_1,\ C_2,\ C_3,\ C_4) \begin{Bmatrix} \mathbf{q}(s) \\ s\mathbf{q}(s) \end{Bmatrix}, \tag{1.5}$$

and $A_1, A_2, a_1, \dots, a_3, b, C_1, \dots, C_4$ are constant aerodynamic parameters (in addition to ρ and U). Equations (1.4) and (1.5) result in the following relationship for the generalized unsteady aerodynamic loads:

$$\mathbf{Q}(s) = \mathsf{G}(s) \begin{Bmatrix} \mathbf{q}(s) \\ s\mathbf{q}(s) \end{Bmatrix}, \tag{1.6}$$

where the aerodynamic transfer matrix is given by

$$\mathsf{G}(s) = \frac{1}{2}\rho U \left(a_1 + a_2 s + \frac{a_3 s}{s+b} \right) \begin{Bmatrix} A_1 \\ A_2 \end{Bmatrix} (C_1,\ C_2,\ C_3,\ C_4). \tag{1.7}$$

The simple aeroelastic model considered here leads to a state-space representation given as follows:

$$\dot{\mathbf{x}} = \mathsf{A}\mathbf{x} + \mathsf{B}u, \tag{1.8}$$

$$y = \alpha = \mathsf{C}\mathbf{x} + \mathsf{D}u, \tag{1.9}$$

where $\mathbf{x} = (\mathbf{q}^T, \dot{\mathbf{q}}^T, x_a)^T$ is the state vector,

$$\mathsf{A} = \begin{bmatrix} 0 & \mathsf{I} & 0 \\ -\bar{\mathsf{M}}^{-1}\bar{\mathsf{K}} & \bar{\mathsf{M}}^{-1}\bar{\mathsf{C}}_d & -\frac{1}{2}\rho U b a_3 \bar{\mathsf{M}}^{-1} \begin{Bmatrix} A_1 \\ A_2 \end{Bmatrix} \\ (C_1,\ C_2) & (C_3,\ C_4) & -b \end{bmatrix}, \tag{1.10}$$

$$
B = \begin{bmatrix} 0 \\ \bar{M}^{-1}(0,\ 1)^T \\ 0 \end{bmatrix}, \tag{1.11}
$$

$$
C = (1,\ 0,\ 0,\ 0\ 0)\ ; \qquad D = 0, \tag{1.12}
$$

$$
\bar{M} = M - \frac{1}{2}\rho U a_2 \begin{Bmatrix} A_1 \\ A_2 \end{Bmatrix} (C_3,\ C_4), \tag{1.13}
$$

$$
\bar{C}_d = \frac{1}{2}\rho U \begin{Bmatrix} A_1 \\ A_2 \end{Bmatrix} [a_2(C_1,\ C_2) + (a_1 + a_3)(C_3,\ C_4)], \tag{1.14}
$$

$$
\bar{K} = K - \frac{1}{2}\rho U (a_1 + a_3) \begin{Bmatrix} A_1 \\ A_2 \end{Bmatrix} (C_1,\ C_2). \tag{1.15}
$$

The matrices $\bar{K}, \bar{C}_d, \bar{M}$ are the generalized stiffness, damping, and mass matrices of the aeroelastic system, which reduce to $K, 0, M$, respectively, for the in vacuo case ($\rho = 0$). It is to be noted that the order of the aeroelastic system is increased by one due to the aerodynamic state, $x_a(t)$, arising out of the lag term, $a_3 s/(s + b)$, which augments the (last row and column of the) dynamics matrix A. The given aeroelastic plant is controllable with the motor torque input, which can be verified from the rank of the controllability test matrix for the pair (A, B) [168].

The aeroelastic stability is determined from the eigenvalues of A:

$$
\mid sI - A \mid = 0
$$

at a certain flight condition. Let $I_\alpha = 10, I_\beta = 1$ kg m^2, $k_\alpha = 40, k_\beta = 9$ N m/rad, $m_c = 0.1$ kg, $x_c = 0.3$ m , $x_\beta = 0.1$ m, and the aerodynamic parameters be the following:

$$
\rho = 1.225 \text{ kg/m}^3\ ; \quad A_1 = 0.5\ ; \quad A_2 = -0.0268,
$$

$$
C_1 = 1.0\ ; \quad C_2 = 0.7067\ ; \quad C_3 = 0.5\ ; \quad C_4 = 0.3387,
$$

$$
a_1 = 1.0\ ; \quad a_2 = 0\ ; \quad a_3 = -2.0\ ; \quad b = 0.05.
$$

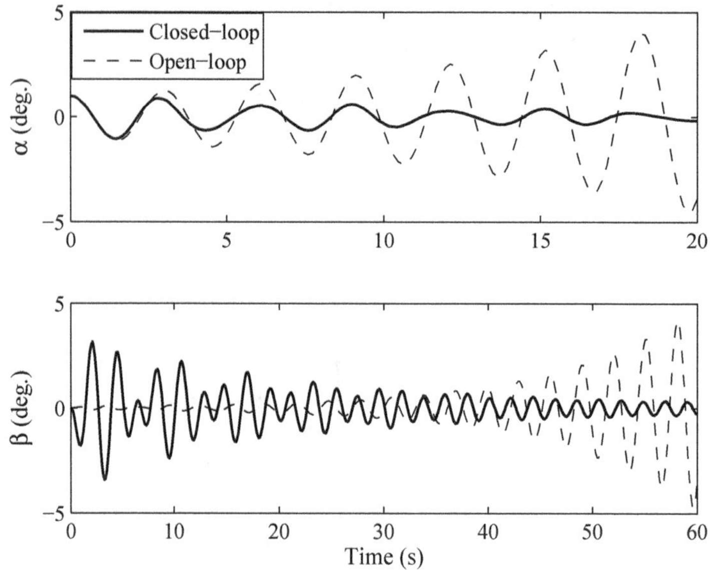

Fig. 1.3 Initial reponse of a pitching airfoil with trailing-edge control surface and a linear feedback control law

It can be easily shown that the aeroelastic plant is unstable at any speed U due to the unsteady aerodynamic characteristics associated with the wake (modeled here as a simple time lag).

In order to stabilize the plant, the following linear feedback control law is tried for $U = 10\,\mathrm{m/s}$:

$$u = -(8.2703,\ 0.4301,\ -3.2839,\ 0.3131) \left\{ \begin{array}{c} \alpha \\ \beta \\ \dot{\alpha} \\ \dot{\beta} \end{array} \right\}.$$

This feedback control law stabilizes the closed-loop system, which is shown by the initial response to a unit angle-of-attack perturbation (caused by a vertical gust) plotted in Fig. 1.3. For a practical implementation of such a control law, there must be angle and rate sensors mounted on the airfoil pitch axis and the control surface hinge line, whose combined electrical signals are fed to a multichannel amplifier for driving the control surface motor. Alternatively, a single output y can be selected, which results in a linear combination of all the state variables. Such an output could be the normal acceleration measured by an accelerometer placed at a point on the control surface (the reader can show such a plant will be observable [168]). However, the derivation of the control law in that case will require the design of an observer, which can reconstruct the information about the state variables from the knowledge of the input u and the output y. The reader is advised at this point to refresh the definitions of stability, controllability, and observability of linear systems [84].

The gains of the amplifier—indicated above for $U = 10\,\text{m/s}$—can be individually adjusted with a changing flight speed U (or atmospheric density ρ), until the desired closed-loop response is obtained. An adjustment of the controller gains with a changing operating condition is called adaptation. Instead of a human operator, a separate control system—called adaptation mechanism—can be devised such that the gains are automatically adjusted to the correct values with changing flight speed and density. Such a control system which has the capability of self-modification in order to always achieve the desired closed-loop behavior is called an adaptive control system. A modern flight control system has an inbuilt adaptive mechanism, which can sense a change in the flight condition by pressure and temperature sensors, and applies a correction to the feedback gains.

Chapter 2
Structural Modeling

2.1   Introduction

Aircraft have thin-walled, built-up structures for high strength-to-weight ratio and stiffness. The transverse and longitudinal members share the external loads with the outer skin panels in a semimonocoque construction. It would appear that an analysis of such a structure requires a detailed model of each component, which specifies the exact manner in which it is connected to the other members. Such a modeling would be a daunting task, requiring enormous computational resources. Fortunately, although a detailed analysis of individual structural components is indeed necessary for structural design, it is not required for aeroelastic purposes where some simplifying approximations can be made. Many aircraft components—such as the wings and the fuselage—are designed to be slender and streamlined for a high lift-to-drag ratio. The associated structures thus have small thicknesses and can be often idealized as either solid beams or plates. Furthermore, the necessity of preserving aerodynamic shape results in a much higher bending stiffness in the transverse (chordwise or radial) direction, compared with that in the longitudinal (lengthwise or spanwise) direction, which is achieved in practice by closely spaced ribs and frames. The resulting assumption of chordwise rigidity is quite valuable in reducing the degrees of freedom for a structural model. However, such a model would be inaccurate for wings with very small aspect ratio where chordwise and spanwise bending stiffnesses would be comparable. Another major simplification is the fact that any inelastic deformation leading to buckling of skin panels under design loads is unacceptable from aerodynamic design viewpoint. Therefore, an elastic stress–strain behavior is necessary, and results in a linear load–displacement relationship—a valuable model from aeroelastic perspective. However, post-buckling behavior of skin panels requires nonlinear structural modeling, which is excluded from usual aeroelastic design and analysis.

The main emphasis of an aeroelastic model is upon wing-like structures, which are quite thin in comparison with the chord and span. This offers a valuable modeling simplicity, which combined with the chordwise rigidity of high aspect-ratio wings, results in plane cross-sections remaining essentially plane due to a free warping of the structure under twisting loads. Conversely, a bending load would not produce any twisting deformation due to the same reason. Hence by Saint-Venant's theory [68],

© Springer Science+Business Media, LLC 2015
A. Tewari, *Aeroservoelasticity,* Control Engineering,
DOI 10.1007/978-1-4939-2368-7_2

one can decouple bending and torsion, thereby leading to the very useful concepts of shear center and elastic axis. Of course, such a decoupling is not possible for either short beams, or plate-like structures where sectional warping invariably causes some normal (bending) stresses. Furthermore, if shear deformations can be neglected due to an essentially thin beam, the bending deformations can be treated by a simple Euler–Bernoulli beam theory.

2.2 Static Load Deflection Model

Consider an elastic wing with an unloaded and undeformed mean surface defined by $z = s(x, y)$ and generated by smoothly joining the wing's chord lines, camber lines, or any other centroidal features of the cross-sections. If a concentrated load, $\mathbf{P}$, is now applied at a point, (ξ, η), located on the original mean surface, it will cause a structural deflection, $\delta(x, y)$ such that a new equilibrium is achieved in the deformed configuration given by the deformed surface, $z = s'(x, y)$. The deflection vector, δ, can be regarded as a change of location of an original point on the surface, (x, y, z), to its new position on the deformed surface, (x', y', z'), and is given by

$$\delta(x, y) = \left\{ \begin{array}{c} x' - x \\ y' - y \\ s'(x', y') - s(x, y) \end{array} \right\} \tag{2.1}$$

The deflection vector is thus based upon a one-to-one mapping of all points in the closed set constituting the original surface to those on the deformed surface:

$$\{(x, y, z) : x_1 \le x \le x_2; y_1 \le y \le y_2; z = s(x, y)\}$$
$$\rightarrow \{(x', y', z') : x'_1 \le x' \le x'_2; y'_1 \le y' \le y'_2; z' = s(x', y')\} \tag{2.2}$$

Such a map can be geometrically represented by the transformation

$$\left\{ \begin{array}{c} x' \\ y' \\ s'(x', y') \end{array} \right\} = \mathbf{T}(x', y', z' : x, y, z) \left\{ \begin{array}{c} x \\ y \\ s(x, y) \end{array} \right\} \tag{2.3}$$

Since the mean surface of most wing-like structures is essentially flat, the deflection at each point can be approximated by the displacement normal to the original surface,

$$\delta(x, y) = z' - s(x, y), \tag{2.4}$$

as shown in Fig. 2.1. In such a case, the deformed mean surface may not turn out to be flat, but can have a local curvature due to shear, twist, and spanwise and chordwise bending.

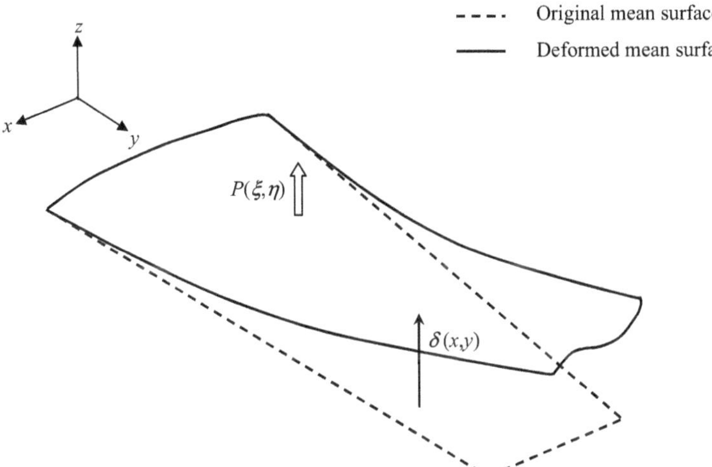

Fig. 2.1 Deformation of a wing-like structure under a concentrated load

The transformation matrix, **T**, in Eq.(2.3) produced by a general loading must obey the material properties called *constitutive relationships*, and is also subject to the geometric constraints (also called kinematical relationships, compatibility requirements, or boundary conditions) on the structure. For a linearly elastic structure, the constitutive relationships take the form of a linear stress–strain behavior, such as

$$\tau = \mathsf{C}\epsilon, \tag{2.5}$$

where τ and ϵ are the stress and strain vectors, respectively, experienced by an infinitesimal structural element, and C denotes the matrix comprising the material properties. The linear stress–strain behavior of the material produces a linear relationship between the load and displacement of the structure, given by:

$$\delta(x, y) = \mathbf{R}(x, y : \xi, \eta)\mathbf{P}(\xi, \eta), \tag{2.6}$$

where $\mathbf{R}(x, y : \xi, \eta)$ is the matrix of *flexibility influence-coefficient* functions (also called *Green's functions*). By applying linear superposition, Eq. (2.6) can be extended for the case of a continuously distributed load per unit area, $\mathbf{p}(\xi, \eta)$, as follows:

$$\delta(x, y) = \iint_s \mathbf{R}(x, y : \xi, \eta)\mathbf{p}(\xi, \eta)\mathrm{d}\xi\,\mathrm{d}\eta. \tag{2.7}$$

Unfortunately, the flexibility influence-coefficient functions are rarely available in a closed form for any but the most simple structures. Therefore, numerical approximations must be made for the integral relationship given by Eq. (2.7). Such approximations are based upon a discretization of Eq. (2.7), whereby a continuous structure with infinitely many degrees of freedom is converted into an equivalent finite dimensional form. For example, the mean surface can be approximated by

n flat elemental panels of individual dimensions, $(\Delta\xi_i, \Delta\eta_i), i = 1\ldots n$. The load distribution on the jth panel is then approximated by an average generalized load, $\mathbf{P}_j = \mathbf{p}(\xi, \eta)\Delta\xi_j\,\Delta\eta_j$ acting at a given *load point* (such as the panel centroid) in each panel. Similarly, the displacement, $\boldsymbol{\delta}(x, y)$, averaged over the ith panel is taken as the average deflection vector, $\boldsymbol{\delta}_i$, at a given *collocation point*, $(x_i, y_i), i = 1\ldots n$. The discretized load–displacement relationship is then given for the ith panel as follows:

$$\boldsymbol{\delta}_i = \sum_{j=1}^{n} \mathbf{R}_{ij}\mathbf{P}_j \; ; i = 1\ldots n \,. \tag{2.8}$$

Often, the individual elements of the displacement and load vectors are identified as *generalized displacements*,

$$\boldsymbol{\delta}_i = \begin{Bmatrix} q_{1i} \\ q_{2i} \\ q_{3i} \end{Bmatrix}, \tag{2.9}$$

and *generalized loads*,

$$\mathbf{P}_i = \begin{Bmatrix} Q_{1i} \\ Q_{2i} \\ Q_{3i} \end{Bmatrix}, \tag{2.10}$$

respectively, corresponding to the individual degrees of freedom at each point. Then Eq. (2.8) collected for all points takes the following vector-matrix form:

$$\mathbf{q} = \mathsf{R}\mathbf{Q}, \tag{2.11}$$

where the $3n \times 3n$ influence-coefficient matrix, R, consists of $\mathbf{R}_{ij}$ as its elements. Here, we note that $\mathbf{P}_i$ denotes the vector of all generalized loads (Q_{1i}, Q_{2i}, Q_{3i}) acting on the ith panel, while $\mathbf{Q}$ is the generalized loads vector for the entire structure derived by collecting all the generalized loads acting on all the panels. Similarly, the generalized displacement vector on the ith panel, $\boldsymbol{\delta}_i$, is to be distinguished from the overall generalized displacement vector of the structure, $\mathbf{q}$.

By making simplifying assumptions for a typical aircraft, the number of generalized coordinates can be significantly reduced. Such idealizations include chordwise rigidity, plate or beam-shaft approximations, and negligible streamwise loading on the structure. Due to their smaller dimensions and high stiffnesses, control surfaces are modeled simply by rotating angles about rigid hinge axes. Various definitions of the generalized loads and generalized displacements are possible, depending upon the idealizations and constitutive relationships used in deriving Eq. (2.11). There are also various alternative techniques for deriving the load–displacement relationship of Eq. (2.11) from constitutive relations and structural constraints. These include the *lumped parameters approximation*, the *finite-element method* (FEM) (or *Galerkin* method), the *assumed modes* (or *Rayleigh–Ritz*) method, and the *boundary-element method*. Of these, the FEM is the most commonly employed due to its ease of implementation and modeling efficiency. We shall have the occasion to consider some examples of the FEM modeling a little later.

The influence-coefficient matrix,

$$
\mathsf{R} = \begin{pmatrix}
R_{11} & R_{12} & \cdots & R_{1N} \\
R_{21} & R_{22} & \cdots & R_{2N} \\
\vdots & \vdots & \vdots & \vdots \\
R_{N1} & R_{N2} & \cdots & R_{NN}
\end{pmatrix},
\tag{2.12}
$$

with $N = 3n$, consists of influence coefficients, R_{ij}, which are defined as the ith generalized *virtual coordinate*, δq_{ij}, produced by an isolated generalized load at a given point, Q_j,

$$
\delta q_{ij} = R_{ij} P_j.
\tag{2.13}
$$

A virtual coordinate is an arbitrary, infinitesimal deflection in any of the three possible directions at a given point due to an isolated generalized load, and must be compatible with any kinematical constraints of the structure. The actual generalized coordinate, q_i, is a sum of all the virtual coordinates, δq_{ij}, caused by the generalized load, Q_j, and is given by the ith row of Eq. (2.11),

$$
q_i = \sum_{j=1}^{N} \delta q_{ij} = \sum_{j=1}^{N} R_{ij} Q_j.
\tag{2.14}
$$

Therefore, the discrete influence coefficient, R_{ij}, can be understood as the ith virtual generalized coordinate due to the jth *unit* generalized load. The *reciprocal principle* of a linear structure requires that the ith virtual coordinate due to the jth unit load is the same as the jth virtual coordinate caused by the ith generalized load, i.e.,

$$
R_{ij} = R_{ji},
\tag{2.15}
$$

which implies that the matrix R is symmetric.

In order to determine the generalized loads from the generalized displacements they actually produce an inversion of Eq. (2.11) is required as follows:

$$
\mathbf{P} = \mathbf{Kq},
\tag{2.16}
$$

where $\mathsf{K} = \mathsf{R}^{-1}$ is the generalized *stiffness matrix* of the structure. Both R and K must be nonsingular and symmetric matrices. The element of $[K]$, k_{ij}—called the *stiffness coefficient*—is the ith generalized *virtual load* due to the jth *unit* generalized displacement. The work done by a generalized virtual load,

$$
\delta Q_i = k_{ij} q_j,
\tag{2.17}
$$

in producing a generalized displacement, q_j, is given by

$$U_{ij} = \int \delta Q_i \mathrm{d}q_j = \int k_{ij} q_j \mathrm{d}q_j = \frac{1}{2} k_{ij} q_j^2 = \frac{1}{2} \delta Q_i q_j. \qquad (2.18)$$

When summed over all points on the structure, the net work done by all the static forces is the total *strain energy* stored in the structure, given by

$$U = \sum_{i=1}^{N} \sum_{j=1}^{N} U_{ij} = \frac{1}{2} \mathbf{Q}^T \mathbf{q} = \frac{1}{2} \mathbf{q}^T \mathbf{K} \mathbf{q}. \qquad (2.19)$$

The strain energy is the potential energy responsible for restoring the structure to its original shape once the loading is removed, and its quadratic form given by Eq. (2.19) is an important consequence of the linear elastic behavior. Since the external forces must be balanced by equal and opposite internal forces for a static equilibrium, one can regard U as the net work done by the internal, restoring (or conservative) forces.

2.3 Beam-Shaft Idealization

Consider a thin, high aspect-ratio wing with an essentially flat mean surface. Let (x, y, z) be Cartesian coordinates, such that x is in the chordwise direction measured from the elastic axis, y in the spanwise direction along the elastic axis on the mean plane, and z is normal to the mean plane. Saint-Venant's theory [68] postulates that a point exists at each cross-section of a slender beam about which a twisting load will produce a pure twist and a free warping, but no bending deformation. Such a point is called the *shear center*. Conversely, if a vertical load P is applied directly at the shear center, it will only cause pure bending deformation without any twisting or warping of the beam. The *elastic axis* is the line joining the shear centers at all spanwise locations. Assuming there is no bending in the chordwise (x) direction, and that plane cross-sections remain plane in the deformed configuration, the resulting structural displacement at any given point from the original (undeformed) shape can be represented by a combination of the normal deflection of the elastic axis at the given spanwise station, $w(y)$, the twist angle of the section about the elastic axis, $\theta(y)$, and the in-plane warp angle, $\phi(y)$, as depicted in Fig. 2.2. The net vertical deflection at location (x, y) is thus given by

$$\delta(x, y) = w(y) + x \tan \theta(y), \qquad (2.20)$$

whereas the in-plane deformation due to warping is merely $x \tan \phi(y)$. Since the angles θ, ϕ are small, one can apply the approximation $\tan \theta \simeq \theta$ and $\tan \phi \simeq \phi$, leading to the linear relationship

$$\delta(x, y) = w(y) + x \theta(y). \qquad (2.21)$$

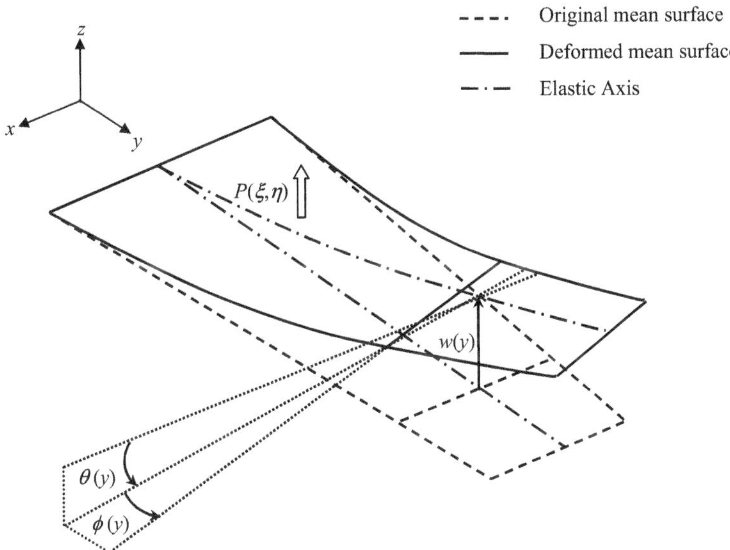

Fig. 2.2 Deformation of a thin wing of large aspect ratio under a concentrated normal load

The warp angle ϕ is inconsequential for aerodynamic loading, thus we have no need to model it any further. Since the structure is assumed to be linearly elastic, the load and displacement are linearly related by

$$\delta(x, y) = R(x, y : \xi, \eta)P(\xi, \eta), \tag{2.22}$$

where $R(x, y : \xi, \eta)$ is the flexibility influence-coefficient function (Green's function). By applying linear superposition, Eq. (2.22) can be extended for the case of a continuously distributed, normal load per unit area (pressure), $p(\xi, \eta)$, as follows:

$$\delta(x, y) = \iint_s R(x, y : \xi, \eta)p(\xi, \eta)\mathrm{d}\xi\,\mathrm{d}\eta. \tag{2.23}$$

The constitutive relations of the bending and twisting deformations are separately derived by considering a segment of the structure. For this purpose, the spanwise direction y is taken along the elastic axis, and bending deflection, $w(y)$ measured normal to the mean surface (called *neutral axis*) as shown in Fig. 2.3a. The kinematics (or compatibility) of the bending and shearing deformations is based upon the assumption that an originally plane section normal to the neutral axis must remain plane after deformation. However, this section can undergo a rotation due to shear deformation (shearing strain), $\beta(y)$, such that the section is no longer normal to the neutral axis. The bending slope, $w'(y)$, and the rotation angle due to shear, $\beta(y)$, thus collectively produce the net rotation, $\alpha(y)$, according to the following kinematical relationship:

$$\alpha = w' - \beta. \tag{2.24}$$

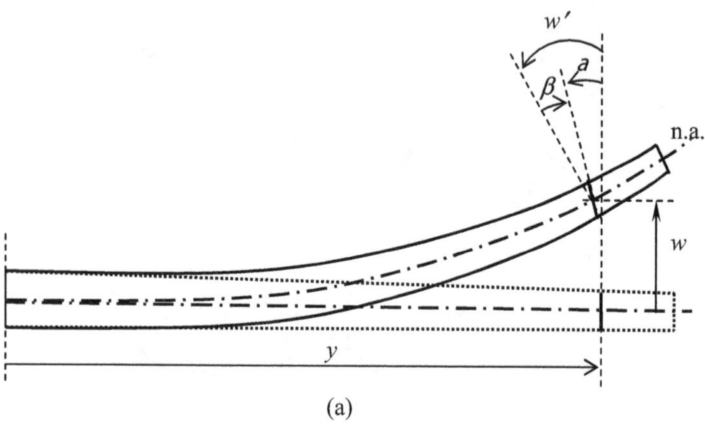

(a)

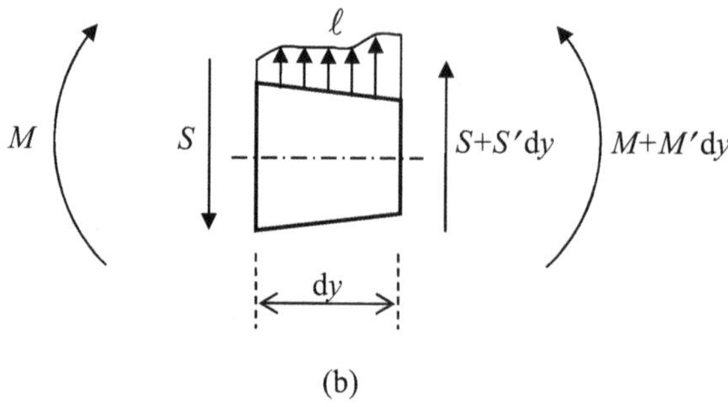

(b)

Fig. 2.3 Beam deflection geometry and static equilibrium of a beam segment

Equilibrium of a beam segment of infinitesimal length, dy, with a static lift load per unit span, $\ell(y)$, requires an internal shear force, $S(y)$, and bending moment, $M(y)$, in order to balance the external load, $\ell(y)$, as shown in Fig. 2.3b. Neglecting second and higher order terms of dy results in the following linear equilibrium equations:

$$-S' = \ell, \tag{2.25}$$

$$S + M' = 0. \tag{2.26}$$

The bending and shear constitutive relationships of a material with Young's modulus of elasticity, E, and shear modulus, G, are the following:

$$\alpha' = \frac{M}{EI}, \tag{2.27}$$

$$\beta = \frac{S}{GK}, \tag{2.28}$$

Fig. 2.4 Twisting of a slender shaft at a spanwise station, y

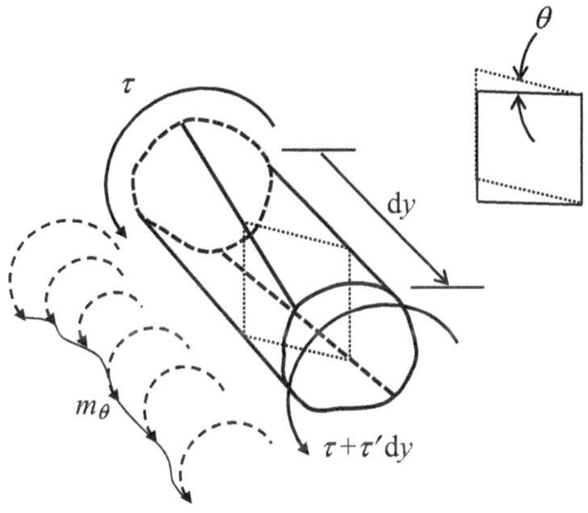

where $I(y)$ is the area moment of inertia and $K(y)$ the shearing constant of the beam cross-section. The quantity $EI(y)$ is called the bending stiffness and $GK(y)$ the shearing stiffness of the local beam cross-section. Substitution of Eqs. (2.27), (2.28), and (2.24) into Eqs. (2.25) and (2.26) results in the following differential equations for the beam:

$$\left(EI\alpha'\right)'' = \ell, \tag{2.29}$$

$$\left(EI\alpha'\right)' + GK\left(\alpha - w'\right) = 0, \tag{2.30}$$

which must be solved for $\alpha(y)$ and $w(y)$, subject to the boundary conditions of the structure. The net strain energy of the wing semispan idealized as a beam of length $b/2$ is then given by

$$U = \frac{1}{2}\int_0^{b/2} EI\left(\alpha'\right)^2 dy + \frac{1}{2}\int_0^{b/2} GK\left(\alpha - w'\right)^2 dy. \tag{2.31}$$

For thin, slender structures, the shear deformation, $\beta(y)$, can be neglected in comparison with the bending slope, $w'(y)$, leading to the following *Euler–Bernoulli* beam equation:

$$\left(EIw''\right)'' = \ell, \tag{2.32}$$

which must be solved for bending deflection, $w(y)$, subject to the boundary conditions. The net bending strain energy is now simply the following:

$$U = \frac{1}{2}\int_0^{b/2} EI\left(w''\right)^2 dy. \tag{2.33}$$

Wherever possible to apply, Euler–Bernoulli assumptions are extremely valuable due to the simplicity of the resulting model.

For a slender, shaft-like structure (Fig. 2.4), the twisting deformation, $\theta(y)$, by shear of an originally straight edged element, is related to the local twisting moment, $\tau(y)$, about the elastic axis by Saint-Venant's theory [68] as follows:

$$\tau = GJ\theta', \tag{2.34}$$

where G is the shear modulus and $J(y)$ the local torsional constant (polar moment of inertia) of the cross-section. Due to the linear proportionality of the local torque, $\tau(y)$, with the twisting slope, $\theta'(y)$, the factor $GJ(y)$ is termed torsional stiffness of the structure. Equilibrium of a segment of infinitesimal length, dy, under a distributed torque (pitching moment) load per unit length, $m_\theta(y)$, results in the following differential equation:

$$\left(GJ\theta'\right)' + m_\theta = 0, \tag{2.35}$$

which must be solved for twist angle, $\theta(y)$, subject to the boundary conditions on the shaft. The net twisting strain energy of the wing semispan is then given by

$$U = \frac{1}{2}\int_0^{b/2} GJ\left(\theta'\right)^2 dy. \tag{2.36}$$

When Euler–Bernoulli assumption of negligible shear deformation is applied along with that of a slender shaft, the net vertical deflection at a point, $\delta(x, y)$, is given by Eq. (2.21), which implies

$$\frac{\partial \delta}{\partial x} = \theta(y). \tag{2.37}$$

2.4   Dynamics

Aerodynamic loads on a vibrating structure are time dependent. Therefore, in order to construct a dynamic model, it is necessary to consider not only the strain (potential) energy, U, but also the net *kinetic energy*, T, and the work done upon the structure by nonconservative forces, W_n. For a thin wing, the linear load–displacement relationship can be used to derive the strain energy in terms of the vertical deflection, $\delta(x, y, t)$,

$$U = \frac{1}{2}\iint_s \delta(x, y, t)\iint_s k(x, y : \xi, \eta)\delta(\xi, \eta, t)d\xi d\eta dx dy, \tag{2.38}$$

where $k(x, y : \xi, \eta)$ is the stiffness influence function of the displacement produced at point (ξ, η) by a load applied at the point (x, y). Similarly, the net kinetic energy of the structure is given by

$$T = \frac{1}{2}\iint_s \rho(x, y)\dot{\delta}^2(x, y, t)dx dy, \tag{2.39}$$

where $\rho(x, y)$ is the structural mass per unit area and $\dot{\delta}(x, y, t)$, the local vertical velocity.

An aircraft wing is a thin structure essentially cantilevered at the root, with the span (b) generally much larger than the average chord (Fig. 2.5). For an ASE (aeroservoelasticity) model, one is primarily concerned with wings of large aspect ratio that

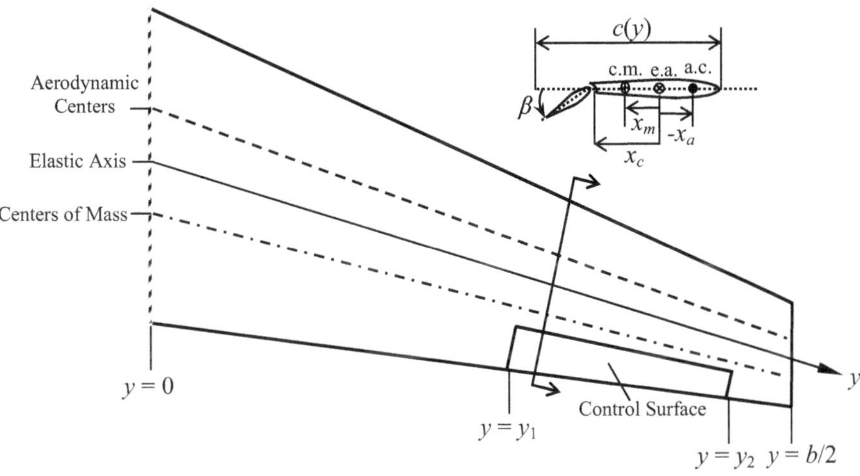

Fig. 2.5 Schematic diagram of a wing semispan and a chordwise section for dynamic modeling

are typically more prone to aeroelastic instabilities. Such a wing has a well-defined elastic axis (e.a.), defined as the line joining the shear centers of all cross-sections. The line of the centers of mass (c.m.) at each chordwise section affects the natural frequencies of the structural modes. The net aerodynamic loads at each chordwise section act at a point called the *aerodynamic center* (a.c.), which is defined as the location about which the pitching moment is invariant with the angle of attack. The net aerodynamic loading per unit span of the wing is thus a concentrated pitching moment, a lift force, and a drag force, all applied at the a.c. For a thin wing, the drag is very much smaller than the lift, and therefore has a negligible aeroelastic contribution. Due to the offset of the line joining the aerodynamic centers from the elastic axis (see Fig. 2.5), there is a net spanwise bending moment, a vertical shear load, and a torsional (or twisting) moment at each spanwise location. The bending and twisting produced at each spanwise location cause a vertical translation, $w(y)$, of the elastic axis (heave), a rotation, $\theta(y)$, (pitch) about the elastic axis (as shown in Fig. 2.6), and a negligible shear deformation. If a control surface is also present, it is usually modeled as being rigid due to its small dimensions. In such a case, the control surface rotation, $\beta(y)$, relative to the wing is the third degree of freedom at a given cross-section. A rotational spring mounted on the hinge line models the torsional stiffness pertaining to the control surface rotation.

For a slender, thin wing of semispan $b/2$ with negligible shear, the Euler–Bernoulli beam assumptions are applicable, and the net strain energy is given by

$$U = \frac{1}{2} \int_0^{b/2} EI \left(w'' \right)^2 dy + \frac{1}{2} \int_0^{b/2} GJ \left(\theta' \right)^2 dy + \frac{1}{2} \int_{y_1}^{y_2} k_\beta \left(\beta' \right)^2 dy , \quad (2.40)$$

where $EI(y)$ is the wing's bending stiffness, $GJ(y)$ its torsional stiffness, and $k_\beta(y)$ the torsional spring stiffness at the control surface hinge line. For modeling

Fig. 2.6 Degrees of freedom, forces, and moments at a given chordwise section

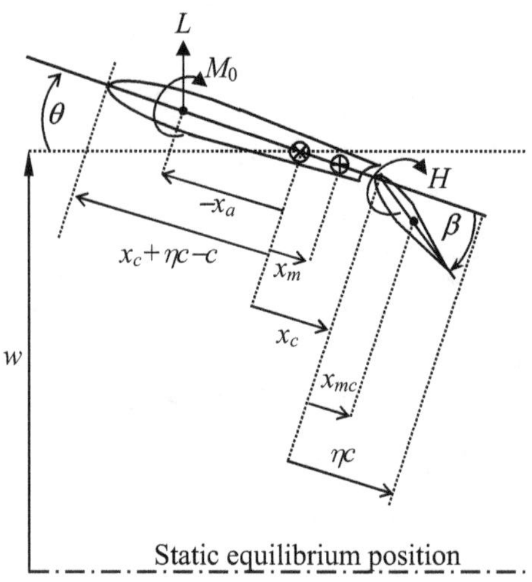

the kinetic energy of the structure, the wing's mass of density $\rho(x, y)$ per unit area is assumed to be distributed along the span such that mass per unit span is $m(y)$ with the local center of mass located at $x_m(y)$ aft of the elastic axis (Fig. 2.6), the rotary inertia under shear is negligible, and the local moment of inertia about the elastic axis is $I_\theta(y)$. The control surface has a spanwise mass distribution per unit span, $m_c(y)$, with the local center of mass at $x_{mc}(y)$ aft of the hinge line, and the net moment of inertia about the hinge line, I_β. If the control surface chord is a fraction, $\eta(y)$, of the local wing chord, then the trailing edge is located a distance $x_c + \eta c$ aft of the elastic axis. Then the net kinetic energy is given by

$$T = \frac{1}{2} \int_0^{b/2} \int_{x_c+\eta c-c}^{x_c} \left(\dot{w} + x\dot\theta\right)^2 \rho \mathrm{d}x\mathrm{d}y$$

$$+ \frac{1}{2} \int_{y_1}^{y_2} \int_{x_c}^{x_c+\eta c} \left[\dot{w} + x_c\dot\theta + (x - x_c)\dot\beta\right]^2 \rho \mathrm{d}x\mathrm{d}y, \tag{2.41}$$

or

$$T = \frac{1}{2} \int_0^{b/2} \int_{x_c+\eta c-c}^{x_c+\eta c} \left(\dot{w}^2 + 2x\dot{w}\dot\theta + x^2\dot\theta^2\right) \rho \mathrm{d}x\mathrm{d}y$$

$$+ \frac{1}{2} \int_{y_1}^{y_2} \int_{x_c}^{x_c+\eta c} \left[x^2\dot\beta^2 + 2(x - x_c)\dot{w}\dot\beta + 2x_c(x - x_c)\dot\theta\dot\beta\right] \rho \mathrm{d}x\mathrm{d}y. \tag{2.42}$$

Given the definitions of the center of mass and the moment of inertia, we have

$$\int_0^{b/2} \int_{x_c+\eta c-c}^{x_c+\eta c} \dot{w}^2\rho \mathrm{d}x\mathrm{d}y = \int_0^{b/2} m\dot{w}^2\mathrm{d}y, \tag{2.43}$$

$$\int_0^{b/2} \int_{x_c+\eta c-c}^{x_c+\eta c} x\dot{w}\dot\theta\rho \mathrm{d}x\mathrm{d}y = \int_0^{b/2} mx_m\dot{w}\dot\theta\mathrm{d}y, \tag{2.44}$$

$$\int_0^{b/2} \int_{x_c+\eta c-c}^{x_c+\eta c} x^2 \dot{\theta}^2 \rho \mathrm{d}x\mathrm{d}y = \int_0^{b/2} I_\theta \dot{\theta}^2 \mathrm{d}y, \tag{2.45}$$

$$\int_{y_1}^{y_2} \int_{x_c}^{x_c+\eta c} x^2 \dot{\beta}^2 \rho \mathrm{d}x\mathrm{d}y = \int_{y_1}^{y_2} I_\beta \dot{\beta}^2 \mathrm{d}y, \tag{2.46}$$

$$\int_{y_1}^{y_2} \int_{x_c}^{x_c+\eta c} (x-x_c)\dot{w}\dot{\beta}\rho \mathrm{d}x\mathrm{d}y = \int_{y_1}^{y_2} m_c x_{mc} \dot{w}\dot{\beta}\mathrm{d}y, \tag{2.47}$$

and

$$\int_{y_1}^{y_2} \int_{x_c}^{x_c+\eta c} x_c(x-x_c)\dot{\theta}\dot{\beta}\rho \mathrm{d}x\mathrm{d}y = \int_{y_1}^{y_2} m_c x_c x_{mc} \dot{\theta}\dot{\beta}\mathrm{d}y. \tag{2.48}$$

Consequently, the net kinetic energy is the following:

$$T = \frac{1}{2} \int_0^{b/2} \left(m\dot{w}^2 + I_\theta \dot{\theta}^2 + 2mx_m \dot{w}\dot{\theta} \right) \mathrm{d}y$$

$$+ \frac{1}{2} \int_{y_1}^{y_2} \left[I_\beta \dot{\beta}^2 + 2m_c x_{mc} \dot{\beta}(\dot{w} + x_c\dot{\theta}) \right] \mathrm{d}y. \tag{2.49}$$

The motion at a spanwise location is thus completely described by the generalized displacement vector, $\mathbf{q}(y,t) = (w,\theta,\beta)^T$, measured in an inertial frame, and its time derivative, $\dot{\mathbf{q}}$.

The generalized coordinates now represent the *degrees of freedom* of the dynamic motion. Before proceeding further, it is necessary to split the generalized force vector into conservative (elastic) and nonconservative parts, $\mathbf{Q}_c(t)$ and $\mathbf{Q}_n(t)$, respectively:

$$\mathbf{Q} = \mathbf{Q_c}(\mathbf{q}) + \mathbf{Q_n}(\mathbf{q},\dot{\mathbf{q}}). \tag{2.50}$$

By definition, the conservative force is a function only of the generalized coordinates, while the nonconservative force can also depend upon the time derivatives of the generalized coordinates. Since the elastic stiffness creates a restoring internal force, it is a conservative force. For a linear structure, the generalized forces created by viscous (Rayleigh) damping effects are proportional to $\dot{\mathbf{q}}$, and are therefore nonconservative forces. Finally, the generalized, unsteady aerodynamic forces given by the vector $\mathbf{Q_a}$ are nonconservative in nature. Thus we write

$$\mathbf{Q_c} = -\mathbf{K}\mathbf{q}$$

$$\mathbf{Q_n} = -\mathbf{C}\dot{\mathbf{q}} + \mathbf{Q_a}, \tag{2.51}$$

where the negative sign indicates an internal (opposing) force, $\mathbf{C}$ is the *generalized damping matrix* comprising the viscous damping coefficients, and $\mathbf{Q_a}$ can have a nonlinear relationship with the generalized coordinates and their time derivatives. By Newton's second law of motion applied to the discretized structure, we have

$$\mathbf{M}\ddot{\mathbf{q}} = \mathbf{Q} = -\mathbf{K}\mathbf{q} - \mathbf{C}\dot{\mathbf{q}} + \mathbf{Q_a}, \tag{2.52}$$

or

$$\mathbf{M}\ddot{\mathbf{q}} + \mathbf{C}\dot{\mathbf{q}} + \mathbf{K}\mathbf{q} = \mathbf{Q_a}, \tag{2.53}$$

where $\mathbf{M}$ is the *generalized mass matrix* representing the individual masses and moments of inertia corresponding to the various degrees of freedom.

One can alternatively adopt an energy approach to derive Eq. (2.53). For achieving an arbitrary, infinitesimal generalized displacement, $\delta q_i(t)$, Hamilton's principle requires that of all possible trajectories, the correct one is that which minimizes the net mechanical energy, $(T - U + W_n)$. This statement is given in the variational form by the necessary condition,

$$\delta \int (T - U + W_n)dt = \int \delta(T - U)dt + \int \delta W_n dt = 0, \tag{2.54}$$

where $\delta(.)$ represents the *variational operator*,

$$\delta T = \sum_{i=1}^{N} \frac{\partial T}{\partial q_i}\delta q_i + \frac{\partial T}{\partial \dot{q}_i}\delta \dot{q}_i, \tag{2.55}$$

$$\delta U = \sum_{i=1}^{N} \frac{\partial U}{\partial \dot{q}_i}\delta \dot{q}_i, \tag{2.56}$$

and

$$\delta W_n = \sum_{i=1}^{N} Q_{ni}\delta \dot{q}_i. \tag{2.57}$$

Since the initial virtual displacement must be zero, $\delta q_i(0) = 0$, integrating the first term of kinetic energy variation, Eq. (2.55), by parts yields the following expression:

$$\int \frac{\partial T}{\partial \dot{q}_i}\delta \dot{q}_i dt = -\int \frac{d}{dt}\left(\frac{\partial T}{\partial \dot{q}_i}\right)\delta q_i dt, \tag{2.58}$$

which substituted into Eq. (2.54) produces the well-known *Lagrange's equations* :

$$\frac{d}{dt}\left(\frac{\partial T}{\partial \dot{q}_i}\right) - \frac{\partial T}{\partial q_i} + \frac{\partial U}{\partial q_i} = Q_{ni} \quad (i = 1,\ldots,N). \tag{2.59}$$

For a linearly elastic structure, the potential and kinetic energies are expressed as follows:

$$U = \frac{1}{2}\mathbf{q}^T\mathbf{K}\mathbf{q}$$

$$T = \frac{1}{2}\dot{\mathbf{q}}^T\mathbf{M}\dot{\mathbf{q}}, \tag{2.60}$$

thereby yielding

$$\frac{\mathrm{d}}{\mathrm{d}t}\left(\frac{\partial T}{\partial \dot{\mathbf{q}}}\right) = \mathsf{M}\ddot{\mathbf{q}}$$

$$\frac{\partial T}{\partial \mathbf{q}} = \mathbf{0} \tag{2.61}$$

$$\frac{\partial U}{\partial \mathbf{q}} = \mathsf{K}\mathbf{q}.$$

Thus, substituting Eq. (2.61) along with the second of Eq. (2.51) into Eq. (2.59) yields the equation of motion, Eq. (2.53).

For a simple harmonic vibration in vacuum, we have $\mathbf{Q_a}(t) = \mathbf{0}$, and

$$\mathbf{q}(t) = \bar{\mathbf{q}}e^{i\omega t}, \tag{2.62}$$

where ω is the in-vacuo natural frequency and $\bar{\mathbf{q}}$ the mode shape of the structure. In this case, Eq. (2.53) results into the following linear eigenvalue problem:

$$(\mathsf{M}\omega^2 + \mathsf{C}i\omega + \mathsf{K})\bar{\mathbf{q}} = \mathbf{0} \tag{2.63}$$

Much of the effort in structural modeling involves the derivation of the generalized mass, stiffness, and viscous damping matrices by an appropriate discretization scheme, and the separation of the generalized coordinates into spatial and time-dependent parts.

2.5   Lumped Parameters Method

The simplest way to discretize the structural dynamics equations of motion is to assume that the structure consists of several piecewise rigid elements, connected together by discrete springs. Each rigid element has well-defined degrees of freedom with corresponding inertial, stiffness, and damping parameters that are considered concentrated (or lumped) at a point in each element. For example, a plate-like structure is approximated by n rigid elements capable of only vertical displacement, δ_i, measured at the centroid under an external load distribution discretized as a vertical load, P_i, acting at each elemental centroid. The discretized load–displacement relationship at each elemental point is then given by

$$P_i = \sum_{j=1}^{n} k_{ij}\delta_j , \quad (i = 1,\ldots,n) \tag{2.64}$$

where k_{ij} is the discrete stiffness associated with the jth load and ith displacement. Similarly, the dynamic equation of each element can be derived to be the following:

$$m_i\ddot{\delta}_i + \sum_{j=1}^{n} k_{ij}\delta_j = Q_i , \quad (i = 1,\ldots,n) \tag{2.65}$$

with m_i being the elemental mass and Q_i the concentrated load at the centroid. The assembled form of Eq. (2.65) for the whole structure is thus the following:

$$\mathsf{M}\ddot{\delta} + \mathsf{K}\delta = \mathbf{Q} \,, \tag{2.66}$$

where $\delta = (\delta_1, \delta_2, \ldots, \delta_n)^T$ is the displacement vector, $\mathbf{Q} = (Q_1, Q_2, \ldots, Q_n)^T$, the external load vector, and M and K are the mass and stiffness matrices. It is to be noted that due to the absence of dynamic coupling between the elements, M is a diagonal matrix.

Another important example of the lumped parameters method is a slender wing, which can be approximated as having rigid chordwise sections attached to one another by linear (bending) and rotational (torsion) springs, and free to deflect vertically as well as to rotate relative to one another. Each section has a discrete center of mass and moment of inertia about the torsion axis. The strain energy and kinetic energy, as well as the work done by nonconservative forces of all elements are merely summed up in order to yield the mass, stiffness, and damping matrices, and the generalized force vector by Lagrange's equations.

Consider a chordwise rigid section at spanwise station $y = y_i$, as shown in Fig. 2.7. The three degrees of freedom (DOFs) of a rigid elemental section are denoted by the generalized coordinates, h_i (plunge), θ_i (pitch), and β_i (control surface angle). While h_i is the vertically downward displacement of the elastic axis (e.a.) from its static equilibrium position, θ_i gives the rotation of the section about elastic axis (considered positive nose-up as shown in Fig. 2.7), whereas β_i is the angle between the chord lines of the wing and the control surface (positive with the trailing edge down). The chordwise locations of the elastic axis, the aerodynamic center, the center of mass (c.m.), and the control surface hinge line are $x = x_{hi}$, $x = x_{ai}$, $x = x_{mi}$, and $x = x_{ci}$, respectively, measured from the leading edge.

The elastic motion is assumed to have linear stiffnesses corresponding to each degree of freedom. Thus k_{hi} is the stiffness in plunge, $k_{\theta i}$, stiffness in pitch, and $k_{\beta i}$ that of the control rotation. The structural damping is considered negligible in comparison with aerodynamic damping. The generalized forces corresponding to the degrees of freedom must be calculated from the aerodynamic loads shown in Fig. 2.7 for the individual rigid sections. These are L_i (lift), M_{0i} (zero-lift pitching moment) concentrated at the a.c., and H_i (hinge moment) acting at the hinge line. The equations of motion are derived by the Lagrange's equations as follows, using the generalized coordinates per section, $\mathbf{q}_i = (h_i, \theta_i, \beta_i)^T$. The potential (strain) energy and kinetic energy are given for the ith discrete element as follows:

$$U_i = \frac{1}{2}\left(k_{hi}h_i^2 + k_{\theta i}\theta_i^2 + k_{\beta i}\beta_i^2\right)$$

$$T_i = \frac{1}{2}\int_0^{c_i} \dot{h}_i^2\,dm + \frac{1}{2}\int_0^{x_{ci}} \left[\dot{h}_i + (x - x_{hi})\dot{\theta}_i\right]^2 dm$$

$$+ \frac{1}{2}\int_{x_{ci}}^{c_i} \left[\dot{h}_i + (x_{ci} - x_{hi})\dot{\theta}_i + (x - x_{ci})\dot{\beta}_i\right]^2 dm$$

$$= \frac{1}{2}\left[m_i\dot{h}_i^2 + I_{\theta i}\dot{\theta}_i^2 + I_{\beta i}\dot{\beta}_i^2\right.$$

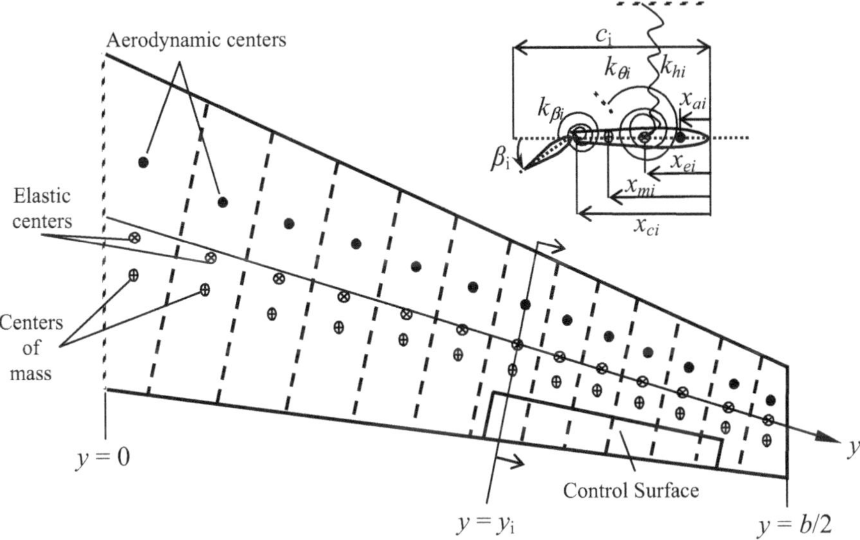

Fig. 2.7 Schematic diagram of a wing semispan approximated as chordwise rigid sections attached by springs

$$+ 2S_{\theta i}\dot{\theta}_i\dot{h}_i + 2S_{\beta i}\dot{\beta}_i\dot{h}_i + 2(x_{ci} - x_{hi})S_{\beta i}\dot{\theta}_i\dot{\beta}_i\Big], \tag{2.67}$$

where

$$I_{\theta i} = \int_0^{c_i} (x - x_{hi})^2\, dm \tag{2.68}$$

is the moment of inertia of the ith rigid element about the elastic axis,

$$I_{\beta i} = \int_{x_{ci}}^{c_i} (x - x_{ci})^2\, dm \tag{2.69}$$

is the moment of inertia of the control surface section about its hinge line,

$$S_{\theta i} = \int_0^{c_i} (x - x_{hi})\, dm = m_i\,(x_{mi} - x_{hi})$$

$$S_{\beta i} = \int_0^{x_{ci}} (x - x_{ci})\, dm = m_{ci}\,(x_{m_{ci}} - x_{ci}) \tag{2.70}$$

with m_{ci} being the mass of the control surface and $x_{m_{ci}}$ its center of mass. By substituting Eq. (2.67) into Eq. (2.26), we have

$$\frac{\mathrm{d}}{\mathrm{d}t}\left(\frac{\partial T_i}{\partial \dot{\mathbf{q}}_i}\right) = \mathsf{M}_i \ddot{\mathbf{q}}_i = \begin{pmatrix} m_i & S_{\theta i} & S_{\beta i} \\ S_{\theta i} & I_{\theta i} & (x_{ci} - x_{hi})S_{\beta i} \\ S_{\beta i} & (x_{ci} - x_{hi})S_{\beta i} & I_{\beta i} \end{pmatrix} \begin{Bmatrix} \ddot{h}_i \\ \ddot{\theta}_i \\ \ddot{\beta}_i \end{Bmatrix}$$

$$\frac{\partial T_i}{\partial \mathbf{q}_i} = \mathbf{0} \tag{2.71}$$

$$\frac{\partial U}{\partial \mathbf{q}_i} = \mathbf{K}_i \mathbf{q}_i = \begin{pmatrix} k_{hi} & 0 & 0 \\ 0 & k_{\theta i} & 0 \\ 0 & 0 & k_{\beta i} \end{pmatrix} \begin{Bmatrix} h_i \\ \theta_i \\ \beta_i \end{Bmatrix}.$$

The generalized loads vector of each section is due to unsteady aerodynamics, and can be expressed as follows:

$$\mathbf{Q}_{\mathbf{a}i}(t) = \begin{Bmatrix} -L_i(t) \\ M_{0i}(t) + (x_{hi} - x_{ai})L_i(t) \\ H_i(t) \end{Bmatrix}. \tag{2.72}$$

Since structural damping has been neglected, $\mathbf{Q}_{\mathbf{a}i}$ is the only nonconservative force acting on the structure. The aerodynamic loading (L_i, M_{0i}, H_i) is a function of the flow speed and density, as well as the control surface deflection, β_i. In addition, the lift, L_i, and hinge moment, H_i, also depend upon the flow incidence (called geometric *angle of attack*), which is defined as the angle made by the chord with the air flow far upstream of the wing (*freestream*). The geometric angle of attack at a given point is related to the freestream flow component seen normal to the wing, w_i (called *upwash*), which, in turn, is a function of the generalized coordinates, h_i, θ_i, β_i. For example, the local angle of attack at the aerodynamic center is given by[1]

$$\alpha_i \simeq \frac{w_i}{U_\infty} = \theta_i + \frac{\dot{h}_i + (x_{ai} - x_{hi})\dot{\theta}_i}{U_\infty}, \tag{2.73}$$

where U_∞ is the speed of the uniform, relative air flow far upstream of the wing (called *freestream speed*). Furthermore, the location of the aerodynamic center can be a function of freestream flow properties. For a thin airfoil, the wing's aerodynamic center in subsonic freestream is near the quarter-chord location ($x_{ai} = c_i/4$), and moves to the mid-chord point ($x_{ai} = c_i/2$) for a supersonic free stream. It may thus be appreciated that there is a complicated relationship between the generalized coordinates and the generalized loads on the structure. The next chapter is concerned with how such a relationship can be modeled by the use of aerodynamic concepts.

The mass and stiffness matrices of each section are symmetric, and the structural dynamical system is linear. The sectional degrees of freedom are also the degrees of freedom of the entire structure, therefore the mass and stiffness matrices, as well as the generalized loads, can be readily assembled into those of the structure. The ease of assembly of elemental vectors and matrices into global ones is the main benefit of

---

[1] In a steady flow, we have $\alpha_i \simeq \theta_i$, which is why most textbooks on aeroelasticity use the two angles interchangeably. However, we will distinguish the angles α_i and θ_i here, because we are mainly concerned with unsteady flow.

using the lumped parameters method, as opposed to more sophisticated discretization schemes given below. However, for any reasonable modeling accuracy, the lumped parameter approach requires a large number of discrete elements, thus a much larger modeling dimension.

2.6 Rayleigh–Ritz Method

The Rayleigh–Ritz (or assumed modes method) uses the Lagrange's equations to formulate a series expression for the wing's vertical deformation, $\delta(x, y, t)$, in terms of a finite number of continuous mode shapes, $\gamma_i(x, y)$, and the associated generalized coordinates, $q_i(t)$, where $i = 1, \ldots, n$:

$$\delta(x, y, t) = \sum_{i=1}^{n} \gamma_i(x, y) q_i(t). \tag{2.74}$$

Here the assumed modes, $\gamma_i(x, y)$, must satisfy the geometric boundary conditions, such as those at the cantilevered root,

$$\gamma_i(x, 0) = 0, \qquad \frac{\partial \gamma_i}{\partial y}(x, 0) = 0, \tag{2.75}$$

and possibly also the natural boundary conditions at the free tip (although this is not required by the principle of virtual work) :

$$\frac{\partial^2 \gamma_i}{\partial y^2}(x, b/2) = 0, \qquad \frac{\partial^3 \gamma_i}{\partial y^3}(x, b/2) = 0. \tag{2.76}$$

The generalized coordinates, $q_i(t), i = 1, \ldots, n$, represent the contribution of each mode to the overall displacement, and satisfy the Lagrange's equations, Eq. (2.26), with total kinetic energy given by

$$T = \frac{1}{2} \iint_s \rho \dot{\delta}^2 dx dy = \frac{1}{2} \sum_{i=1}^{n} \sum_{j=1}^{n} M_{ij} \dot{q}_i \dot{q}_j, \tag{2.77}$$

where $\rho(x, y)$ is the density per unit area, and the modal masses, M_{ij}, are given by

$$M_{ij} = \iint_s \rho \gamma_i \gamma_j dx dy. \tag{2.78}$$

The derivation of strain energy is in terms of modal stiffness parameters (generalized stiffness influence coefficients), K_{ij}:

$$U = \frac{1}{2} \iint_s \delta(x, y, t) \iint_s k(x, y : \xi, \eta) \delta(\xi, \eta, t) d\xi d\eta dx dy$$

$$\tag{2.79}$$

$$= \frac{1}{2} \sum_{i=1}^{n} \sum_{j=1}^{n} K_{ij} q_i q_j, \tag{2.80}$$

where

$$K_{ij} = \iint_s \gamma_i(x,y) \iint_s k(x,y:\xi,\eta)\gamma_j(\xi,\eta)d\xi d\eta dxdy, \tag{2.81}$$

and $k(x,y:\xi,\eta)$ is the stiffness influence function per unit area. The substitution of kinetic and strain energy expressions into Lagrange's equations results in the usual dynamic equations of motion:

$$\frac{1}{2} \sum_{j=1}^{n} \left(M_{ij}\ddot{q}_j + K_{ij}q_j \right) = Q_i, \qquad (i = 1,\ldots,n), \tag{2.82}$$

where

$$Q_i = \iint_s F(x,y,t)\gamma_i(x,y)dxdy \tag{2.83}$$

is the generalized air force with $F(x,y,t)$ being the normal aerodynamic force (lift) acting on the wing. Equation (2.82) is expressed in the following matrix form:

$$\mathsf{M}\ddot{\mathbf{q}} + \mathsf{K}\mathbf{q} = \mathbf{Q}, \tag{2.84}$$

where M, K are the generalized mass and stiffness matrices, respectively, and $\mathbf{q}, \mathbf{Q}$ are the generalized coordinates and forces vectors, respectively.

The primary modeling effort requires the determination of the generalized stiffness matrix. For an arbitrary-shaped wing planform where the stiffness influence function, $k(x,y:\xi,\eta)$, is defined at N discrete points, k_{pq}, it is common to apply the following numerical approximation to Eq. (2.81):

$$\mathsf{K} = \Gamma^T \Lambda \begin{pmatrix} k_{11} & k_{12} & \cdots & k_{1N} \\ k_{21} & k_{22} & \cdots & k_{2N} \\ \vdots & \vdots & \vdots & \vdots \\ k_{N1} & k_{N2} & \cdots & k_{NN} \end{pmatrix} \Lambda\Gamma, \tag{2.85}$$

where

$$\Gamma = \begin{pmatrix} \gamma_{11} & \gamma_{21} & \cdots & \gamma_{n1} \\ \gamma_{12} & \gamma_{22} & \cdots & \gamma_{n2} \\ \vdots & \vdots & \vdots & \vdots \\ \gamma_{1N} & \gamma_{2N} & \cdots & \gamma_{nN} \end{pmatrix}$$

is the modal matrix consisting of the assumed mode shapes evaluated at the N discrete points as its columns, and Λ is an $(N \times N)$ diagonal matrix of integration parameters, called the weighting matrix. Depending upon the choice of the numerical integration (quadrature) scheme, the weighting matrix can be suitably selected.

The high aspect-ratio wing, which is of main interest to us, is relatively easily modeled by Rayleigh–Ritz method. Herein, the assumption of chordwise rigidity of the sections gives rise to the following expression of strain energy [21]:

$$U = \frac{1}{2} \int_0^{b/2} \left[EI(y) \left(\frac{\partial^2 \delta}{\partial y^2} \right)^2 + GJ(y) \left(\frac{\partial \theta}{\partial y} \right)^2 \right] dy, \qquad (2.86)$$

where $\theta(y) = \partial \delta / \partial x$ is the local twist angle at a given spanwise location. By substituting Eq. (2.74) into Eq. (2.86), the stiffness coefficients are obtained as follows:

$$K_{ij} = \int_0^{b/2} \left[EI(y) \left(\frac{\partial^2 \gamma_i}{\partial y^2} \frac{\partial^2 \gamma_j}{\partial y^2} \right) + GJ(y) \left(\frac{\partial^2 \gamma_i}{\partial y \partial x} \frac{\partial^2 \gamma_j}{\partial y \partial x} \right) \right] dy. \qquad (2.87)$$

2.7   Finite-Element Method

The FEM is based upon the approximation that a continuous structure can be represented by a finite number of discrete elements, over each of which the displacement due to a given loading is given by smooth interpolation (or shape) functions. The shape functions are simple polynomials whose coefficients are determined by satisfying the load or displacement conditions at certain specific points, called the nodes of each element. The mass and stiffness matrices are derived for each element from energy considerations, and are then assembled into global mass and stiffness matrices, respectively, while taking into account the geometric continuity (compatibility) and natural boundary conditions (force and moment balance) of the structure by a process called global assemblage. The utility of the FEM lies in modeling a complex structure by a large number of simple structural elements, while working out the global assemblage and connectivity through automated computational procedures. In fact, FEM is the most commonly used structural modeling and analysis tool in most engineering applications.

2.7.1   Weak Formulation and Galerkin's Approximation

The FEM is based upon the approximate solution of a boundary-value problem formulated in a *weak form* [75]. While the FEM is applicable to any engineering problem governed by partial differential equations, we will confine our discussion here to structural problems only. Consider for example a structure governed by the following spatio-temporal partial differential equation,

$$f(u) + \ell = 0, \qquad (x, y, z) \in \Omega, \ 0 \le t < \infty, \tag{2.88}$$

where $u(x, y, z, t) : \mathcal{R}^4 \to \mathcal{R}$ is the unknown deformation, $f(.)$ is a partial differential operator, and $\ell(x, y, z, t) : \mathcal{R}^4 \to \mathcal{R}$ is a smooth functional called the *Lagrangian*, specified over a closed spatial domain, $(x, y, z) \in \Omega$, and for the time interval, $0 \le t < \infty$. The problem is closed by specifying a set of m boundary conditions to be satisfied by the solution, $u(x, y, z)$, expressed as follows:

$$g_k(u) = h_k, \qquad (x, y, z) \in \Gamma, \ 0 \le t < \infty, k = 1, 2, \ldots, m, \tag{2.89}$$

where $g_k(.), k = 1, 2, \ldots, m$ are partial differential operators, and

$$h_k(x, y, z, t) : \mathcal{R}^4 \to \mathcal{R}, k = 1, 2, \ldots, m$$

are smooth functionals specified on the boundary Γ of the domain Ω for all times [2] $0 \le t < \infty$. For aeroelastic problems of interest to us, the boundary-value problem posed by Eqs. (2.88) and (2.89) are linearized in both space and time, such that the operators $f(.), g_k(.)$ are linear. The functional $f(u)$ then represents linear constitutive behavior, and the dynamic boundary-value problem can be expressed as follows:

$$\mu \frac{\partial^2 u}{\partial t^2} + \kappa \mathcal{S}(u) + \ell = 0, \tag{2.90}$$

where $\mu(x, y, z)$ and $\kappa(x, y, z)$ denote the inertial and stiffness properties of the material, respectively, $\mathcal{S}(.)$ is a spatial partial differential operator, and $\ell(x, y, z, t)$ is a loading function. The m spatial boundary conditions specified on the boundary Γ of the domain Ω are given by

$$D_k(u) = h_k, \qquad (x, y, z) \in \Gamma, 0 \le t < \infty, k = 1, 2, \ldots, m, \tag{2.91}$$

where $D_k(.)$ denotes a linear partial differential operator. Often we are only interested in the undamped, free structural vibration models, where the deformation vector has a simple harmonic behavior given by,

$$u(x, y, z, t) = \bar{u}(x, y, z)e^{i\omega t}, \tag{2.92}$$

where $\bar{u}$ is the vibration amplitude. Then the dynamic boundary-value problem can be expressed as the following linear eigenvalue problem,

$$\mu \omega^2 \bar{u} + \kappa \mathcal{S}(\bar{u}) = 0 \qquad (x, y, z) \in \Omega. \tag{2.93}$$

---

[2] A more general boundary-value problem has time-varying governing equations and boundary conditions specified at discrete times, which implies explicit dependence of the partial differential operators on time, $f(u, t), g_k(u, t)$. While such a problem can be posed for the general aeroservoelastic case in which the control inputs produce a certain desired behavior in time, it is not relevant in structural modeling applications.

Obtaining the solution $u(x, y, z, t)$, given the material properties $\mu(x, y, z)$, $\kappa(x, y, z)$, and loading $\ell(x, y, z, t)$, over a closed spatial domain, $(x, y, z) \in \Omega$, as well as the boundary condition functions, $h_k(x, y, z, t), k = 1, 2, \ldots, m$ specified on the boundary Γ of the domain Ω is called the *strong form* of the boundary-value problem posed by Eqs. (2.90) and (2.91). Only rarely can the strong form lead to a closed-form solution. Therefore, an approximation called the *weak form* (or the variational form) is almost always necessary for the boundary-value problem. This requires a series approximation for the solution, such as

$$u(x, y, z, t) \simeq \sum_{i=1}^{n} u_i(t) N_i(x, y, z), \tag{2.94}$$

where $u_i(t), i = 1, \ldots, N$ are the unknown deflections at n discrete points interpolated by smooth, prescribed functions $N_i(x, y, z)$. These interpolation (or shape) functions are selected such that they automatically satisfy the following *natural boundary conditions* associated with the physical domain, Ω:

$$D_k^{(p)}(u) = h_k^{(p)}, \qquad (x, y, z) \in \Gamma, 0 \le t < \infty, k = 1, 2, \ldots, m_1 \tag{2.95}$$

The remaining $m - m_1$ boundary conditions, called the *geometric boundary conditions*, are enforced on Γ,

$$D_k^{(g)}(u) = h_k^{(g)}, \quad (x, y, z) \in \Gamma, \ 0 \le t < \infty, k = m_1 + 1, m_1 + 2, \ldots, m, \tag{2.96}$$

in order to provide constraint equations for determining the discrete deflections.

The weak formulation requires the difference between the correct solution and its approximation must be minimized. This requires the *residual* of the approximate governing equation, given by

$$\epsilon = f\left(\sum_{i=1}^{n} u_i N_i\right) + \ell = f(\mathbf{N}\mathbf{U}) + \ell, \tag{2.97}$$

should be as close to zero as possible over the entire domain Ω. Here,

$$\mathbf{U} = \left\{ \begin{array}{c} u_1 \\ u_2 \\ \vdots \\ u_n \end{array} \right\}, \qquad \mathbf{N} = (N_1, N_2, \ldots, N_n)$$

For the linear structure, the residual is given by

$$\epsilon = \left(\mu \frac{\partial^2}{\partial t^2} + \kappa \mathcal{S}(.)\right)\left(\sum_{i=1}^{n} u_i N_i\right) + \ell, \tag{2.98}$$

or

$$\epsilon = \mu \mathbf{N} \ddot{\mathbf{U}} + \kappa \mathcal{S}\left(\mathbf{N}\right)\mathbf{U} + \ell, \qquad (2.99)$$

where overdot represents the time derivative.

The solution of the weak formulation requires a discretization of the domain Ω. In a computational approach called the *collocation method*, the domain is discretized into a set of n points at each of which the residual is made to vanish:

$$\epsilon(x_j, y_j, z_j, t) = 0, \qquad j = 1, \ldots, N \qquad (2.100)$$

The collocation procedure is a simple approach, but results in a large number of linear algebraic equations to be solved for a given computational accuracy.

An alternative procedure is the *weighted-residual method* in which the residual, $\epsilon(x, y, z)$, weighted by elements of a desired functional, $\boldsymbol{\Phi}(x, y, z)$,

$$\boldsymbol{\Phi} = (\phi_1, \phi_2, \ldots, \phi_n),$$

is brought to zero when integrated over the discretized domain Ω:

$$\int_\Omega \epsilon \phi_j \, d\Omega = 0, \qquad j = 1, \ldots, n. \qquad (2.101)$$

Substituting Eq. (2.95) into Eq. (2.101), we have

$$\int_\Omega \left[f\left(\mathbf{NU}\right) + \ell\right] \phi_j \, d\Omega = 0, \qquad j = 1, \ldots, n, \qquad (2.102)$$

which, for the linear structure, becomes

$$\int_\Omega \left[\mu \mathbf{N}\ddot{\mathbf{U}} + \kappa \mathcal{S}\left(\mathbf{N}\right)\mathbf{U} + \ell\right] \phi_j \, d\Omega = 0, \qquad j = 1, \ldots, n. \qquad (2.103)$$

By taking variation of Eq. (2.101) with respect to the unknown deflections, we have

$$\delta \int_\Omega \epsilon \phi_j \, d\Omega = \int_\Omega \phi_j \delta \epsilon \, d\Omega$$

$$= \int_\Omega \phi_j \left(\partial f / \partial \mathbf{U}\right) \delta \mathbf{U} \, d\Omega$$

$$= 0, \qquad j = 1, \ldots, n, \qquad (2.104)$$

or

$$\int_\Omega \mathbf{N}^T \left(\partial f / \partial u\right) \boldsymbol{\Phi} \delta u \, d\Omega = 0. \qquad (2.105)$$

For the linear structure, the weighted-residual method results in the following:

$$\int_\Omega \mathbf{N}^T \mu \boldsymbol{\Phi} \delta \ddot{u} + \mathcal{S}\left(\mathbf{N}\right)^T \kappa \mathcal{S}\left(\boldsymbol{\Phi}\right) \delta u \, d\Omega = 0. \qquad (2.106)$$

Since the weighting functions, $\phi_i, i = 1,\ldots,n$ multiplied with the variation δu represent the virtual displacement vector in the weighted-residual integral, they are also known as *variations*, or *virtual coordinates*. A practical and efficient procedure in which the weighted-residual method can be implemented is *Galerkin's method* [75], wherein the computational domain is discretized into a finite number n_e of elements, over each of which the weights are taken to be the same as the interpolation functions, $\phi_i = N_i, i = 1,\ldots,k$. This discretization procedure, called *FEM*, is schematically depicted in Fig. 2.8. The weighting functions $N_i(x,y,z), i = 1,\ldots,k$ satisfying the natural boundary conditions are called the *shape functions*, and the points at which the integrated weighted residual is made to vanish are called the *nodes*. The subdivision of the domain into n_e finite elements interconnected at the nodes is represented as follows:

$$\Omega = \bigcup_{i=1}^{n_e} \Omega_i, \tag{2.107}$$

where Ω_i is the subdomain comprising the ith element. An integration performed over the domain is then the summation of integrations carried out over the individual elements,

$$\int_\Omega F(x,y,z)\,\mathrm{d}\Omega = \sum_{i=1}^{n_e}\int_{\Omega_i} F(x,y,z)\,\mathrm{d}\Omega_i \tag{2.108}$$

For a linear structure, the FEM yields:

$$\int_\Omega \mathbf{N}^T \mu \mathbf{N}\delta\ddot{u} + \mathcal{S}\,(\mathbf{N})^T \kappa \mathcal{S}\,(\mathbf{N})\,\delta u\,\mathrm{d}\Omega = 0, \tag{2.109}$$

which immediately lets us identify the mass and stiffness matrices of the structure as follows:

$$\mathsf{M} = \int_\Omega \mathbf{N}^T \mu \mathbf{N}\,\mathrm{d}\Omega$$

$$\mathsf{K} = \int_\Omega \mathcal{S}\,(\mathbf{N})^T \kappa \mathcal{S}\,(\mathbf{N})\,\mathrm{d}\Omega. \tag{2.110}$$

It is to be noted however that the FEM carries out the integration for the mass and stiffness matrices in a piecewise manner, wherein integrations are first performed over each element, and then assembled into global matrices using the kinematical compatibility of the elemental and global degrees of freedom, while enforcing the geometric boundary conditions, Eq. (2.96). The finite-element approach offers a choice of a large range of possible shape functions for a given domain, and is thus quite flexible. A higher order of interpolation function is expected to produce a better numerical accuracy with a given number of elements. A variation of the FEM is the Rayleigh–Ritz (or assumed modes) method of the previous section, where the interpolation functions are applied to the global structure as mode shapes, each of which is made to satisfy the geometric boundary conditions. Both the methods use the interpolation functions as the weighting functions in the weighted-residual formulation, and are thus essentially Galerkin's methods. For this purpose, the equivalence between the energy (or Lagrange's equations) formulation given by

$$\delta \int (T - U + W_n)dt = 0, \qquad (2.111)$$

and the weighted-residual formulation by Galerkin's variational approach (Eq. 2.106) must be carefully noted. Here T, U, W_n are the total kinetic energy, strain energy, and external work, respectively, of the structure.

The *least-squares method* is another weighted-residual procedure alternative to the FEM, and uses the minimization of the square of the residual at the given discrete points,

$$\frac{\partial}{\partial u_i} \int_\Omega \epsilon^2 \, d\Omega = 0, \qquad i = 1, \ldots, n.$$

This results in n linear algebraic equations for the unknown deflections, $u_i(t)$.

2.7.2 *Euler–Bernoulli Beam and Shaft Elements*

As discussed earlier in this chapter, the wings of high-aspect ratio are of main interest in aeroelastic models, because their structural modes are capable of significant interaction with unsteady air loads. Consider a high-aspect ratio wing idealized as Euler–Bernoulli beam in bending deformation, and a slender shaft in torsion Fig. 2.9. For simplicity, the structure is approximated as consisting of n beam elements with a node at the either end, and an equal number of two-noded shaft elements. At a given spanwise station, y, the bending displacement is given by $w(y, t)$, and the torsional displacement (twist angle) by $\theta(y, t)$. If n_c control surfaces are present, they are approximated to be rigid in both chordwise and spanwise directions, and their deflections, $\beta_i(t), i = 1, \ldots, n_c$, are considered to be invariant with y. The control surface degrees of freedom are then added to the mass and stiffness matrices in the same manner as carried out by the lumped parameters method. At a given spanwise station, y, the equations of motion of the wing are expressed as follows:

$$\frac{\partial^2}{\partial y^2} \left(EI \frac{\partial^2 w}{\partial y^2} \right) + m \frac{\partial^2 w}{\partial t^2} - mx_\theta \frac{\partial^2 \theta}{\partial t^2} = \ell$$

$$-\frac{\partial}{\partial y} \left(GJ \frac{\partial \theta}{\partial y} \right) - mx_\theta \frac{\partial^2 w}{\partial t^2} + I_\theta \frac{\partial^2 \theta}{\partial t^2} = \tau, \qquad (2.112)$$

where $EI(y)$ and $GJ(y)$ are the bending and torsional stiffnesses, respectively, at the local spanwise station, $x_\theta(y)$ is the aft displacement of the local center of mass

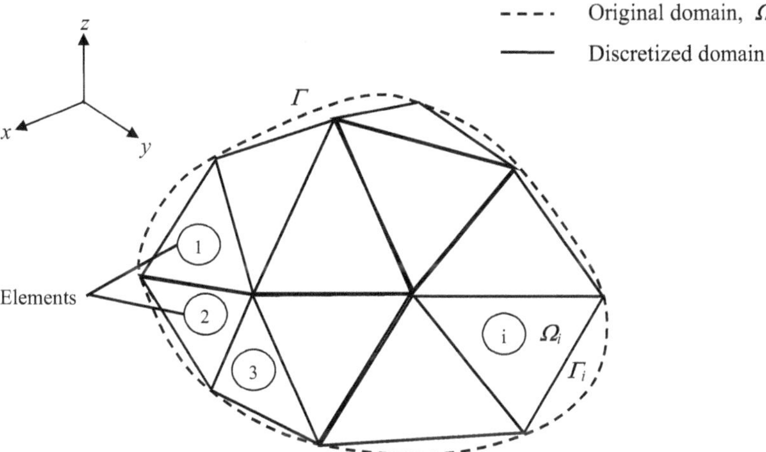

Fig. 2.8 Finite-element approximation of a domain, Ω, with boundary Γ, by a number of triangular elements, Ω_i, each of boundary Γ_i

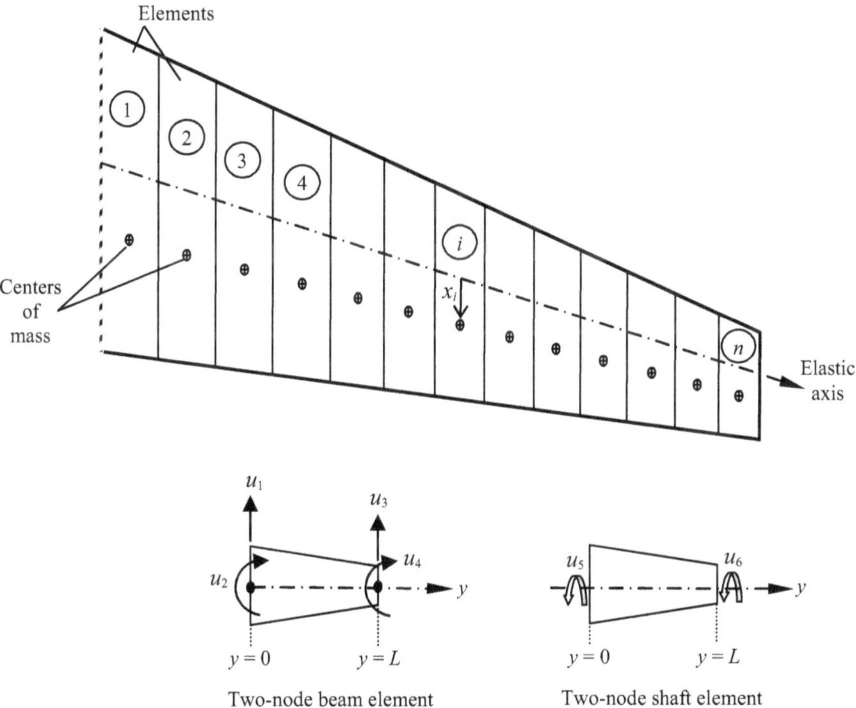

Fig. 2.9 Schematic diagram of a high-aspect ratio wing approximated by a finite number of beam and shaft elements

from the elastic axis (Fig. 2.9), $m(y)$ is the mass per unit span, $I_\theta(y)$ is the local mass moment of inertia per unit span about the elastic axis, $\ell(y, t)$ is the lift loading per unit span, and $\tau(y, t)$ is the twisting moment per unit span about the elastic axis. These partial differential equations must satisfy the boundary conditions at the either end of the wing span. The constraint arising out of a rigid attachment of the wing to the fuselage requires that the following *geometric boundary conditions* must hold at the root ($y = 0$):

$$w(0) = \left. \frac{\partial w}{\partial y} \right|_{y=0} = 0$$

$$\theta(0) = 0. \tag{2.113}$$

At the free end (the wing tip, $y = b/2$), the shear force, $(EIw'')'$, (where prime denotes $\partial/\partial y$), the bending moment, EIw'', and the twisting moment, $GJ\theta'$, must all vanish due to force and moment equilibrium. This results in the following *natural boundary conditions*:

$$\left. \frac{\partial^2 w}{\partial y^2} \right|_{y=b/2} = \left. \frac{\partial^3 w}{\partial y^3} \right|_{y=b/2} = 0$$

$$\left. \frac{\partial \theta}{\partial y} \right|_{y=b/2} = 0. \tag{2.114}$$

The solution of the strong form given by Eq. (2.112) subject to the boundary conditions of Eqs. (2.113)–(2.114) cannot be carried out in a closed form, but requires weak formulation by Galerkin's method. Here, the deformation is a two-dimensional vector, $\mathbf{u} = (w, \theta)^T$ governed by a vector differential equation equivalent of Eq. (2.90), given by

$$\mu\ddot{\mathbf{u}} + \mathbf{S}(\mathbf{u}) + \Lambda = 0, \tag{2.115}$$

where overdot denotes $\partial/\partial t$,

$$\ddot{\mathbf{u}} = \begin{Bmatrix} \ddot{w} \\ \ddot{\theta} \end{Bmatrix}, \qquad \mu = \begin{pmatrix} m & -mx_\theta \\ -mx_\theta & I_\theta \end{pmatrix}$$

$$\mathbf{S}(\mathbf{u}) = \begin{Bmatrix} \frac{\partial^2}{\partial y^2}\left(EI \frac{\partial^2 w}{\partial y^2} \right) \\ -\frac{\partial}{\partial y}\left(GJ \frac{\partial \theta}{\partial y} \right) \end{Bmatrix}, \qquad \Lambda = \begin{Bmatrix} \ell \\ \tau \end{Bmatrix}$$

The finite-element discretization is carried out by unidimensional n elements of length L each arranged in the spanwise direction (Fig. 2.9). Lagrange's equations formulation of the variational approach is derived in the form of Eq. (2.111) for the dynamically coupled case. Since bending and torsion deformations are decoupled for the static case ($\ddot{w} = \ddot{\theta} = 0$) by Saint-Venant's principle, the stiffness matrix for each element can be computed either from the weighted-residual integral of $\mathbf{S}(\mathbf{u})$, or from the strain energy,

$$U_e = \frac{1}{2}\int_0^L EI(w'')^2 dy + \frac{1}{2}\int_0^L GJ(\theta')^2 dy. \tag{2.116}$$

This offers a major simplification wherein the static deflections are treated separately for bending and torsion by using beam and shaft elements, respectively, each with different shape functions. The governing equation and natural boundary conditions can be satisfied by taking cubic shape functions, $N_i, i = 1, 2, \ldots, 4$, for the two-node beam elements shown in Fig. 2.9:

$$w(y,t) \simeq \sum_{i=1}^{4} u_i(t)N_i(y), \tag{2.117}$$

and linear shape functions, $N_i, i = 5, 6$, for shaft elements,

$$\theta(y,t) \simeq \sum_{i=5}^{6} u_i(t)N_i(y). \tag{2.118}$$

Since the two motions are coupled dynamically, the elemental mass matrix is derived from the kinetic energy,

$$T_e = \frac{1}{2}\int_0^L m\dot{w}^2 dy + \frac{1}{2}\int_0^L I_\theta \dot{\theta}^2 dy + \int_0^L mx_\theta \dot{w}\dot{\theta}\,dy, \tag{2.119}$$

and does not have a diagonal structure.

Elemental Matrices

Consider a two-node beam element shown in Fig. 2.9. In order to satisfy the natural boundary conditions, the bending deflection at a point y on the element is assumed to be given by the following Hermite cubic polynomial:

$$w(y,t) = a_0 + a_1 y + a_2 y^2 + a_3 y^3, \tag{2.120}$$

where the coefficients (a_0, a_1, a_2, a_3) are to be expressed in terms of the elemental degrees of freedom, (u_1, u_2, u_3, u_4), and the following kinematical constraints at the nodes:

$$w(0) = a_0 = u_1$$

$$w'(0) = a_1 = u_2$$

$$w(L) = a_0 + a_1 L + a_2 L^2 + a_3 L^3 = u_3 \tag{2.121}$$

$$w'(L) = a_1 + 2a_2 L + 3a_3 L^2 = u_4.$$

This results in the following expression for the interpolation coefficients:

$$\begin{Bmatrix} a_0 \\ a_1 \\ a_2 \\ a_3 \end{Bmatrix} = A \begin{Bmatrix} u_1 \\ u_2 \\ u_3 \\ u_4 \end{Bmatrix}, \tag{2.122}$$

where

$$A = \begin{pmatrix} 1 & 0 & 0 & 0 \\ 0 & 1 & 0 & 0 \\ -3/L^2 & -2/L & 3/L^2 & -1/L \\ 2/L^3 & 1/L^2 & -2/L^3 & 1/L^2 \end{pmatrix}. \tag{2.123}$$

As mentioned above, the FEM consists of interpolating the nodal deflections by smooth shape functions, $N_1(y)$, $N_2(y)$, $N_3(y)$, $N_4(y)$, in order to find the deflection at a particular point:

$$w(y,t) = N_1(y)u_1(t) + N_2(y)u_2(t) + N_3(y)u_3(t) + N_4(y)u_4(t) \tag{2.124}$$

Since the shape functions must satisfy the kinematical constraints, they can be determined as follows from Eqs. (2.120)–(2.124):

$$[N_1(y),\ N_2(y),\ N_3(y),\ N_4(y)] = (1,\ y,\ y^2,\ y^3)A, \tag{2.125}$$

resulting in

$$\begin{aligned} N_1(y) &= 1 - 3(y/L)^2 + 2(y/L)^3 \\ N_2(y) &= y - 2y^2/L + y^3/L^2 \\ N_3(y) &= 3(y/L)^2 - 2(y/L)^3 \\ N_4(y) &= -y^2/L + y^3/L^2. \end{aligned} \tag{2.126}$$

For a two-node shaft element (Fig. 2.9), the twisting deformation at a point y is approximated by

$$\theta(y,t) = b_0 + b_1 y = N_5(y)u_5(t) + N_6(y)u_6(t), \tag{2.127}$$

where

$$\begin{Bmatrix} b_0 \\ b_1 \end{Bmatrix} = \begin{pmatrix} 1 & 0 \\ -1/L & 1/L \end{pmatrix} \begin{Bmatrix} u_5 \\ u_6 \end{Bmatrix}, \tag{2.128}$$

which results in the following shape functions :

$$N_5(y) = 1 - y/L$$

$$N_6(y) = y/L. \tag{2.129}$$

When the interpolated bending and twisting deformations, $w(y)$ and $\theta(y)$, respectively, are substituted in the strain–energy of each element, we have

$$U_e = \frac{1}{2} \int_0^L E I \mathbf{u}_b^T \mathbf{N}_b'' \left(\mathbf{N}_b''\right)^T \mathbf{u}_b dy$$

$$+ \frac{1}{2} \int_0^L G J (\mathbf{u}_t^T \mathbf{N}_t' \left(\mathbf{N}_t'\right)^T \mathbf{u}_t dy \tag{2.130}$$

$$= \frac{1}{2} \mathbf{u}_e^T \mathsf{K}_\mathsf{e} \mathbf{u}_e, \tag{2.131}$$

where

$$\mathbf{u}_e = \left\{ \begin{matrix} \mathbf{u}_b \\ \mathbf{u}_t \end{matrix} \right\}, \qquad \mathbf{u}_b = \left\{ \begin{matrix} u_1 \\ u_2 \\ u_3 \\ u_4 \end{matrix} \right\}, \qquad \mathbf{u}_t = \left\{ \begin{matrix} u_5 \\ u_6 \end{matrix} \right\}, \tag{2.132}$$

$$\mathbf{N}_b = \left\{ \begin{matrix} N_1(y) \\ N_2(y) \\ N_3(y) \\ N_4(y) \end{matrix} \right\}, \qquad \mathbf{N}_t = \left\{ \begin{matrix} N_5(y) \\ N_6(y) \end{matrix} \right\}, \tag{2.133}$$

and the elemental stiffness matrix is given by

$$\mathsf{K}_\mathsf{e} = \begin{pmatrix} \mathsf{K}_\mathsf{eb} & 0 \\ 0 & \mathsf{K}_\mathsf{et} \end{pmatrix}, \tag{2.134}$$

with the following bending stiffness matrix:

$$\mathsf{K}_\mathsf{eb} = \int_0^L E I \mathbf{N}_b'' \left(\mathbf{N}_b''\right)^T dy$$

$$= \int_0^L E I \left\{ \begin{matrix} N_1''(y) \\ N_2''(y) \\ N_3''(y) \\ N_4''(y) \end{matrix} \right\} \left(N_1''(y), N_2''(y), N_3''(y), N_4''(y) \right) dy , \tag{2.135}$$

and the following torsional stiffness matrix:

$$\mathsf{K}_{\mathrm{et}} = \int_0^L GJ\mathbf{N}_t' \left(\mathbf{N}_t'\right)^T dy = \int_0^L GJ \left\{ \begin{matrix} N_5'(y) \\ N_6'(y) \end{matrix} \right\} \left(N_5'(y), N_6'(y)\right) dy. \qquad (2.136)$$

If the material properties EI and GJ are assumed to be constants over each element for simplicity [3] and when the spanwise integrals are carried out, they result in the following matrices:

$$\mathsf{K}_{\mathrm{eb}} = \frac{EI}{L^3} \begin{pmatrix} 12 & 6L & -12 & 6L \\ 6L & 4L^2 & -6L & 2L^2 \\ -12 & -6L & 12 & -6L \\ 6L & 2L^2 & -6L & 4L^2 \end{pmatrix}, \qquad (2.137)$$

$$\mathsf{K}_{\mathrm{et}} = \frac{GJ}{L} \begin{pmatrix} 1 & -1 \\ -1 & 1 \end{pmatrix}. \qquad (2.138)$$

For determining the elemental mass matrix, the shape functions are substituted into the expression for kinetic energy as follows:

$$T_e = \frac{1}{2} \int_0^L m\dot{\mathbf{u}}_b^T \mathbf{N}_b \mathbf{N}_b^T \dot{\mathbf{u}}_b dy + \frac{1}{2} \int_0^L I_\theta \dot{\mathbf{u}}_t^T \mathbf{N}_t \mathbf{N}_t^T \dot{\mathbf{u}}_t dy$$

$$+ \int_0^L m x_\theta \dot{\mathbf{u}}_b^T \mathbf{N}_b \mathbf{N}_t^T \dot{\mathbf{u}}_t dy$$

$$= \frac{1}{2} \dot{\mathbf{u}}_e^T \mathsf{M}_e \dot{\mathbf{u}}_e, \qquad (2.139)$$

where the elemental mass matrix is given by

$$\mathsf{M}_e = \begin{pmatrix} \mathsf{M}_{\mathrm{eb}} & \mathsf{M}_{e_s} \\ \mathsf{M}_{e_s}^T & \mathsf{M}_{\mathrm{et}} \end{pmatrix}, \qquad (2.140)$$

with

$$\mathsf{M}_{\mathrm{eb}} = \int_0^L m\mathbf{N}_b \mathbf{N}_b^T dy = \int_0^L m \left\{ \begin{matrix} N_1(y) \\ N_2(y) \\ N_3(y) \\ N_4(y) \end{matrix} \right\} \left(N_1(y), N_2(y), N_3(y), N_4(y)\right) dy,$$

$$(2.141)$$

---

[3] One can alternatively specify a variation of the material properties by smooth functions, $EI(y)$ and $GJ(y)$, which are then included in the integration over each element.

$$\mathbf{M_{es}} = 2 \int_0^L mx_\theta \mathbf{N}_b \mathbf{N}_t^T \, dy = 2 \int_0^L mx_\theta \begin{Bmatrix} N_1(y) \\ N_2(y) \\ N_3(y) \\ N_4(y) \end{Bmatrix} (N_5(y), N_6(y)) \, dy, \quad (2.142)$$

and

$$\mathbf{M_{et}} = \int_0^L I_\theta \mathbf{N}_t \mathbf{N}_t^T \, dy = \int_0^L I_\theta \begin{Bmatrix} N_5(y) \\ N_6(y) \end{Bmatrix} (N_5(y), \ N_6(y)) \, dy. \quad (2.143)$$

If the material properties m, I_θ, and x_θ are assumed to be constants over each element, the following matrices are derived by integration:

$$\mathbf{M_{eb}} = m \begin{pmatrix} 13L/35 & 11L^2/210 & 9L/70 & -13L^2/420 \\ 11L^2/210 & L^3/105 & 13L^2/420 & -L^3/140 \\ 9L/70 & -13L^2/420 & 13L/35 & -11L^2/210 \\ -13L^2/420 & -L^3/140 & -11L^2/210 & L^3/105 \end{pmatrix}, \quad (2.144)$$

$$\mathbf{M_{e_s}} = 2mx_\theta \begin{pmatrix} 7L/20 & 3L/20 \\ L^2/20 & L^2/30 \\ -3L/20 & 7L/20 \\ -L^2/30 & -L^2/20 \end{pmatrix}, \quad (2.145)$$

$$\mathbf{M_{et}} = I_\theta \begin{pmatrix} L/3 & L/6 \\ L/6 & L/3 \end{pmatrix}. \quad (2.146)$$

Global Assembly

For assembling the elemental matrices into global stiffness and mass matrices, a mapping is necessary from the elemental onto global degrees of freedom. Consider the global degrees of freedom numbered in the way shown in Fig. 2.10 for a wing discretized into n finite elements.

The connectivity array, Γ, showing the contribution of each elemental degree of freedom to that of the complete structure is given as follows:

$$\Gamma = \begin{pmatrix} 1 & 2 & 3 & 4 & (2n+3) & (2n+4) \\ 3 & 4 & 5 & 6 & (2n+4) & (2n+5) \\ 5 & 6 & 7 & 8 & (2n+5) & (2n+6) \\ \cdots & \cdots & \cdots & \cdots & \cdots & \cdots \\ (2n-1) & (2n) & (2n+1) & (2n+2) & (3n+2) & (3n+3) \end{pmatrix}, \quad (2.147)$$

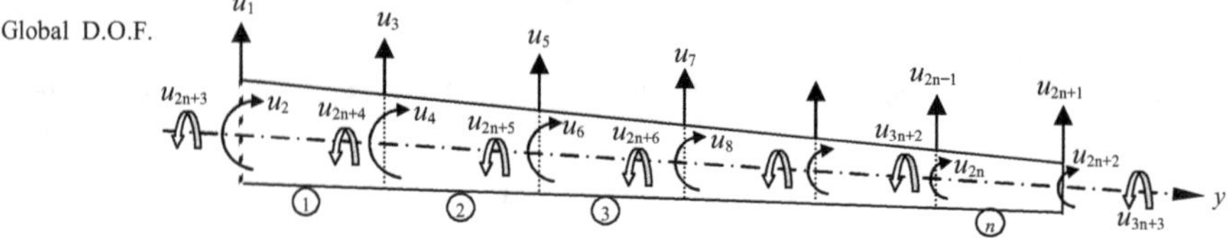

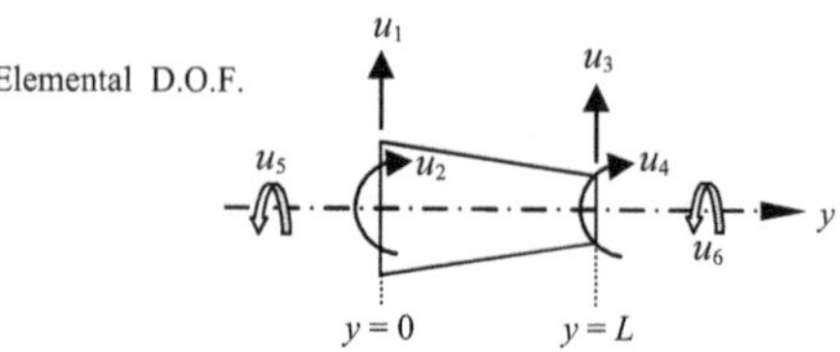

Fig. 2.10 Elemental and global degrees of freedom ($DOFs$) for a high-aspect ratio wing discretized by n, two-noded, beam-shaft finite elements

whose element Γ_{ij} identifies the location of the global degree of freedom to which the jth degree of freedom of the ith finite element contributes. In order to assemble the global matrices, the elemental stiffness and mass matrices are first expressed as follows:

$$
\mathbf{K_e} = \begin{pmatrix}
12\frac{EI}{L^3} & 6\frac{EI}{L^2} & -12\frac{EI}{L^3} & 6\frac{EI}{L^2} & 0 & 0 \\
6\frac{EI}{L^2} & 4\frac{EI}{L} & -6\frac{EI}{L^2} & 2\frac{EI}{L} & 0 & 0 \\
-12\frac{EI}{L^3} & -6\frac{EI}{L^2} & 12\frac{EI}{L^3} & -6\frac{EI}{L^2} & 0 & 0 \\
6\frac{EI}{L^2} & 2\frac{EI}{L} & -6\frac{EI}{L^2} & 4\frac{EI}{L} & 0 & 0 \\
0 & 0 & 0 & 0 & \frac{GJ}{L} & -\frac{GJ}{L} \\
0 & 0 & 0 & 0 & -\frac{GJ}{L} & \frac{GJ}{L}
\end{pmatrix},
$$

$$
\mathbf{M_e} =
$$

$$
\begin{pmatrix}
13mL/35 & 11mL^2/210 & 9mL/70 & -13mL^2/420 & 7mx_\theta L/10 & 3mx_\theta L/10 \\
11mL^2/210 & mL^3/105 & 13mL^2/420 & -mL^3/140 & mx_\theta L^2/10 & mx_\theta L^2/15 \\
9mL/70 & -13mL^2/420 & 13mL/35 & -11mL^2/210 & -3mx_\theta L/10 & 7mx_\theta L/10 \\
-13mL^2/420 & -mL^3/140 & -11mL^2/210 & mL^3/105 & -mx_\theta L^2/15 & -mx_\theta L^2/10 \\
7mx_\theta L/10 & mx_\theta L^2/10 & -3mx_\theta L/10 & -mx_\theta L^2/15 & I_\theta L/3 & I_\theta L/6 \\
3mx_\theta L/10 & mx_\theta L^2/15 & 7mx_\theta L/10 & -mx_\theta L^2/10 & I_\theta L/6 & I_\theta L/3
\end{pmatrix}.
$$

Next, the elements of the elemental matrices are added in order to yield the corresponding elements of the respective global matrices at the locations specified by the connectivity array:

$$u_{\Gamma_{ij}} = u_j^{(i)} \tag{2.148}$$

$$\mathsf{K}(\Gamma_{ij}, \Gamma_{ik}) = \sum_{i=\ell}^{\ell+1} \mathsf{K}_e^{(i)}(j,k)$$

$$\mathsf{M}(\Gamma_{ij}, \Gamma_{ik}) = \sum_{i=\ell}^{\ell+1} \mathsf{M}_e^{(i)}(j,k), \qquad \ell = 1,\ldots,n \tag{2.149}$$

The indices (j,k) locate an element of the elemental matrix with a shared degree of freedom between two adjacent elements [4], for which the summation over the corresponding elements of the elemental matrices is carried out as indicated in Eq. (2.149). Of course, if there is no shared degree of freedom, the summation is not carried out. The location of each elemental term in the assembled matrix is determined by the connectivity matrix. For the high aspect-ratio wing with Euler–Bernoulli beam-shaft approximation considered above, the assembly procedure is schematically depicted in Figs. 2.11 and 2.12 for the bending and torsional degrees of freedom, respectively. The nonzero elements are seen to fall in a diagonal band structure. Finally, the geometric boundary conditions at the root ($y = 0$) are applied to the assembled matrices by imposing the following restriction on the global degrees of freedom:

$$u_1 = u_2 = u_{2n+3} = 0. \tag{2.150}$$

This is carried out by removing the corresponding rows and columns from the assembled stiffness and mass matrices. An example of the FEM for deriving the stiffness and mass matrices is given in the following section.

2.7.3 Illustrative Example

Consider a wing planform shown in Fig. 2.13 discretized by taking 20 finite elements per semispan of equal width in the spanwise direction. The stiffness and mass parameters of the wing are the following:

$$EI = 10^6[-5(y/b) + 6.7] \text{ kg.m}^2$$

$$GJ = 10^6[-6.25(y/b) + 9.38] \text{ kg.m}^2$$

---

[4] Since the elements are arranged sequentially in a spanwise direction, only the adjacent elements can have a shared degree of freedom.

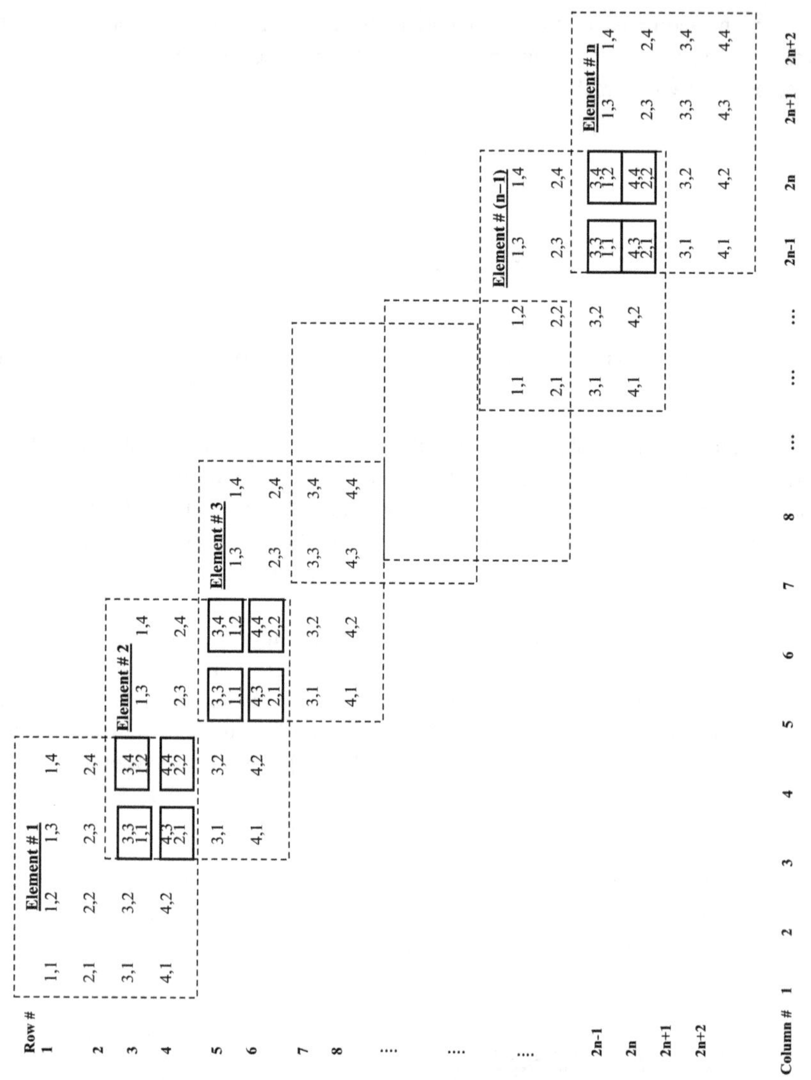

Fig. 2.11 Assembly procedure for the bending degrees of freedom for a high-aspect ratio wing discretized by n, two-noded, beam-shaft finite elements

$$m = -23(y/b) + 68 \text{ kg/m}$$

$$I_\theta = -51(y/b) + 160 \text{ kg.s}^2$$

$$x_\theta = -0.25(y/b) + 0.38 \text{ m}$$

The global stiffness and mass matrices derived by the FEM with 20, two-noded, beam-shaft finite elements are each of dimension 60×60, and are listed in Appendix A. Table 2.1 shows the in-vacuo natural frequencies of the first 12 structural vibration

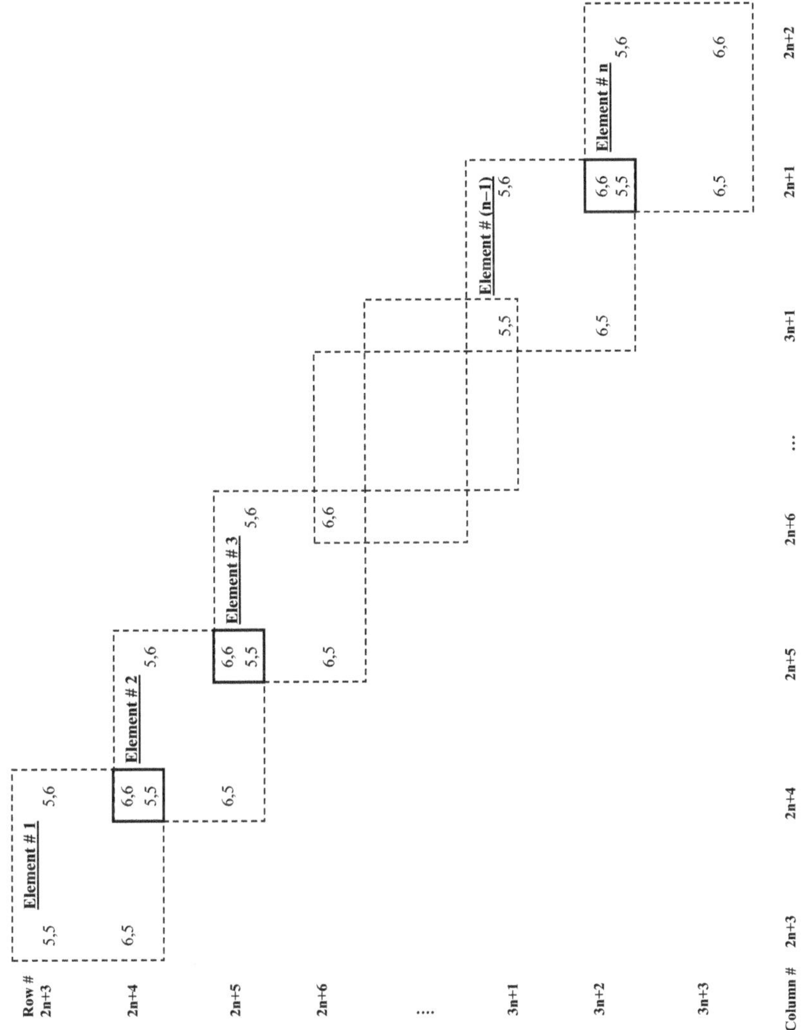

Fig. 2.12 Assembly procedure for the torsional degrees of freedom for a high-aspect ratio wing discretized by n, two-noded, beam-shaft finite elements

modes ranging from 9.7845 to 457.76 rad/s, which are computed by solving the linear eigenvalue problem (Eq. (2.63)) of the assembled structure. The bending deflection mode shapes for the first six structural modes (also computed from the eigenvalue problem) are plotted in in Fig. 2.14, while the bending slopes of the same modes are shown in Fig. 2.15. The twist angle modes shapes or the first six structural modes are plotted in Fig. 2.16. The relative magnitudes of the bending displacement and twist angle in each mode shape allow one to label it as either a bending, torsion, or mixed (bending/torsion) mode (Table 2.1).

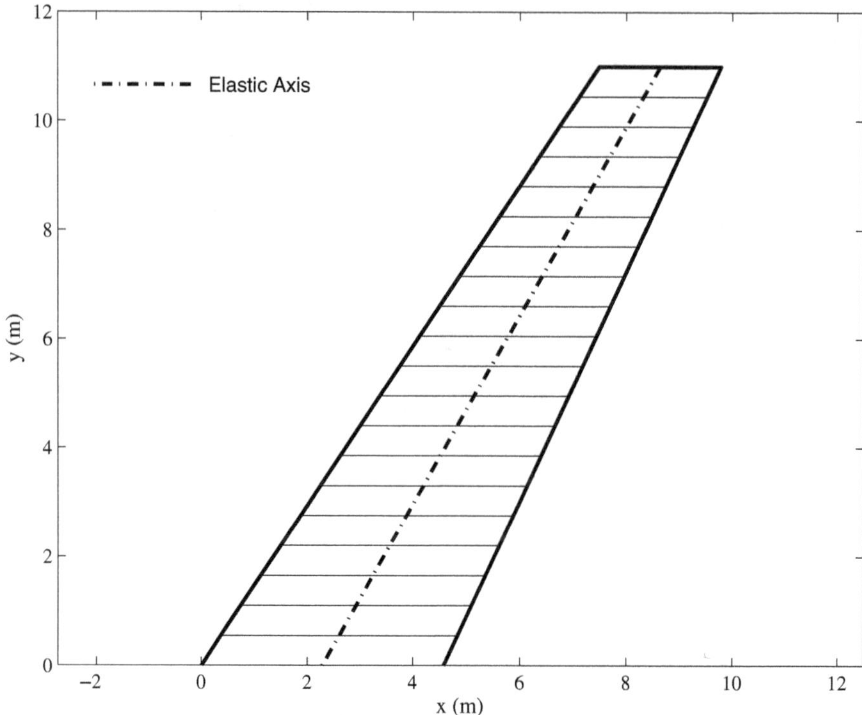

Fig. 2.13 Planform geometry of a high-aspect ratio wing discretized by 20, two-noded, beam-shaft finite elements

Table 2.1 Structural vibration modes

No.	Type	Natural frequency (rad/s)
1	Bending	9.7845
2	Torsion	35.036
3	Bending	52.371
4	Torsion	93.797
5	Bending/torsion	134.23
6	Bending/torsion	161.78
7	Torsion	210.02
8	Bending/torsion	266.09
9	Bending/torsion	291.19
10	Bending/torsion	337.59
11	Torsion	405.05
12	Bending	457.76

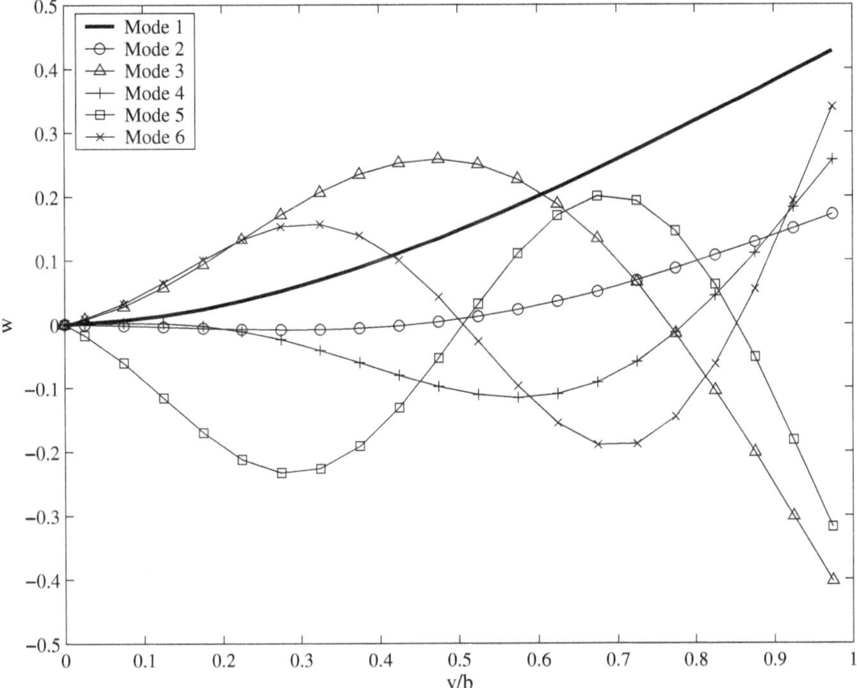

Fig. 2.14 Bending deflection mode shapes for the first six structural modes for the high-aspect ratio wing of Fig. 2.13

2.7.4   Plate Bending Elements

The Euler–Bernoulli beam-shaft model given above cannot be applied to wings, tails, and fins of small aspect ratio where chordwise and spanwise bendings are of similar magnitudes. This means that approximating the deformation by a linear combination of spanwise bending and chordwise rigid sections twisting about an elastic axis is no longer valid. In such a case, resort has to be made to a more rigorous approach of a plate theory. Thus while the Euler–Bernoulli model involves unidimensional finite elements, a plate model is essentially two-dimensional in nature, and, therefore, requires a more sophisticated set of shape functions.

Plates are thin, solid structures bounded by nearly parallel planes, and are mathematically represented as follows:

$$\Omega = \{(x, y, z) \in \mathcal{R}^3 | z \in [-d/2, d/2], (x, y) \in A \subset \mathcal{R}^2, \tag{2.151}$$

where $d(x, y)$ denotes the plate's local thickness and A its area. For convenience of notation, the deflections in the Cartesian directions are denoted by the individual component's subscript, i.e., $(\delta_x, \delta_y, \delta_z)$. The angular rotations of fibers, initially normal to the plate's mid surface about the $-y$ and x axes, are denoted by α and β,

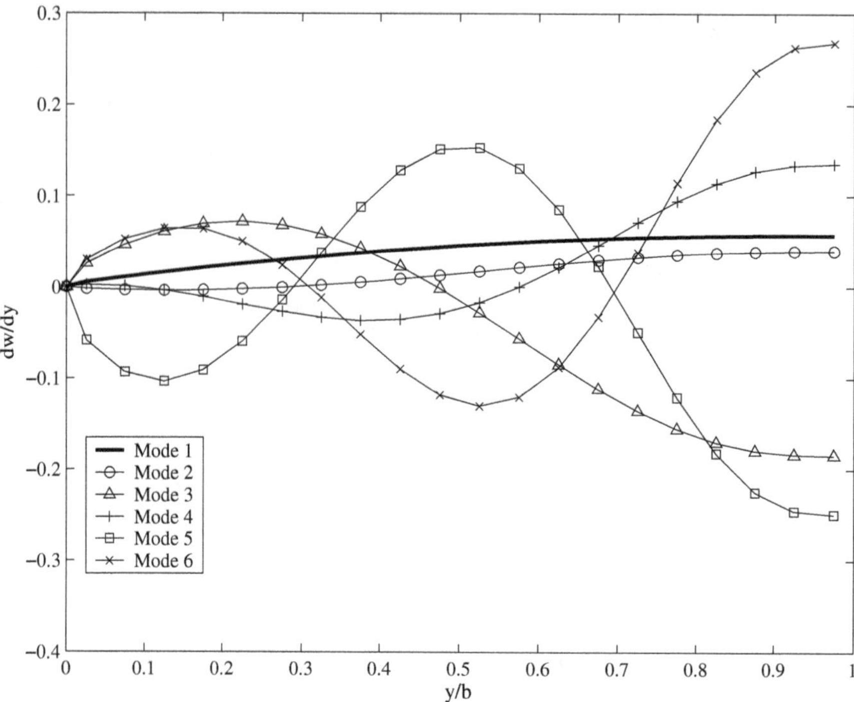

Fig. 2.15 Bending slope mode shapes for the first six structural modes for the high-aspect ratio wing of Fig. 2.13

respectively, and shear strains in the corresponding directions are denoted by γ_α and γ_β, respectively (as indicated in Fig. 2.17a for $-y$ rotation).

There are two types of plate theories in vogue [138]: (a) Poisson–Kirchhoff (or classical) plate theory (CPT), and (b) Reissner–Mindlin plate theory. While (a) is an extension of the Euler–Bernoulli model from two to three dimensions, and does not allow for any shear deformations, (b) is a more elaborate model that includes transverse shear deformations. Since most wing planforms of small aspect ratio have significant shearing stresses, the Reissner–Mindlin model is expected to produce a greater accuracy in such cases.

The relevant assumptions of the Reissner–Mindlin [75] plate theory are the following:

1. The stress normal to the plate mid surface is zero ($\sigma_{zz} = 0$), which implies a *plain stress* hypothesis.
2. The bending deflection is given by the deflection normal to the plate, which is taken to be constant across the thickness, $w(x, y, t) = \delta_z(x, y, z, t)$.
3. Plane sections remain plane after deformation. This implies that the fibers initially normal to the plate's mid surface remain straight, but are rotated about the $-y$ and x axes by angles α and β, respectively (Fig. 2.17a).

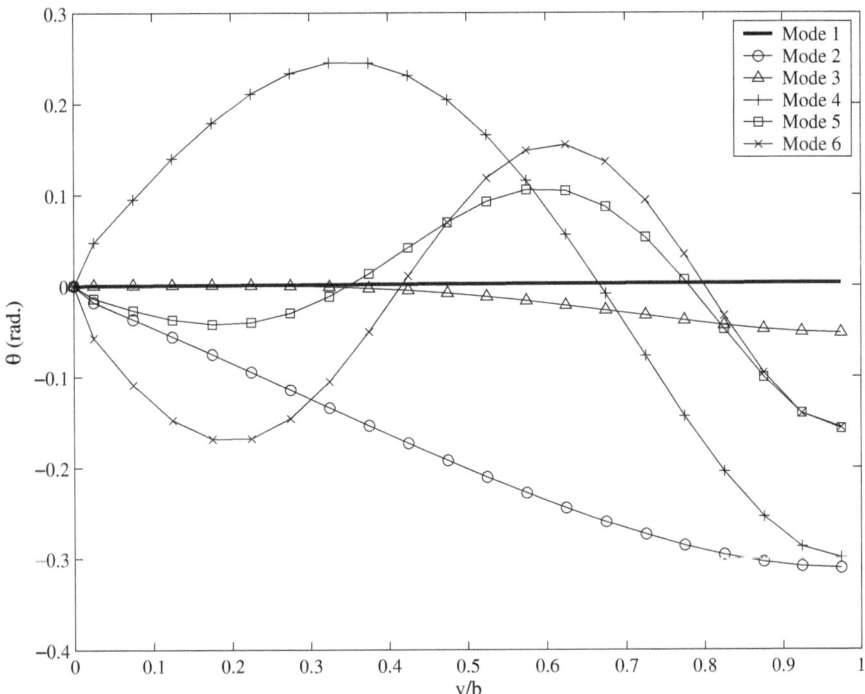

Fig. 2.16 Twist angle mode shapes for the first six structural modes for the high-aspect ratio wing of Fig. 2.13

While the first assumption contradicts the second, it offers a major simplification which is accurate for thin plates. It is to be noted that plate theories cannot be entirely consistent with the exact three-dimensional models, but are only useful approximations. Of course, there is no allowance for any shear deformation in the CPT theory, and thus there is no difference between α and w'_α (or between β and w'_β) shown in Fig. 2.17a. Consequently, the CPT approach is easier to model, and the discussion here will be confined to it as a logical extension of the Euler–Bernoulli beam theory. For the Reissner–Mindlin plate theory, the reader is referred to Chap. 5 of Hughes [75], or to Reddy [138].

The bending displacements for CPT are given by

$$\delta_x(x, y, z, t) = u_x(x, y, t) - z\frac{\partial w}{\partial x}$$

$$\delta_y(x, y, z, t) = u_y(x, y, t) - z\frac{\partial w}{\partial y} \qquad (2.152)$$

$$\delta_z(x, y, z, t) = u_z(x, y, t) = w(x, y, t),$$

and result in the following strain field:

$$\epsilon_{zz} = \epsilon_{xz} = \epsilon_{yz} = 0.$$

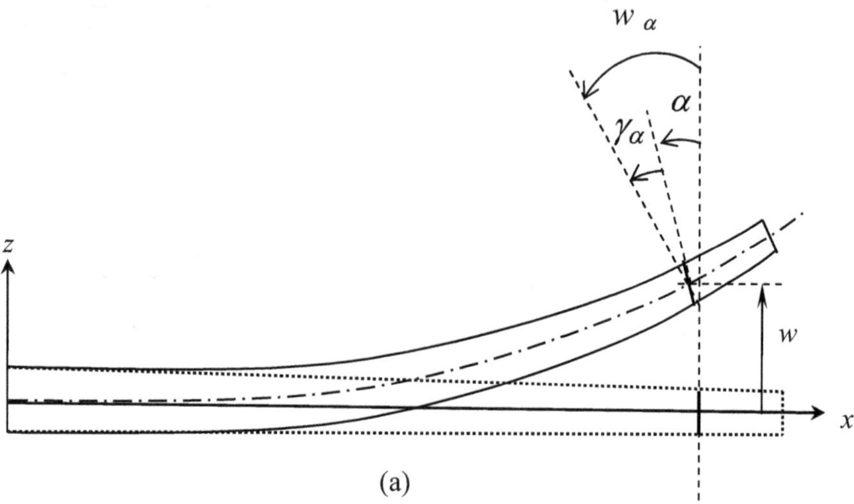

(a)

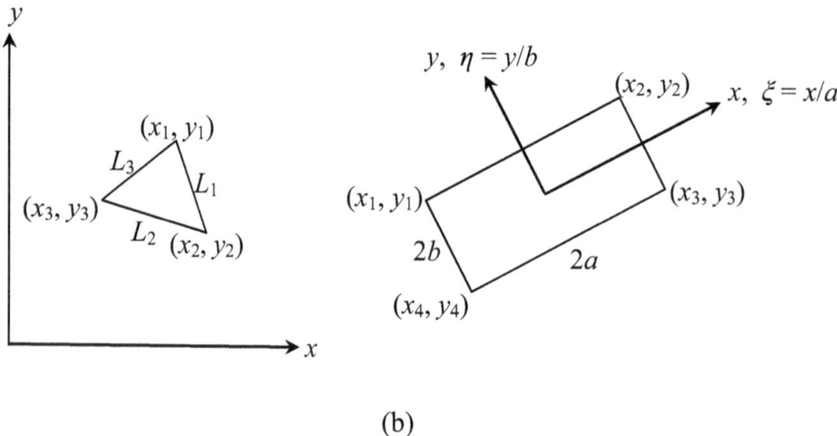

(b)

Fig. 2.17 Schematic diagram showing **a** plate bending and shear deformation geometry, and **b** plate bending elements for the CPT theory

The solution of the plate problem is to find the displacements, u_x, u_y, w, which are defined for the mid plane ($z = 0$). However, since we are primarily interested in solving for the bending deformation $w(x, y, t)$, we only consider the following strain field produced by the bending displacement:

$$\left\{ \begin{array}{c} \epsilon_{xx} \\ \epsilon_{yy} \\ \epsilon_{xy} \end{array} \right\} = -z \left\{ \begin{array}{c} \frac{\partial^2 w}{\partial x^2} \\ \frac{\partial^2 w}{\partial y^2} \\ \frac{\partial^2 w}{\partial x \partial y} \end{array} \right\} \tag{2.153}$$

The principle of virtual work results in the weak form of the bending problem, and the following FEM [138]:

$$\mathbf{M}^e \ddot{\mathbf{q}}^e + \mathbf{K}^e \mathbf{q}^e = \mathbf{F}^e + \mathbf{Q}^e, \tag{2.154}$$

where $\mathbf{q}^e(t) = (w, \partial w/\partial x, \partial w/\partial y)^T$ denotes the generalized displacement vector evaluated at elemental nodes,

$$w(x, y, t) \simeq \sum_{i=1}^{n} q_i(t) N_i(x, y), \tag{2.155}$$

$N_i(x, y)$ are shape interpolation functions for the plate bending element, and the corresponding elemental matrices and force vectors are given by their following respective elements:

Mass matrix:

$$M_{ij}^e = \int_{\Omega_e} \left[J_0 N_i N_j + J_2 \left(\frac{\partial N_i}{\partial x} \frac{\partial N_j}{\partial x} + \frac{\partial N_i}{\partial y} \frac{\partial N_j}{\partial y} \right) \right] dx \, dy \tag{2.156}$$

Stiffness matrix:

$$K_{ij}^e = \int_{\Omega_e} \left[D_{11} \frac{\partial^2 N_i}{\partial x^2} \frac{\partial^2 N_j}{\partial x^2} + D_{12} \left(\frac{\partial^2 N_i}{\partial x^2} \frac{\partial^2 N_j}{\partial y^2} + \frac{\partial^2 N_i}{\partial y^2} \frac{\partial^2 N_j}{\partial x^2} \right) \right.$$
$$\left. + D_{22} \frac{\partial^2 N_i}{\partial y^2} \frac{\partial^2 N_j}{\partial y^2} + 4 D_{66} \frac{\partial^2 N_i}{\partial x \partial y} \frac{\partial^2 N_j}{\partial x \partial y} \right] dx \, dy \tag{2.157}$$

Transverse force due to distributed load $\ell(x, y, t)$:

$$F_i^e = \int_{\Omega_e} N_i \ell dx \, dy \tag{2.158}$$

Transverse force due to boundary loads (shear force V_n, and bending moment M_n applied normal to the elemental edge boundary Γ_e with local normal $\mathbf{n}$):

$$Q_i^e = \int_{\Gamma_e} \left(-M_n \frac{\partial N_i}{\partial n} + V_n N_i \right) ds \tag{2.159}$$

Here, the constitutive (material) constants of the plate are the following:

$$J_0 = \int_{d/2}^{d/2} \rho dz \; ; \qquad J_2 = \int_{d/2}^{d/2} \rho z^2 dz \tag{2.160}$$

$$D_{11} = \frac{E_1 d^3}{12(1 - \nu_{12}\nu_{21})}$$

$$D_{22} = \frac{E_2 d^3}{12(1 - \nu_{12}\nu_{21})}$$

$$D_{12} = \frac{\nu_{12}E_2 d^3}{12(1 - \nu_{12}\nu_{21})} \tag{2.161}$$

$$D_{66} = \frac{G_{12}d^3}{12},$$

where $\rho(x, y)$ is the mass per unit area, $E_i, i = 1, 2$ are the Young's moduli in the respective directions (x, y) indicated by the subscript i, G_{12} is the shear modulus in the xy plane, and ν_{ij} is the Poisson ratio representing the negative ratio of the transverse strain in the j direction to the strain in the i direction, caused by a stress in the j direction. The elasticity coefficients are related as follows:

$$\nu_{21} = \nu_{12}\frac{E_2}{E_1}$$

The plate bending elements commonly used in CPT theory are the three-noded triangular and four-noded rectangular elements, as depicted in Fig. 2.17b. These elements have 3 degrees of freedom at each node, $(w, -\partial w/\partial y, \partial w/\partial x)$, and therefore the triangular element has a total of 9 degrees of freedom while the rectangular element has 12. For example, the interpolation for the triangular element is given by

$$w = a_1 + a_2 x + a_3 y + a_4 xy + a_5 x^2 + a_6 y^2$$
$$+ a_7(x^2 y + y^2 x) + a_8 x^3 + a_9 y^3, \tag{2.162}$$

which is regarded as an incomplete polynomial, because of the shared coefficient of the terms $x^2 y$ and $y^2 x$. Due to the cubic variation, the slope $\partial w/\partial x$ can be discontinuous across the boundaries of two consecutive elements. Furthermore, the derivative $\partial^2 w/(\partial x \partial y)$ can be multivalued at the corner points. Such an interpolation, which does not strictly satisfy the geometric compatibility conditions, is said to be *nonconformal*. This is also the case with the rectangular, four-noded elements with 3 degrees of freedom per node. However, if sufficiently large number of elements are employed, nonconformal elements produce accurate results. An example of the shape interpolation functions for the triangular CPT element is the following:

$$N_1 = L_1 + L_1^2 L_2 + L_1^2 L_3 - L_1 L_2^2 - L_1 L_3^2$$
$$N_2 = x_{31}(L_3 L_1^2 - L_1 L_2 L_3/2) - x_{12}(L_2 L_1^2 + L_1 L_2 L_3/2)$$
$$N_3 = y_{31}(L_3 L_1^2 + L_1 L_2 L_3/2) - y_{12}(L_2 L_1^2 + L_1 L_2 L_3/2)$$
$$N_4 = L_2 + L_2^2 L_3 + L_2^2 L_1 - L_2 L_3^2 - L_2 L_1^2 \tag{2.163}$$
$$N_5 = x_{12}(L_1 L_2^2 - L_1 L_2 L_3/2) - x_{23}(L_3 L_2^2 + L_1 L_2 L_3/2)$$
$$N_6 = y_{12}(L_1 L_2^2 + L_1 L_2 L_3/2) - y_{23}(L_3 L_2^2 + L_1 L_2 L_3/2)$$
$$N_7 = L_3 + L_3^2 L_1 + L_3^2 L_2 - L_3 L_1^2 - L_3 L_2^2$$
$$N_8 = x_{23}(L_2 L_3^2 - L_1 L_2 L_3/2) - x_{31}(L_1 L_3^2 + L_1 L_2 L_3/2)$$
$$N_9 = y_{23}(L_2 L_3^2 + L_1 L_2 L_3/2) - y_{31}(L_1 L_3^2 + L_1 L_2 L_3/2)$$

Here $L_i, i = 1, 2, 3$ are the respective lengths of the three sides, $(x_i, y_i), i = 1, 2, 3$ are the nodal (corner) coordinates, $x_{ij} = x_i - x_j$, and $y_{ij} = y_i - y_j$.

The use of triangular CPT elements can give accurate results for most problems [138], except for a highly curved planform shape, for which convergence issues could arise. A combination of triangular and rectangular elements (with appropriate shape functions) can also be employed for complicated geometries. A better accuracy can be achieved by employing the *conforming* rectangular CPT elements, which have the following 4 degrees of freedom per node (total 16 degrees of freedom):

$$w, \frac{\partial w}{\partial x}, \frac{\partial w}{\partial y}, \frac{\partial^2 w}{\partial x \partial y}$$

The shape functions of the conforming rectangular element are given by

$$\begin{aligned}
N_i &= g_{i1} \quad (i = 1, 5, 9, 13) \\
N_i &= g_{i2} \quad (i = 2, 6, 10, 14) \\
N_i &= g_{i3} \quad (i = 3, 7, 11, 15) \\
N_i &= g_{i4} \quad (i = 4, 8, 12, 16),
\end{aligned} \tag{2.164}$$

where

$$\begin{aligned}
g_{i1} &= \frac{1}{16}(\xi + \xi_i)^2(\xi_0 - 2)(\eta + \eta_i)^2(\eta_0 - 2) \\
g_{i2} &= \frac{1}{16}\xi_i(\xi + \xi_i)^2(1 - \xi_0)(\eta + \eta_i)^2(\eta_0 - 2) \\
g_{i3} &= \frac{1}{16}\eta_i(\xi + \xi_i)^2(\xi_0 - 2)(\eta + \eta_i)^2(1 - \eta_0) \\
g_{i4} &= \frac{1}{16}\xi_i\eta_i(\xi + \xi_i)^2(1 - \xi_0)(\eta + \eta_i)^2(1 - \eta_0).
\end{aligned} \tag{2.165}$$

The coordinates are rendered nondimensional by dividing by half side of the rectangle in each direction (Fig. 2.17b), $\xi = (x - x_c)/a$, $\eta = (y - y_c)/b$, where (x_c, y_c) is the centroid, the subscript i refers to the corner (node) points, $\xi_0 = \xi\xi_i$, and $\eta_0 = \eta\eta_i$.

However, it is not expected that anything beyond triangular elements will be necessary for modeling low-aspect ratio wing planforms. Assemblage of the elemental matrices and nodal vectors into corresponding global ones must account for the boundary conditions, as in the case of a beam-shaft idealization. A further discussion of plate elements is beyond our present scope, and the reader is referred to a specialized textbook on this topic [138].

Chapter 3
Unsteady Aerodynamic Modeling

3.1 Introduction

An aeroservoelastic (ASE) model requires the representation of the generalized unsteady aerodynamic forces, $\mathbf{Q}(t)$, as functions of the generalized structural displacements, $\mathbf{q}(t)$, which describe the structural dynamics. For an arbitrary structural motion, derivation of unsteady aerodynamic forces and moments often requires a *computational fluid dynamics* (CFD) model. However, when dealing with small displacements of a thin lifting surface (which is the objective of ASE modeling), one would like to employ a linear operational relationship between the approximate unsteady airloads and the motion variables, such as

$$\mathbf{Q} = \mathbf{D}\,(\mathbf{q}), \tag{3.1}$$

where $\mathbf{D}(.)$ is an *aerodynamic differential operator matrix*, which depends upon the geometry and flow properties. A linear aerodynamic representation exemplified by Eq. (3.1) not only enables a systematic ASE analysis, but also breaks down much more complex geometries and flow patterns into a set of much simpler problems by linear superposition. The present chapter is concerned with devising appropriate aerodynamic models for typical flow conditions that are representative of modern airplane flight, namely subsonic, transonic, and supersonic flight regimes.

The most useful form of Eq. (3.1) is derived for a wing-like surface from a linear superposition of elementary flat-plate (or flat-panel) solutions to the governing partial differential equation of unsteady aerodynamics, resulting in an integral equation of the following form:

$$w(x, y, t) = \int_S K[(x, y : \xi, \eta), t] \Delta p(\xi, \eta, t) \mathrm{d}\xi \, \mathrm{d}\eta, \tag{3.2}$$

where Δp is the pressure difference between the upper and lower faces of the wing's mean surface, $S(x, y) = 0$, at a given point, (ξ, η), and w is the flow component normal to the mean surface (z-component) called the *upwash* (Fig. 3.1). The *kernel function* (or Green's function), $K[(x, y : \xi, \eta), t]$, represents the discrete influence coefficient of upwash induced at *collocation point* (x, y) due to a unit concentrated

© Springer Science+Business Media, LLC 2015

A. Tewari, *Aeroservoelasticity*, Control Engineering,

DOI 10.1007/978-1-4939-2368-7_3

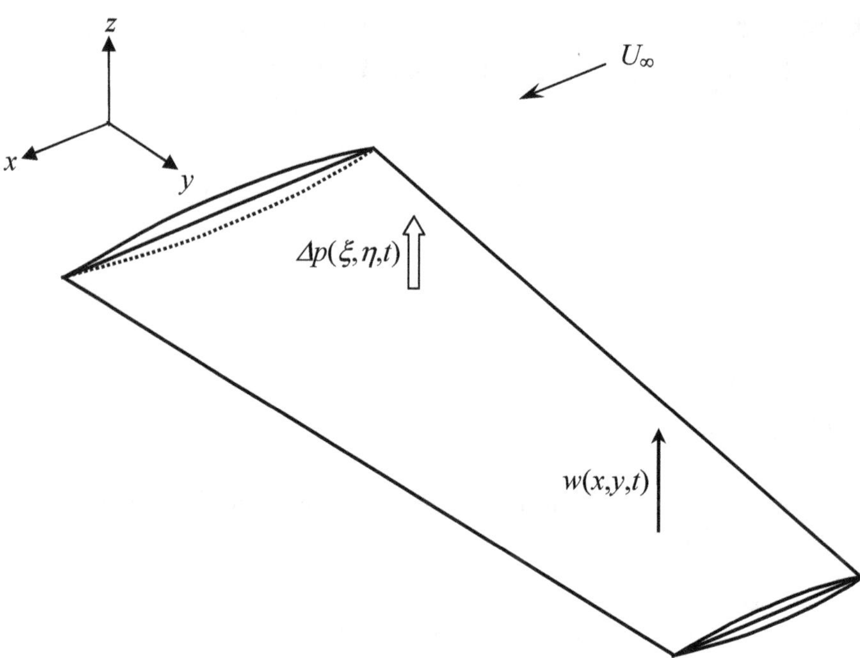

Fig. 3.1 A thin wing placed in a uniform flow of freestream speed U_∞

pressure load, $\Delta p\,\mathrm{d}\xi\,\mathrm{d}\eta$, applied at the *load point* (ξ,η). The pressure-upwash integral equation, Eq. (3.2), representing the linear aerodynamic model is analogous to the linear load-deflection model of the structure derived in Chap. 2, and taken together, the two approximations produce a linear aeroelastic model. The aerodynamic integral equation must be solved for the unsteady pressure distribution, yielding the net loads, $\mathbf{Q}(t)$, on the structure, due to a prescribed upwash distribution resulting from the structural motion, $\mathbf{q(t)}$. If the structural deflections are measured from a steady state (or static equilibrium) condition, the linear aerodynamic model allows a superimposition of the unsteady solution on that for the steady flow condition. Consequently, the unsteady flow past a wing with camber and thickness can be modeled as the unsteady flow due to time-dependent displacements of the mean surface, superimposed steady positions. This is the main advantage of having a linear aerodynamic model.

Solution to the integral equation, Eq. (3.2), is unavailable in a closed form, and thus requires a numerical approximation based upon the discretization of the surface integral into a number of panels. The pressure and upwash in each panel is represented by elementary solutions to the governing flowfield equation. For example, in an inviscid subsonic or supersonic flow past an oscillating surface, elementary solutions take the form of either velocity or acceleration potential sources, sinks, and doublets, which can be distributed in each panel in a variety of ways. Such a discretization of the flowfield is quite valuable in ASE modeling and is called a *panel method*. Many panel methods have been devised for determining the unsteady aerodynamic loads on oscillating lifting surfaces.

3.2   Governing Equations

It is necessary to begin from first principles while trying to understand the aerodynamic modeling procedure. The reader is advised to consult textbooks on fluid mechanics [153] and gas dynamics [155] for fundamental concepts. Fluid mechanics models derive the governing equations in two possible ways: (a) Lagrangian model, which considers the change in the properties of fluid elements as they pass through fixed points in space, and (b) Eulerian model, in which fluid elements are continuously tracked as they move through the space. While there is no basic difference in the two models, they follow different terminologies, and hence the resulting equations are in different forms.

3.2.1   Viscous Flow

Navier–Stokes (N–S) equations govern the flow of a viscous fluid [153], and are derived using the principles of conservation of mass (continuity), momentum, and energy, given in a body-fixed coordinate frame, (x, y, z). These are respectively the following:

- *Continuity*

$$\frac{\partial \rho}{\partial t} + \frac{\partial (\rho u)}{\partial x} + \frac{\partial (\rho v)}{\partial y} + \frac{\partial (\rho w)}{\partial z} = 0 \tag{3.3}$$

- *x-Momentum*

$$\frac{\partial (\rho u)}{\partial t} + \frac{\partial (\rho u^2)}{\partial x} + \frac{\partial (\rho uv)}{\partial y} + \frac{\partial (\rho uw)}{\partial z} = -\frac{\partial p}{\partial x} + \frac{\partial \tau_{xx}}{\partial x} + \frac{\partial \tau_{xy}}{\partial y} + \frac{\partial \tau_{xz}}{\partial z} \tag{3.4}$$

- *y-Momentum*

$$\frac{\partial (\rho v)}{\partial t} + \frac{\partial (\rho uv)}{\partial x} + \frac{\partial (\rho v^2)}{\partial y} + \frac{\partial (\rho vw)}{\partial z} = -\frac{\partial p}{\partial y} + \frac{\partial \tau_{yy}}{\partial y} + \frac{\partial \tau_{xy}}{\partial x} + \frac{\partial \tau_{yz}}{\partial z} \tag{3.5}$$

- *z-Momentum*

$$\frac{\partial (\rho w)}{\partial t} + \frac{\partial (\rho uw)}{\partial x} + \frac{\partial (\rho vw)}{\partial y} + \frac{\partial (\rho w^2)}{\partial z} = -\frac{\partial p}{\partial z} + \frac{\partial \tau_{yz}}{\partial y} + \frac{\partial \tau_{xz}}{\partial x} + \frac{\partial \tau_{zz}}{\partial z} \tag{3.6}$$

- *Energy*

$$\frac{\partial (\rho h)}{\partial t} + \frac{\partial (\rho uh)}{\partial x} + \frac{\partial (\rho vh)}{\partial y} + \frac{\partial (\rho wh)}{\partial z} = \frac{\partial p}{\partial t} + u\frac{\partial p}{\partial x} + v\frac{\partial p}{\partial y} + w\frac{\partial p}{\partial z}$$

$$
\begin{aligned}
&+ \tau_{xx}\frac{\partial u}{\partial x} + \tau_{yy}\frac{v}{y} + \tau_{xy}\left(\frac{\partial u}{\partial y} + \frac{\partial v}{\partial x}\right) \\
&+ \tau_{xz}\left(\frac{\partial u}{\partial z} + \frac{\partial w}{\partial x}\right) \\
&+ \tau_{yz}\left(\frac{\partial v}{\partial z} + \frac{\partial w}{\partial y}\right) \\
&+ \tau_{zz}\frac{\partial w}{\partial z} \\
&- \frac{\partial q_x}{\partial y} - \frac{\partial q_x}{\partial z} - \frac{\partial q_y}{\partial x} - \frac{\partial q_y}{\partial z} \\
&- \frac{\partial q_z}{\partial x} - \frac{\partial q_z}{\partial y}.
\end{aligned}
\tag{3.7}
$$

Here (u, v, w) denote the flow velocity components in the (x, y, z) directions, respectively, ρ is the density, p the static pressure, h the specific enthalpy, $(\tau_{xx}, \tau_{xy}, \tau_{xz}, \tau_{yy}, \tau_{yz}, \tau_{zz})$ the elements of symmetric stress tensor,

$$
\begin{aligned}
\tau_{xx} &= (\lambda + 2\mu)\frac{\partial u}{\partial x} \\
\tau_{xy} &= \mu\left(\frac{\partial u}{\partial y} + \frac{\partial v}{\partial x}\right) \\
\tau_{xz} &= \mu\left(\frac{\partial u}{\partial z} + \frac{\partial w}{\partial x}\right) \\
\tau_{yy} &= (\lambda + 2\mu)\frac{\partial v}{\partial y} \\
\tau_{yz} &= \mu\left(\frac{\partial v}{\partial z} + \frac{\partial w}{\partial y}\right) \\
\tau_{zz} &= (\lambda + 2\mu)\frac{\partial w}{\partial z},
\end{aligned}
\tag{3.8}
$$

and (q_x, q_y, q_z) the heat-flux components depending upon variation of static temperature T,

$$
\begin{aligned}
q_x &= -k\frac{\partial T}{\partial x} \\
q_y &= -k\frac{\partial T}{\partial y} \\
q_z &= -k\frac{\partial T}{\partial z}.
\end{aligned}
\tag{3.9}
$$

The basic variables governing the flow are thus (u, v, w, ρ, p, h, T), and are called flow variables. The fluid properties are defined by the coefficient of dynamic viscosity, μ, the coefficient of bulk viscosity, $\lambda = -\frac{2}{3}\mu$, and the thermal conductivity coefficient, k. These could depend upon temperature, T, if the temperature is high. If the flow is turbulent, there is a variation of the viscosity and thermal conductivity coefficients with the flow variables, which can be described by statistical equations called *turbulence models*.

Since there are seven flow variables and only five conservation equations, other relationships are necessary before a solution of the fluid flow can be derived. These additional relationships are obtained by assuming the flowing gas to be thermo-dynamically perfect (which is a reasonable assumption for air under normal flight applications), and expressed as follows:

$$p = \rho R T \tag{3.10}$$

$$h = c_p T, \tag{3.11}$$

where R is the specific gas constant (287 J/(kg.K) for air) and c_p the constant-pressure specific heat.

N–S equations must be solved for the unsteady flowfield defined by the flow variables, whenever an accurate model of the viscous flow effects is necessary, such as in large amplitude oscillations involving unsteady flow separation, turbulence, and shock-wave/boundary-layer interaction. However, N–S solutions are computationally expensive, requiring memory-intensive, iterative numerical techniques, even for simple geometries. Their application to ASE applications involving a rapidly changing flowfield due to a dynamically deforming boundary is currently infeasible, but is an area of active research. Hence, in a practical situation, a reasonable approximation to N–S equations is quite desirable.

With regard to obtaining a simplified flow model, the following categorization of viscous flows is useful:

1. Separated flow, which is governed by essentially nonlinear aerodynamic relationships, thereby requiring N–S solutions.
2. Partially separated flow, which can be handled by a *boundary-layer* approximation consisting of an outer inviscid region and an inner viscous (or shear) layer. The inner viscous region, although nonlinear, is governed by ordinary (rather than partial) differential equations, which are simpler to solve than N-S equations. The outer inviscid region for a thin, slender body is governed by linear aerodynamic relationships. Flows of this type are inherently unsteady, and thus capable of causing nonlinear aeroelastic effects, such as buffet and aileron buzz.
3. Attached flow, where the effects of viscosity can be neglected while calculating the pressure distributions required for aeroelastic analyses. In such flows, the relationships between generalized aerodynamic forces and small deflections of a thin, slender body due to rigid and structural deformations are essentialy linear.

3.2.2   *Inviscid Flow*

The most common approximation for aeroelastic applications is that of attached flow (Type 3 above) with small perturbations. Since viscous effects are of secondary importance in such cases, one can often drop viscosity and thermal conductivity from the N–S equations, leading to the following *Euler equations*, which govern inviscid flows:

$$
\frac{\partial}{\partial t}
\begin{Bmatrix} \rho \\ \rho u \\ \rho v \\ \rho w \\ e \end{Bmatrix}
+ \frac{\partial}{\partial x}
\begin{Bmatrix} \rho u \\ \rho u^2 + p \\ \rho uv \\ \rho uw \\ u(e + p) \end{Bmatrix}
+ \frac{\partial}{\partial y}
\begin{Bmatrix} \rho v \\ \rho uv \\ \rho v^2 + p \\ \rho vw \\ v(e + p) \end{Bmatrix}
$$

$$
+ \frac{\partial}{\partial z}
\begin{Bmatrix} \rho w \\ \rho uw \\ \rho vw \\ \rho w^2 + p \\ w(e + p) \end{Bmatrix}
= \begin{Bmatrix} 0 \\ 0 \\ 0 \\ 0 \\ 0 \end{Bmatrix}.
\tag{3.12}
$$

Here $e = h - p/\rho$ is the specific internal energy for which the perfect gas relations, Eqs. (3.10) and (3.11), yield the following thermodynamic relationship:

$$
p = (\gamma - 1)\left[e - \frac{1}{2}(u^2 + v^2 + w^2) \right],
\tag{3.13}
$$

where

$$
\gamma = \frac{c_p}{c_v} = \frac{c_p}{c_p - R}
\tag{3.14}
$$

is the ratio of specific heats.

Euler equations are much simpler in form than the N–S equations, but they retain the nonlinear character of the latter. Therefore, Euler equations can capture the essential unsteady aerodynamic phenomena that are necessary for deriving workable ASE models. These include pressure distributions due to moving boundaries and shock waves. The isentropic and shock wave relations are solutions of Euler equations. Thus it is not necessary to model shock waves separately since they arise naturally out of Euler solutions. However, Euler solvers do require certain degree of care and complexity. The typical unsteady Euler solver is based upon a time-marching, finite-difference (or finite-volume) procedure, such as explicit (MacCormack, Lax-Wendroff) or implicit (ADI) methods [166]. The primitive flow variables, u, w, ρ, e, must be solved in an iterative manner over a computational domain discretized into

a large number of grid points for a realistic problem. Furthermore, the convergence
and stability of an Euler solution algorithm are problematic due to the nondissipative
nature of the inviscid momentum flux terms, particularly so, if strong shock waves
are present in the flowfield. This typically requires the introduction of artificial viscosity (or entropy condition) into a solution procedure. Even more importantly, Euler
equations have nonunique solutions, which must be resolved by a proper application
of tangential boundary condition on the solid surface. This is termed *closure* and
typically takes the form of *Kutta condition* at the trailing edge which determines
circulation around an airfoil. Another problem is the generation of body conforming
grids for Euler solver when the solid boundaries and shock waves are moving, as in
an unsteady application. Therefore, much of the advantage of simplified governing
equations is lost in having to devise a sophisticated numerical scheme. The unsteady
Euler equations are thus of limited utility in ASE modeling, and must be simplified
further.

3.2.3   Potential Flow

In order to gain further insight into the physical characteristics of inviscid flow, we
define *specific entropy*, S, as an additional thermodynamic variable by the following
Gibbs relation [155]:

$$T\,\mathrm{d}S = \mathrm{d}h - \frac{1}{\rho}\mathrm{d}p. \tag{3.15}$$

Entropy is a measure of disorder in the flowfield, which by Gibbs relation is seen
to increase with heat transfer (enthalpy gradient) and the presence of strong shock
waves (large pressure gradient). When substituted into the momentum equations,
the Gibbs relation and the continuity equation yield the following important result,
called *unsteady Crocco's equation* :

$$T\nabla S + \mathbf{U} \times \mathbf{\Omega} = \nabla h_0 + \frac{\partial \mathbf{U}}{\partial t}, \tag{3.16}$$

where $\mathbf{U} = (u, v, w)^T$ is the *velocity vector*,

$$\mathbf{\Omega} = -\nabla \times \mathbf{U} \tag{3.17}$$

is the *vorticity vector*, which is a measure of rotation at a given point in the flowfield,

$$\nabla(.) = \left\{ \begin{array}{c} \partial(.)/\partial x \\ \partial(.)/\partial y \\ \partial(.)/\partial z \end{array} \right\} \tag{3.18}$$

is the gradient operator, and

$$h_0 = h + \frac{1}{2}(u^2 + v^2 + w^2) = h + \frac{1}{2}\mathbf{U} \cdot \mathbf{U} \tag{3.19}$$

is the *stagnation enthalpy*. If the flow is steady, adiabatic (h_0 =const.), and irrotational ($\mathbf{\Omega = 0}$) at all points, then Crocco's equation implies a constant entropy flowfield (*isentropic* flow). The condition of irrotational flow requires that the velocity vector must be the gradient of a scalar function, Φ, called the *velocity potential*:

$$\mathbf{U} = \nabla\Phi = \left\{ \begin{array}{c} \partial\Phi/\partial x \\ \\ \partial\Phi/\partial y \\ \\ \partial\Phi/\partial z \end{array} \right\} = \left\{ \begin{array}{c} \Phi_x \\ \\ \Phi_y \\ \\ \Phi_z \end{array} \right\}. \tag{3.20}$$

However, even if the flow is unsteady, it can still be isentropic as long as it is irrotational and the following condition is satisfied by the velocity potential:

$$h_0 = -\frac{\partial\Phi}{\partial t} = -\Phi_t. \tag{3.21}$$

This can be verified by substituting Eqs. (3.19)–(3.21) into Eq. (3.16). For an isentropic flow of a perfect gas, Eqs. (3.12)–(3.15) imply that

$$\frac{p}{\rho^\gamma} = \text{const.} \tag{3.22}$$

Isentropic flow is a special case of *barotropic flow* in which the pressure is a function of only the density. The momentum equation of such a potential flow can be directly integrated in order to yield the following *unsteady Bernoulli* equation:

$$\Phi_t + \frac{U^2}{2} + \int \frac{dp}{\rho} = 0. \tag{3.23}$$

The speed at which infinitesimal pressure disturbances move in an otherwise undisturbed medium is called the *speed of sound*, which for a perfect gas is given by the following isentropic relation :

$$a = \left[\frac{\partial p}{\partial \rho}\right]_{\text{isentropic}} = \sqrt{\gamma RT}. \tag{3.24}$$

The nondimensional flow parameter governing the compressible flow is *Mach number*, M, defined as the ratio of flow speed, U, and the speed of sound,

$$M = \frac{\sqrt{u^2 + v^2 + w^2}}{a} = \frac{U}{a}. \tag{3.25}$$

Aeroelastic problems are concerned with an essentially steady flow far upstream of a dynamically flexing wing. Let the subscript ∞ denote the steady flow conditions far upstream that are unaffected by the unsteady flow in the wing's vicinity. Then the freestream Mach number is given by $M_\infty = U_\infty/a_\infty$, and the isentropic flow conditions produce the following interesting relationship for density variation:

$$\frac{\rho}{\rho_\infty} = \left(\frac{T}{T_\infty}\right)^{1/(\gamma-1)} = \left[\frac{(\gamma-1)}{2a_\infty^2}\left(-2\Phi_t - U^2\right)\right]^{1/(\gamma-1)}$$

$$= \left\{1 + \frac{(\gamma-1)}{2}M_\infty^2\left[1 - \frac{\Phi_t}{U_\infty^2} - \left(\frac{U}{U_\infty}\right)^2\right]\right\}^{1/(\gamma-1)}. \tag{3.26}$$

Of course, Eq. (3.26) requires that the flow far upstream should be steady, that is $(\Phi_t)_\infty = 0$. By substituting Eq. (3.20) into Eq. (3.26) we have the following form of unsteady Bernoulli equation :

$$\frac{\rho}{\rho_\infty} = \left\{1 + \frac{(\gamma-1)}{2}M_\infty^2\left[1 - \frac{1}{U_\infty^2}\left(\Phi_t + \Phi_x^2 + \Phi_y^2 + \Phi_z^2\right)\right]\right\}^{1/(\gamma-1)}. \tag{3.27}$$

Substitution of Eq. (3.20) into the continuity equation, Eq. (3.3), yields

$$\rho_t + \rho_x\Phi_x + \rho\Phi_{xx} + \rho_y\Phi_y + \rho\Phi_{yy} + \rho_z\Phi_z + \rho\Phi_{zz} = 0. \tag{3.28}$$

Then by substituting Eq. (3.27) into Eq. (3.28) and carrying out the partial differentiations of ρ, the following *full-potential equation* (FPE) governing the inviscid, isentropic flow can be derived:

$$\begin{aligned}
\left(a^2 - \Phi_x^2\right)\Phi_{xx} + \left(a^2 - \Phi_y^2\right)\Phi_{yy} + \left(a^2 - \Phi_z^2\right)\Phi_{zz} ={}& \Phi_{tt} + 2\left(\Phi_x\Phi_{xt} + \Phi_y\Phi_{yt}\right.\\
&+ \Phi_z\Phi_{zt}) + 2\left(\Phi_x\Phi_y\Phi_{xy}\right.\\
&+ \Phi_x\Phi_z\Phi_{xz}), \tag{3.29}
\end{aligned}$$

where a is the local speed of sound. It is interesting to note that the FPE, Eq. (3.29), can be alternatively expressed as follows [58]:

$$\nabla^2\Phi = \frac{1}{a^2}\frac{D^2\Phi}{Dt^2}, \tag{3.30}$$

where

$$\nabla^2(.) = \nabla \cdot \nabla(.) = \frac{\partial^2}{\partial x^2}(.) + \frac{\partial^2}{\partial y^2}(.) + \frac{\partial^2}{\partial z^2}(.) \tag{3.31}$$

is the Laplacian operator and

$$\frac{D}{Dt}(.) = \frac{\partial}{\partial t}(.) + \mathbf{U} \cdot \nabla(.) \tag{3.32}$$

is the Eulerian (or material) derivative representing the time derivative seen by a fluid particle convecting with the flow at the local velocity $\mathbf{U}$. Equation (3.30) is the governing *wave equation* of acoustics, with the rate of wave propagation a. Hence, merely by transforming the spatial coordinates[1] from a body-fixed frame to a frame convecting with the flow, we have established an equivalence between potential unsteady aerodynamics and acoustics. This is an important result, which is quite useful in analyzing and solving the FPE.

Another interesting aspect of potential flow is the *acceleration potential* (also called *pressure potential*), ψ, related to the velocity potential as follows:

$$\psi = \frac{D\Phi}{Dt} = \frac{\partial \Phi}{\partial t} + \mathbf{U} \cdot \nabla \Phi, \tag{3.33}$$

which, by the integrated form of Euler momentum equation (unsteady Bernoulli equation) can also be expressed as

$$\psi = -\int \frac{dp}{\rho}. \tag{3.34}$$

Clearly, the acceleration potential is directly related to pressure difference, and provides an alternative model of potential flowfield. We shall return to acceleration potential later in the chapter.

The full-potential formulation is a practical alternative to Euler equations because, in order to completely determine the flowfield, one has to solve only for the velocity potential, $\Phi(x, y, z, t)$, rather than each of the primitive variables, u, v, w, ρ, e, of Euler equations. However, whereas Euler equations can be applied to nonisentropic flow caused by strong shock waves, the FPE is valid only for isentropic flow. The nonlinear nature of the FPE makes it almost as formidable to solve as Euler equations, although the number of dependent variables is reduced to one. The absence of viscous dissipation calls for artificial viscosity and tangential flow conditions for closure, as in the case of Euler formulation. The main utility of the FPE formulation is for nearly isentropic, transonic flows where weak, normal shock waves are present, and for which Euler equations are unnecessary. As discussed below, a major simplification of the FPE is possible for transonic small-disturbance (TSD) problems, where the boundary conditions can be applied on a mean surface rather than the actual moving boundary. Furthermore, in subsonic and supersonic regimes, the FPE can be effectively linearized for thin wings undergoing small amplitude motions.

---

[1] It can be verified that Eq. (3.30) is alternatively expressed as follows:

$$\Phi_{\xi\xi} + \Phi_{\eta\eta} + \Phi_{\zeta\zeta} = \frac{1}{a^2}\Phi_{tt},$$

where the fluid-fixed coordinates (ξ, η, ζ) are obtained from the body-fixed coordinates (x, y, z) by the following *Galilean cum Lorentz transformation*:

$$\left\{\begin{array}{c} \xi \\ \eta \\ \zeta \end{array}\right\} = \left\{\begin{array}{c} x \\ y \\ z \end{array}\right\} - \mathbf{U}t.$$

In order to check the applicability of the FPE to transonic speeds, consider the following normal shock relation for entropy gradient at upstream Mach number close to unity, $M_\infty \approx 1$:

$$\frac{\partial S}{\partial n} \simeq \frac{2\gamma}{3(\gamma + 1)^2}(M_\infty^2 - 1)^3. \tag{3.35}$$

In the limit $M_\infty \to 1$, the entropy variation becomes negligible across the shock wave, and isentropic condition prevails, thereby enabling the application of FPE. But the momentum is not exactly conserved in the presence of a normal shock, however weak it may be. Thus, the validity of FPE for transonic flows with weak shock waves is only approximate, but can give a reasonable model for ASE purposes.

Solution procedures for the FPE are essentially based upon the finite-difference approach. In the unsteady case, the FPE is hyperbolic in nature for all speed regimes, hence a time-marching scheme can be adopted (as in a typical unsteady Euler solver). However, for the steady-state problem, the FPE and Euler equations change their character from being elliptic in the local subsonic region to hyperbolic in the supersonic region. Therefore, in a transonic steady-state application a special treatment of the mixed elliptic/hyperbolic behavior is required when locally supersonic regions (normal shock waves) are present in the flowfield. This is either carried out by a switching (or type-dependent) procedure when the coefficient of the Φ_{xx} term changes sign [120], or by spatial upwinding techniques [66, 73] for density computation in a conservative form. As in the case of Euler computations, there is also the need for the introduction of artificial viscosity/entropy for avoiding physically unrealistic solutions [124]. Furthermore, special treatment of circulation at the trailing edge and the wake is also necessary [162] for closure, as in the case of Euler equations. Fortunately, a typical ASE application involves small amplitude motion of thin lifting surfaces, which quite significantly simplifies the full-potential model.

3.2.4 Transonic Small-Disturbance Flow

Aircraft geometries are streamlined for drag minimization such that the cross-flow (lateral) variations in the flow variables caused by the body thickness are small compared to those along the freestream (longitudinal) direction. The velocity potential is thus only slightly changed from its freestream value. This fact offers a major simplification in the governing equations called the *small disturbance* (or small perturbation) approximation. Consider a thin wing placed in a uniform freestream of speed U_∞, with the x-axis along the freestream direction, and the z-axis normal to the flow direction such that the perpendicular to the essentially flat mean surface lies in the (x, z) plane. The y-axis is in the spanwise direction normal to (x, z) axes, and completes the right-handed triad (x, y, z) as shown in Fig. 3.1. There is no sideslip due to (x, z) plane being parallel to the freestream direction. However, there could be a small, nonzero angle of attack, α, defined as the angle made by the wing's mean surface with the freestream direction. The net velocity potential is then regarded as a linear superposition of the *perturbation velocity potential*, ϕ, over that of the freestream:

$$\Phi = U_\infty x + \phi, \tag{3.36}$$

which results in the velocity components

$$u = \frac{\partial \Phi}{\partial x} = U_\infty + \phi_x; \qquad v = \frac{\partial \Phi}{\partial y} = \phi_y; \qquad w = \frac{\partial \Phi}{\partial z} = \phi_z. \tag{3.37}$$

When Eq. (3.36) is substituted into the FPE, Eq. (3.29), with the small-perturbation assumptions,

$$\phi_x << U_\infty; \qquad \phi_y << U_\infty; \qquad \phi_z << U_\infty, \tag{3.38}$$

and

$$\phi_x << a; \qquad \phi_y << a; \qquad \phi_z << a, \tag{3.39}$$

one can safely neglect second- (and higher) order terms involving ϕ_y, ϕ_z, and ϕ_t, and third-order terms involving ϕ_x. However, the second-order term involving ϕ_x must be retained for accuracy in the transonic limit, $U_\infty + \phi_x \simeq a$, at which weak normal shock waves may be present [95]. With these approximations, the full-potential equation is approximated by the following TSD equation:

$$\left[1 - M_\infty^2 - \frac{(\gamma + 1)M_\infty^2}{U_\infty} \phi_x \right] \phi_{xx} + \phi_{yy} + \phi_{zz} = \frac{2M_\infty^2}{U_\infty} \phi_{xt} + \frac{M_\infty^2}{U_\infty^2} \phi_{tt}. \tag{3.40}$$

The main utility of the TSD equation is its applicability in the unsteady, transonic limit, $M_\infty \simeq 1$, for which its essentially nonlinear character cannot be neglected. However, it is much simpler to solve than the FPE, because the unsteady boundary conditions can be applied on the mean surface (rather than the actual boundary) of the thin wing.

A practical simplification of the TSD equation is called the *low-frequency limit* [95], for which the term involving ϕ_{tt} can be neglected, leading to

$$\left[1 - M_\infty^2 - \frac{(\gamma + 1)M_\infty^2}{U_\infty} \phi_x \right] \phi_{xx} + \phi_{yy} + \phi_{zz} = \frac{2M_\infty^2}{U_\infty} \phi_{xt}. \tag{3.41}$$

The applicability of the low-frequency TSD approximation, Eq. (3.41), requires that the largest characteristic frequency, ω, governing the unsteady motion must be sufficiently small such that

$$\omega << \frac{U_\infty}{b}, \tag{3.42}$$

where b is a characteristic length. Typically, b is taken to be the mean semichord of the wing and ω the largest elastic modal frequency (bending, torsion, or control surface rotation) that can influence the flowfield. In terms of these parameters, Eq. (3.41) can be rendered nondimensional as follows:

$$\left[1 - M_\infty^2 - (\gamma + 1)M_\infty^2 \bar{\phi}_\xi\right] \bar{\phi}_{\xi\xi} + \bar{\phi}_{\eta\eta} + \bar{\phi}_{\zeta\zeta} = 2kM_\infty^2 \bar{\phi}_{\xi\tau}, \qquad (3.43)$$

where the nondimensional variables are given by

$$\xi = x/b, \quad \eta = y/b, \quad \zeta = z/b, \quad \tau = t\omega, \quad \bar{\phi} = \frac{\phi}{U_\infty b}, \quad k = \frac{\omega b}{U_\infty}. \qquad (3.44)$$

The *reduced-frequency*, k, is a similarity parameter of the unsteady flow in addition to the Mach number, M_∞. In the low-frequency limit, we have $k \ll 1$. Solution of the low-frequency TSD equation is carried out by finite-difference techniques that are quite like (but much simpler than) those required for the FPE. Being based upon an iterative solution of the following linearized equation by an *approximate factorization* approach [13]:

$$\left(1 - M_\infty^2\right) \bar{\phi}_{\xi\xi} + \bar{\phi}_{\eta\eta} + \bar{\phi}_{\zeta\zeta} = 2kM_\infty^2 \bar{\phi}_{\xi\tau}. \qquad (3.45)$$

It can be said that the simplest transonic unsteady aerodynamic model (that is, low-frequency TSD) is based upon repeated solution of a linear governing equation.

3.3 Linearized Subsonic and Supersonic Flow

A linear, unsteady aerodynamic model is necessary for designing basic ASE control laws. The governing equation of such a model in terms of the disturbance velocity potential is derived from the TSD equation by neglecting the nonlinear term in Eq. (3.40), resulting in the following linearized equation:

$$\left(1 - M_\infty^2\right) \phi_{xx} + \phi_{yy} + \phi_{zz} = \frac{2M_\infty^2}{U_\infty}\phi_{xt} + \frac{M_\infty^2}{U_\infty^2}\phi_{tt}, \qquad (3.46)$$

which can also be expressed as the following wave equation:

$$\nabla^2\phi = \frac{1}{a_\infty^2}\left(\frac{D^2\phi}{Dt^2}\right)_\infty, \qquad (3.47)$$

where

$$\left(\frac{D}{Dt}\right)_\infty (.) = \frac{\partial}{\partial t}(.) + \left\{\begin{array}{c} U_\infty \\ 0 \\ 0 \end{array}\right\} \cdot \nabla(.) = \frac{\partial}{\partial t}(.) + U_\infty\frac{\partial}{\partial x}(.) \qquad (3.48)$$

is the Eulerian derivative representing the time derivative seen by an observer moving with the freestream velocity $(U_\infty, 0, 0)^T$. It is to be noted that the wave propagation speed in Eq. (3.47) is a_∞, the freestream speed of sound, rather than a, the local speed of sound in the FPE, Eq. (3.30). Evidently, the effect of linearization on the FPE is to make small disturbances due to the body spread out at a constant speed in all directions relative to the moving body. Thus, the Mach number is assumed constant, $M \simeq M_\infty$, at all points in the flowfield.

Derivation of Eq. (3.47) from Eq. (3.46) is equivalent to the following *Galilean cum Lorentz transformation* of the space-time coordinates in the subsonic case ($M < 1$):

$$\frac{x}{\sqrt{1 - M^2}} \Rightarrow x; \quad y \Rightarrow y; \quad z \Rightarrow z, \quad t\sqrt{1 - M^2} + \frac{xM}{a_\infty\sqrt{1 - M^2}} \Rightarrow t$$

which changes to the following in the supersonic case ($M > 1$)

$$\frac{x}{1 - M^2} \Rightarrow x; \quad \frac{y}{\sqrt{1 - M^2}} \Rightarrow y; \quad \frac{z}{\sqrt{1 - M^2}} \Rightarrow z, \quad t + \frac{xM}{a_\infty(1 - M^2)} \Rightarrow t.$$

Equations (3.46) and (3.47) are accurate at subsonic and supersonic speeds, but cannot be applied to the transonic regime, where even for a thin wing and a slender body, the nonlinear TSD equation must be solved. Being a linear equation, Eq. (3.46) possesses the important property that its solution to arbitrary boundary conditions is a linear superposition of elementary solutions corresponding to simpler boundary conditions. Such elementary solutions can be velocity potential sources or doublets distributed over a solid boundary, on which the flow tangency and Kutta condition at the trailing edge are to be satisfied [58].

The boundary conditions in terms of the disturbance velocity potential can be posed as follows:

(a) The flow is uniform (undisturbed) far upstream of the body,

$$\phi(-\infty, y, z, t) = 0, \tag{3.49}$$

and perturbations remain bounded at infinity.

(b) Pressure is continuous across the wake. The small perturbation causes a planar wake which always lies in the mean wing plane. The vorticity of the wake satisfies the Helmholtz theorem [86] along with the bound circulation around the wing.

(c) Viscous effects, though unmodeled, are assumed to be just large enough to cause a smooth flow tangential to the mean wing surface at the trailing-edge. This assumption is called the *Kutta condition*, and allows inviscid flow modeling without the attendant problem of nonunique solutions. Its physical validity, however, is questionable when the reduced frequency, $b\omega/U_\infty$ becomes large. In a typical aeroelastic application with small amplitude, low-frequency oscillations, Kutta condition is widely held to be valid.

(d) The flow is tangential to the impervious, solid boundary. This implies that a flow particle contacting the body must follow the instantaneous surface contour of the dynamic body, which can be defined for the two-dimensional case by the functional,

$$F(x, y, z, t) = z - z_b(x, y, t) = 0.$$

Hence, the solid boundary condition is given by

$$\frac{DF}{Dt} = \frac{\partial F}{\partial t} + \mathbf{U} \cdot \nabla F = 0. \tag{3.50}$$

For a thin wing, this boundary condition is effectively linearized by assuming that the local normal on the body surface is along the z-axis, and the product $w\partial z_b/\partial x$ is negligible. Furthermore, the thickness of the body can be considered to be either negligible, or its effect included in the steady flowfield. Thus the unsteady effect of the body can be represented by the mean surface (i.e., median plane between upper and lower surfaces), $z = 0$, as follows:

$$w(x, y, 0, t) = \left(\frac{\partial \phi}{\partial z}\right)_{z=0} = \frac{\partial z_b}{\partial t} + U_\infty \frac{\partial z_b}{\partial x}. \tag{3.51}$$

Since the upwash, w, cannot be physically discontinuous across the mean surface, it must be the same at the upper and lower surfaces, denoted by $z = 0+$ and $z = 0-$, respectively. However, ϕ must change sign across the wing, that is, be an antisymmetric (odd) function of z, because its z-derivative, w, is a symmetric (even) function. Conventionally, ϕ is taken to be positive on the upper surface and negative on the lower surface. Implementation of the tangential flow condition is useful in determining the unknown strength of a distribution of elementary solution (source, vortex, doublet, etc.) on the solid boundary.

A solid boundary can have a different pressure distribution on the upper surface, p_u, from that on the lower surface, p_ℓ. However, the approximation of the body by a mean surface of zero thickness, $z = 0$, only allows the pressure distribution to change sign, but not the magnitude, across the surface. The linearization of the unsteady Bernoulli equation, Eq. (3.23), written as

$$p(x, y, z, t) - p_\infty = -\rho_\infty \left(\frac{\partial \phi}{\partial t} + U_\infty \frac{\partial \phi}{\partial x}\right), \tag{3.52}$$

results in the following relationship between the pressure difference across the wing, $\Delta p = p_u - p_\ell$, the disturbance velocity potential, ϕ:

$$\Delta p(x, y, t) = p(x, y, 0+, t) - p(x, y, 0-, t) = 2p(x, y, 0, t)$$
$$= -2\rho_\infty \left[\left(\frac{\partial \phi}{\partial t} + U_\infty \frac{\partial \phi}{\partial x}\right)\right]_{z=0}. \tag{3.53}$$

The solution to the unsteady Bernoulli equation (Eq. 3.52) can be expressed as the following integral relationship between the pressure and velocity potential:

$$\phi(x, y, z, t) = -\frac{1}{\rho_\infty U_\infty} \int_{-\infty}^{x} \left[p\left(\xi, y, z, t - \frac{x - \xi}{U_\infty}\right) - p_\infty\right] d\xi, \tag{3.54}$$

where the velocity potential far upstream of the wing ($x \to -\infty$) is assumed to be zero. This is the far-field boundary condition (a), discussed above. Hence the condition of uniform flow far upstream, Eq. (3.49), is implicit in Eq. (3.54).

The integral equation, Eq. (3.54), must be solved for the unknown pressure differ-ence across the wing, while taking into account how the velocity potential evolves in selected regions of the flowfield with both space and time. Needless to say, a closed-form solution does not exist for any realistic problem of this kind, and numerical solutions must be resorted to. One such numerical solution can be derived by Green's divergence theorem [88] applied to the boundary-value problem where either the ve-locity potential, or its normal gradient, are prescribed on selected boundaries of the flowfield. The boundary regions over which the velocity potential (or its gradient) must be specified in order to solve for the unsteady pressure distribution on the wing must necessarily include the wing and the wake. A solution alternative to the velocity potential is possible in terms of the perturbation acceleration potential, ψ, defined by

$$\psi = -\frac{p - p_\infty}{\rho_\infty} = \frac{\partial \phi}{\partial t} + U_\infty \frac{\partial \phi}{\partial x}, \tag{3.55}$$

for which the wake surface (or any region off the solid body) need not be modeled, since ψ cannot physically change across such a surface. This results in a simpler numerical scheme, as will be discussed later.

3.4   Incompressible Flow Solution

A simple but effective method for modeling low-speed, unsteady flows over lifting surfaces is by utilizing elementary solutions to the incompressible, irrotational (ideal) flow problem. In the incompressible limit, $a \rightarrow \infty$, the FPE, Eq. (3.29), reduces to the *Laplace* equation for the perturbation velocity potential, given by

$$\nabla^2 \Phi = 0 \tag{3.56}$$

This equation is the same as that governing the *steady* incompressible and irrotational flow, for which elementary solutions are readily available [86]. The simplest solutions are first- and second-order polynomials in Cartesian coordinates, x, y, z, such as a uniform flow,

$$\Phi = Ux + Vy + Wz, \tag{3.57}$$

and the flow around a corner in the (x, y) plane given by

$$\Phi = C(x^2 - y^2). \tag{3.58}$$

A practical solution is the velocity potential *source* described by

$$\Phi_s = -\frac{\sigma}{4\pi r}, \tag{3.59}$$

where r denotes the distance of the test point, (x, y, z), from the source point, (ξ, η, ζ), and σ is the source strength. The source is a point of singularity in the flowfield where the radial velocity becomes infinite. A *sink* has the velocity potential of opposite sign to that of a source. By taking a pair of source and sink of equal strengths and letting the distance between them vanish, while increasing the strength to infinity, the second elementary solution called the *point doublet* is obtained, given by

$$\Phi_d = -\frac{\mu}{4\pi}\mathbf{n}\cdot\nabla(1/r), \tag{3.60}$$

where μ is the doublet strength, and $\mathbf{n}$ is the unit vector called *doublet axis* representing the direction from the source to the sink. The source and doublet solutions are related by

$$\Phi_d = -\frac{\partial}{\partial n}\Phi_s \tag{3.61}$$

for which $\mu = \sigma\cos\theta$, where θ is the angle made by the doublet axis and the line joining the doublet to the test point. The third important elementary solution of the Laplace equation is the *line vortex*, for which the induced flow velocity at a point located by $\mathbf{r}$ from the infinitesimal vortex segment $d\ell$ of constant strength Γ is given by the *Biot–Savart law* as follows:

$$d\mathbf{U} = \frac{\Gamma}{4\pi r^3}d\ell \times \mathbf{r}. \tag{3.62}$$

For a two-dimensional flow, the source, doublet, and vortex velocity potential solutions, Φ_s, Φ_d, Φ_v, respectively, at test point (x, z) for the singularities located at (ξ, ζ) are expressed as follows:

$$\Phi_s = \frac{\sigma}{2\pi}\log r$$

$$\Phi_d = -\frac{\partial}{\partial n}\Phi_s \tag{3.63}$$

$$\Phi_v = -\frac{\gamma}{2\pi}\tan^{-1}\left(\frac{z-\zeta}{x-\xi}\right),$$

where $r = \sqrt{(x-\xi)^2 + (z-\zeta)^2}$.

The linearity of the Laplace equation allows a linear superposition of its n elementary solutions in order to model a complicated flowfield, such as

$$\Phi = \sum_{k=1}^{n} c_k \Phi_k, \tag{3.64}$$

where the unknown coefficients c_k can be determined from n boundary conditions specified at discrete points in the flowfield. For example, steady flow past a circular cylinder can be modeled by superimposing the velocity potential of a uniform flow along the x axis, $\Phi = U_\infty x$ over that of a point doublet with its axis pointing in the negative x direction. Another useful example is the steady flow past a rotating cylinder, with axis $\mathbf{e}$ modeled by the superposition of uniform flow of velocity $\mathbf{U}_\infty$ with that of a line vortex of strength Γ in $\mathbf{e}$ direction, which yields the lift force per unit span of the cylinder by *Kutta Joukowski* theorem to be the following:

$$\mathbf{L} = \rho \mathbf{U}_\infty \times \Gamma \mathbf{e}. \tag{3.65}$$

Clearly, linear superposition of solutions is a very useful concept and allows breaking-up of a complex flowfield into several simpler ones.

A practical application of linear superposition of solutions to the Laplace equation is the model of the flow past a wing by a distribution of sources, doublets, and/or vortices on the wing surface, and calculating their strengths based upon specific boundary conditions. In this way, one can derive the unknown pressure distribution on the wing. Such an approach was the basis of the early analytical, incompressible flow solutions on oscillating two-dimensional airfoils [63, 173], and is also used by the three-dimensional panel methods based upon Green's integral theorem [86, 114]. The steady-state solutions by Prandtl's lifting-line and the quasi-steady, three-dimensional vortex-lattice methods [72] also fall in this category. We will begin with the steady flow integral formulation for the Laplace equation as the basis of the unsteady panel methods, and consider its extension to the unsteady case by a proper treatment of the unsteady boundary conditions on the wing and the wake.

In the steady flow formulation by Green's integral theorem [88], the velocity potential at a given point in the ideal flowfield can be expressed as follows [86] [2]:

$$\Phi(x, y, z) = \iint_S \left[\frac{1}{4\pi r} \frac{\partial \Phi}{\partial n} - \Phi \frac{\partial}{\partial n} \left(\frac{1}{4\pi r} \right) \right] \mathrm{d}S, \tag{3.66}$$

where the closed double integral is carried out over the boundary of the flowfield, S, comprising the solid boundary, S_b, the wake surface, S_w, and the far-field boundary, S_∞, and $\mathbf{n}$ is the outward normal to S, as shown in Fig. 3.2. The contribution of the far-field boundary can be separately derived by

$$\Phi_\infty = \iint_{S_\infty} \left[\frac{1}{4\pi r} \frac{\partial \Phi}{\partial n} - \Phi \frac{\partial}{\partial n} \left(\frac{1}{4\pi r} \right) \right] \mathrm{d}S_\infty, \tag{3.67}$$

where Φ_∞ is the freestream velocity potential. For a point located interior to the solid body, the interior velocity potential distribution, $\Phi_i(x, y, z)$, is given by

$$0 = \iint_{S_b} \left[\frac{1}{r} \frac{\partial \Phi}{\partial n} - \Phi \frac{\partial}{\partial n} \left(\frac{1}{r} \right) \right] \mathrm{d}S_\infty. \tag{3.68}$$

Finally, as the wake is a free (unloaded) surface, the gradient of velocity potential, $\partial \Phi / \partial n$, is continuous across the wake by Bernoulli's equation. When these conditions are substituted into Eq. (3.66), the following solution is obtained for the Laplace equation :

---

[2] Green's integral formulation for unsteady, potential, compressible flow is discussed later in this chapter.

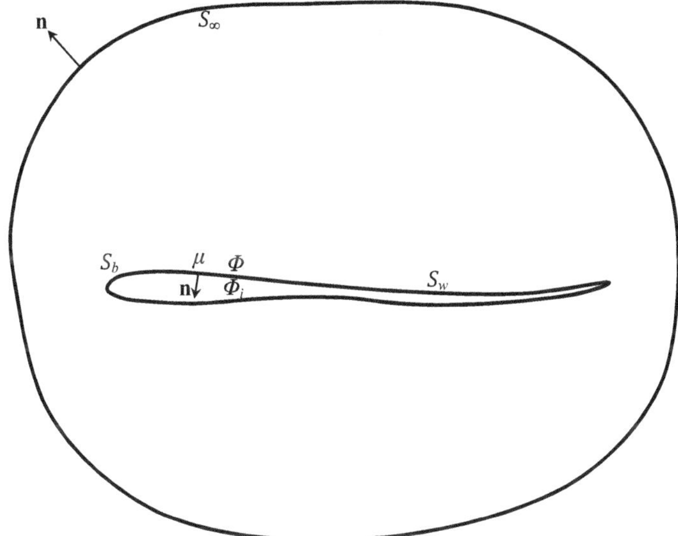

Fig. 3.2 Schematic diagram of potential flowfield for Green's integral formulation

$$\Phi(x,y,z) = \Phi_\infty(x,y,z) + \frac{1}{4\pi} \iint_{\mathcal{S}_b} \left[\frac{1}{r}\left(\frac{\partial \Phi}{\partial n} - \frac{\partial \Phi_i}{\partial n} \right) - (\Phi - \Phi_i)\frac{\partial}{\partial n}\left(\frac{1}{r} \right) \right] d\mathcal{S}_b$$

$$- \frac{1}{4\pi} \iint_{\mathcal{S}_w} \Phi \frac{\partial}{\partial n}\left(\frac{1}{r} \right) d\mathcal{S}_w. \tag{3.69}$$

The integral evaluation requires a distribution of elementary solutions, $\Phi(\xi,\eta,\zeta)$, over the wing and wake boundaries. For example, if one considers an elemental point doublet of strength $\mu(\xi,\eta,\zeta)$ at a point on the external wing surface with the axis pointed along **n**–the outward normal to $\mathcal{S}$, but pointing inward to the wing–then we can select

$$-(\Phi - \Phi_i) = \mu. \tag{3.70}$$

In addition, by choosing an elemental source of strength $\sigma(\xi,\eta,\zeta)$ on the external wing surface given by

$$-\left(\frac{\partial \Phi}{\partial n} - \frac{\partial \Phi_i}{\partial n} \right) = \sigma, \tag{3.71}$$

with the source and doublet strengths related by Eq. (3.61), the integral equation is identically satisfied. Furthermore, since both the solutions decay to zero at the far-field boundary, $r \to \infty$, the far-field boundary condition is also satisfied. However, the unknown source and doublet strength distributions on the wing, $\sigma(\xi,\eta,\zeta), \mu(\xi,\eta,\zeta)$, must be determined by additional boundary conditions to be specified on the wing and the wake, based upon physical considerations. For example, if we define the interior potential by the boundary condition,

$$\Phi = \Phi_i,$$

then the doublet strength μ vanishes at all points, and only the exterior source distribution, σ, has to be computed from the boundary conditions on the external wing and wake surfaces. On the other hand by selecting,

$$\frac{\partial \Phi}{\partial n} = \frac{\partial \Phi_i}{\partial n},$$

we make the source strength vanish, and only require the solution of the unknown doublet strength distribution. If the boundary conditions are only specified for the velocity potential, Φ, the boundary value problem is known as the *Dirichlet problem*. If the boundary conditions are only specified for the gradient of velocity potential, $\partial \Phi / \partial n$, we have a *Neumann problem*. Usually, there is a combination of boundary conditions of either type, resulting in a *mixed formulation*. For example, determining a solution for the unknown source distribution from a specified normal velocity boundary condition at the solid wall is an internal Dirichlet and external Neumann problem.

For example of an application of Green's theorem, consider a two-dimensional, thin airfoil (flat plate) with singularity distributions along the chord plane taken to be the x-axis ($z = 0$). A uniform source distribution, $\sigma(x)$, yields the velocity potential to be the following:

$$\Phi(x, z) = \frac{1}{2\pi} \int_0^x \sigma(\xi) \log \sqrt{(x - \xi)^2 + z^2} d\xi. \tag{3.72}$$

with the normal velocity (upwash) at a point on the wing's upper surface, $z = 0+$, and lower surface, $z = 0-$, given by

$$w(x, 0\pm) = \frac{\partial \Phi}{\partial z}(x, 0\pm) = \lim_{z \to 0\pm} \frac{1}{2\pi} \int_0^x \sigma(\xi) \frac{z}{(x - \xi)^2 + z^2} d\xi$$

$$= \lim_{z \to 0\pm} \frac{\sigma(x)}{2\pi} \int_0^x \frac{z}{(x - \xi)^2 + z^2} d\xi$$

$$= \pm \frac{\sigma(x)}{2}. \tag{3.73}$$

This suggests that the source strength can be directly computed from the upwash distribution on the wing.

Similarly, if a uniform, two-dimensional doublet distribution, $\mu(x)$, with doublet axis pointing in the z direction, is used on the wing chord plane, the velocity potential is the following:

$$\Phi(x, z) = -\frac{1}{2\pi} \int_0^x \mu(\xi) \frac{z}{(x - \xi)^2 + z^2} d\xi, \tag{3.74}$$

and the jump in the velocity potential across the wing is used to calculate the doublet strength distribution:

$$\Phi(x,0+) - \Phi(x,0-) = -\lim_{z\to 0+} \frac{\mu(x)}{2\pi} \int_0^x \frac{z}{(x-\xi)^2 + z^2} d\xi$$

$$+ \lim_{z\to 0-} \frac{\mu(x)}{2\pi} \int_0^x \frac{z}{(x-\xi)^2 + z^2} d\xi$$

$$= -\mu(x). \tag{3.75}$$

The tangential velocity component is, therefore, discontinuous across the wing surface,

$$u(x,0\pm) = \frac{\partial \Phi}{\partial x}(x,0\pm) = \mp \frac{1}{2}\frac{d\mu}{dx}, \tag{3.76}$$

and causes a nonzero circulation about the wing segment $(0, x)$, given by

$$\Gamma(x) = \Phi(x,0+) - \Phi(x,0-) = -\mu(x). \tag{3.77}$$

Hence, the circulation, $\Gamma(x)$, at a given point on the wing is the same as the jump in the velocity potential across the wing, and equals the negative of the doublet strength at that point. A two-dimensional vortex distribution of strength $\gamma(x)$ on the chord plane can be replaced by an equivalent doublet distribution by taking the x derivative. This is easily verified by the result

$$\Phi(x,z) = \frac{1}{2\pi} \int_0^x \gamma(\xi)\tan^{-1}\frac{z}{x-\xi} d\xi, \tag{3.78}$$

or

$$u(x,0\pm) = \frac{\partial \Phi}{\partial x}(x,0\pm) = \lim_{z\to 0\pm} \frac{\gamma(x)}{2\pi} \int_0^x \frac{z}{(x-\xi)^2 + z^2} d\xi$$

$$= \pm \frac{\gamma(x)}{2}, \tag{3.79}$$

thereby implying from Eq. (3.76),

$$\gamma(x) = -\frac{d\mu}{dx}, \tag{3.80}$$

or

$$\mu(x) = -\int_0^x \gamma(\xi)d\xi.$$

Such an equivalence between the vortex and doublet solutions means that either one can be alternatively employed in a numerical solution procedure. The normal velocity component at a point on the airfoil must satisfy the boundary condition of solid wall,

$$w(x,0) = \frac{\partial \Phi}{\partial z}(x,0) = -\frac{1}{2\pi} \int_0^c \frac{\gamma(\xi)}{x-\xi} d\xi$$

$$= U_\infty \alpha \left(\frac{\mathrm{d}F}{\mathrm{d}x} - \alpha \right),$$ (3.81)

where $z = F(x)$ represents the camber line of the airfoil and α the angle of attack. However, the vorticity distribution $\gamma(x)$ is not uniquely determined solely by the normal flow condition, and requires an additional physical consideration to fix it. This is the Kutta condition, which requires the flow should leave smoothly at the trailing-edge $x = c$ of the airfoil, and must have a finite velocity at that point. In order to satisfy the Kutta condition, there should be no discontinuity in the (zero) normal velocity at the trailing-edge as the chord-plane is crossed from the lower to the upper surface,

$$\gamma(c) = 0.$$ (3.82)

The elementary singularity solutions (source, doublet, vortex) are readily extended to the unsteady case, because the governing equation for the unsteady ideal flow is also the Laplace equation. However, since the boundary conditions on the wing are now time dependent, the strengths of the singularity distributions also fluctuate with time, which requires carrying out an additional integration in time over the wing and the wake for a complete solution. In order to illustrate such a strategy, we next consider the unsteady case based upon vortex as the singularity solution.

3.4.1   Unsteady Vortex-Lattice Method

The extension of the steady-flow integral formulation presented above for the unsteady case requires the consideration of Kelvin's theorem, which states that the circulation around a closed curve consisting of a given set of fluid particles is constant with time, as the given particles convect in an irrotational flowfield. This is represented by the equation

$$\frac{D\Gamma_c}{Dt} = \frac{\partial \Gamma}{\partial t} + \mathbf{U} \cdot \nabla(\Gamma)$$ (3.83)

where $\Gamma_c(x, y, z, t)$ is the circulation around the closed curve, and D/Dt denotes the Eulerian derivative representing the time derivative seen by a fluid particle convecting with the flow at the local velocity $\mathbf{U}$.

There are various ways in which Kelvin's theorem of vorticity conservation around a convecting closed curve can be included in the flow model. However, a lattice method based upon the vortex (or a doublet) as the elementary solution is the most direct way to handle circulation, and hence Kelvin's theorem. In order to appreciate this fact, consider a simple model wherein the circulation around the wing is represented by a single vortex concentrated at a point fixed relative to the wing,

$$\Gamma(x) = \int_0^x \gamma(\xi)\,\mathrm{d}\xi.$$

Initially for $t < 0$, the wing is at rest and circulation everywhere is zero. Suddenly, at $t = 0$, the wing is started forward at a constant speed U_∞, with its chord making an angle of attack α. The circulation around the wing, Γ, given by the lumped vortex located at quarter-chord point, is the following line integral around a closed curve, s, surrounding the wing:

$$\Gamma = \int_s \mathbf{U} \cdot \mathrm{d}\mathbf{s}. \tag{3.84}$$

Now, since the integral in Eq. (3.84) is carried out around particles attached to a moving wing, its value is not a constant. To obtain the curve required for Kelvin's theorem, one must consider a coordinate frame attached to the flow rather than the wing, with the closed curve s_c of the same set of fluid particles. Then the circulation around such a curve is given by

$$\begin{aligned}
\Gamma_c &= \int_{s_c} \mathbf{U} \cdot \mathrm{d}\mathbf{s}_c \\
&= \int_s \mathbf{U} \cdot \mathrm{d}\mathbf{s} + \int_{s_w} \mathbf{U} \cdot \mathrm{d}\mathbf{s_w} \\
&= \Gamma + \Gamma_w,
\end{aligned} \tag{3.85}$$

where the second integral on the right-hand side of Eq. (3.85) (denoted Γ_w) is around another closed curve s_w encloses the wake, which is always expanding with time as the wing moves forward. The application of Kelvin's theorem then gives the following:

$$\frac{D\Gamma_c}{Dt} = \frac{D\Gamma}{Dt} + \frac{D\Gamma_w}{Dt} = 0, \tag{3.86}$$

or

$$\frac{D\Gamma}{Dt} = -\frac{D\Gamma_w}{Dt} \tag{3.87}$$

This implies that the circulation around the wing is cancelled by an equal and opposite wake circulation at any given time t. Furthermore, Kelvin's theorem is consistent with the principle of conservation of vorticity in an irrotational flow (Helmholtz theorem), which implies that a vortex tube, line, or surface consisting of the same fluid elements, always has a constant strength.

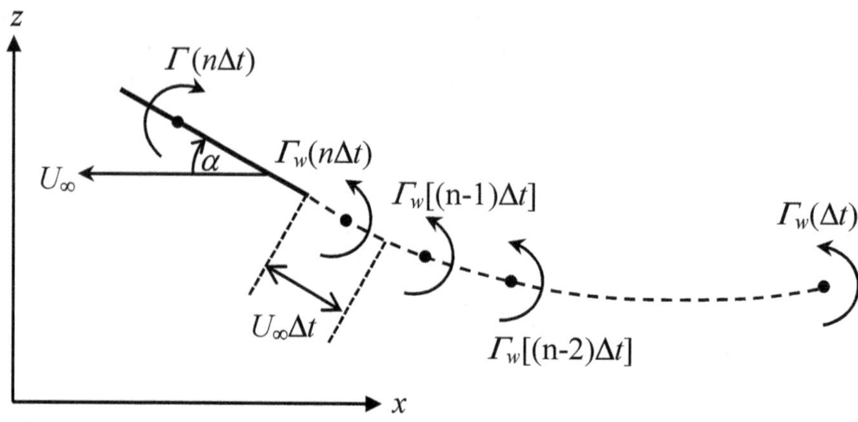

Fig. 3.3 Unsteady circulation, $\Gamma(t)$, of a flat plate at a constant geometric angle of attack, α, suddenly started from rest at $t = 0$, with an evolving wake vorticity, $\Gamma_w(t)$, modeled by the lumped vortex method

The lumped-vortex model for the flat plate suddenly started from rest at $t = 0$, is depicted in Fig. 3.3. Equation (3.87) can be discretized in time by taking finite steps of Δt, and the value of circulation can be easily evolved on the wing and wake accordingly. In order to determine the unknown circulation of the wing, $\Gamma(t)$, and the wake, $\Gamma_w(t)$, boundary condition of zero velocity normal to the wing surface is applied at the three-quarters chord collocation point. The choice of $\xi = c/4$ point for the bound vortex, and $\xi = 3c/4$ for collocation point results in the Kutta condition being instantaneously satisfied at the trailing-edge of the flat plate. This fact can be demonstrated for the steady case by applying Biot–Savart law (Eq. (3.62)) at a point $x = a$ on the wing to determine the net normal velocity due to the bound vortex at $x = c/4$ [86]:

$$w(a,0) = \frac{\partial \Phi}{\partial \zeta}(a,0) = -\frac{\Gamma}{2\pi(a - c/4)} + U_\infty\alpha = 0 \tag{3.88}$$

Substituting the correct value of circulation required for satisfying Kutta condition at the trailing edge (Eq. (3.82)), given by the Kutta–Joukowski theorem [86] as

$$\Gamma = 2\pi\frac{c}{2}U_\infty\alpha, \tag{3.89}$$

we have

$$-\frac{c}{2(a - c/4)} + 1 = 0, \tag{3.90}$$

or $a = 3c/4$.

For the unsteady case, there is an additional contribution of the wake vortices to the induced velocity on the wing's 3/4-chord location. For example, at the first time step, a wake segment of length $U_\infty\Delta t$, with a concentrated vortex in the middle, has been shed off the trailing edge. Thus the first wake vortex reaches a downstream distance $U_\infty \Delta t/2$ behind the wing, thereby (along with the bound vortex) inducing a net upwash at the collocation point,

$$-\frac{\Gamma(\Delta t)}{\pi c} + \frac{\Gamma_w(\Delta t)x_1}{2\pi \left(x_1^2 + z_1^2\right)} = -U_\infty \alpha, \tag{3.91}$$

where

$$x_1 = c/4 + U_\infty \Delta t/2 \tag{3.92}$$

and the vertical location of the first wake vortex, z_1, is determined from the wake roll-up caused by the normal velocity induced by wing vortex on the wake, approximated as follows:

$$z_1 = -\Delta t \frac{\Gamma(\Delta t)}{2\pi(x_1 + c/2)} \tag{3.93}$$

At $t = \Delta t$ (as at all times), the wing and wake vortices must satisfy Kelvin's theorem:

$$\Gamma(\Delta t) + \Gamma_w(\Delta t) = 0. \tag{3.94}$$

While simultaneous solution of Eqs. (3.91)–(3.94), a nonlinear set, requires an iterative approach, an approximate solution can be obtained by utilizing the fact $z_1 << x_1$, which yields $z_1 \simeq 0$ in Eq. (3.91). This allows a linear algebraic solution of the equations for $\Gamma(\Delta t), \Gamma_w(\Delta t)$.

At every fresh instant Δt, a new vortex must be shed into the wake, because by Kelvin's theorem, that is the only way in which the wing's circulation can evolve with time. Thus the following equations must hold at a time $t = n\Delta t$:

$$-\frac{\Gamma(n\Delta t)}{\pi c} + \frac{\Gamma_w(n\Delta t)}{2\pi x_1} + \frac{\Gamma_w[(n-1)\Delta t]x_2}{2\pi \left(x_2^2 + z_2^2\right)}$$

$$+ \frac{\Gamma_w[(n-2)\Delta t]x_3}{\pi \left(x_3^2 + z_3^2\right)} + \cdots + \frac{\Gamma_w(\Delta t)x_n}{\pi \left(x_n^2 + z_n^2\right)}$$

$$= -U_\infty \alpha. \tag{3.95}$$

$$\Gamma(n\Delta t) + \Gamma_w(n\Delta t) + \Gamma_w[(n-1)\Delta t] + \Gamma_w[(n-2)\Delta t] + \cdots + \Gamma_w(\Delta t) = 0. \tag{3.96}$$

The wake roll-up procedure accounts for the evolving wake geometry caused by the upwash due to the wing and wake vortices. This is carried out by applying Biot–Savart law to every new wake segment as follows:

$$w_n = -\frac{\Gamma(n\Delta t)}{2\pi(x_1 + c/2)} + + \frac{\Gamma_w[(n-1)\Delta t](x_2 - x_1)}{2\pi \left[(x_2 - x_1)^2 + (z_2 - z_1)^2\right]}$$

$$+ \frac{\Gamma_w[(n-2)\Delta t](x_3 - x_1)}{2\pi \left[(x_3 - x_1)^2 + (z_3 - z_1)^2\right]} + \cdots$$

$$+ \frac{\Gamma_w(\Delta t)(x_n - x_1)}{2\pi \left[(x_n - x_1)^2 + (z_n - z_1)^2\right]}, \quad n = 2, 3, \ldots, \tag{3.97}$$

and updating the wake vortex coordinates by

$$\begin{Bmatrix} x_n \\ z_n \end{Bmatrix} = \begin{Bmatrix} x_{n-1} \\ z_{n-1} \end{Bmatrix} + \begin{Bmatrix} U_\infty \\ w_n \end{Bmatrix} \Delta t, \quad n = 2, 3, \ldots \tag{3.98}$$

Hence the linear algebraic equations, Eqs. (3.95) and (3.96), must be solved at each instant to determine the unknowns, $\Gamma(n\Delta t), \Gamma_w(n\Delta t), n = 1, 2, \ldots$, using the updated wake coordinates supplied by the wake roll-up procedure.

Once the wing circulation is calculated, lift can be computed by the unsteady Bernoulli equation as follows:

$$L(t) = \int_0^c \Delta p(t)\mathrm{d}\xi = \rho \left[U_\infty \Gamma(t) + c\frac{\partial \Gamma}{\partial t}(t) \right] \tag{3.99}$$

The lumped-vortex procedure applied to the impulsively started flat plate with $c = 1$ m, $U_\infty = 30$ m/s, $\rho = 1.225$ kg/m^3, $5°$ angle of attack, and $\Delta t = 0.25c/U_\infty$ is coded in MATLAB, and results are plotted in Figs. 3.4–3.6 for 10 time steps. As expected, the wing's lift and circulation reach positive, steady-state values, while the wake circulation diminishes to zero in about 1.5 chord lengths of travel. The wake roll-up is evident in the snapshot (Fig. 3.5) taken after 10 time steps (or 2.5 chord lengths). The comparison of computed lift with that provided by the analytical result of Wagner's indicial lift function [189], $k_1(t)$, is shown in Fig. 3.7 for a longer time (120 steps). It is observed that the circulatory (wake-induced) lift predicted by Wagner function [3] reaches the steady-state value, $L(\infty)$, in about 30 chord lengths, whereas the lumped-vortex model—which includes both circulatory and noncirculatory lift— predicts a much faster decay of the transient response. The vortex model also correctly indicates infinite lift at $t = 0$ consistent with the impulsively started airfoil, which corresponds to infinite initial acceleration.

Let us now consider a more sophisticated vortex-lattice model [86] that can be applied to a three-dimensional wing with camber and thickness. The elementary solution used in this case is the vortex-ring consisting of four linear vortices each producing a constant circulation, Γ, linked together in a ring (Fig. 3.8). Each ring is placed on a wing panel such that the leading vortex coincides with the 1/4-chord location of the panel and the trailing vortex is at the 1/4-chord location of the next panel in the downstream direction. The panel edges are parallel to the freestream,

---

[3] Wagner function $k_1(s)$ models the indicial lift of a flat plate airfoil impulsively started from rest by

$$L(t)/L(\infty) = k_1(s) + \frac{b}{2}\delta(s),$$

where $s = tU_\infty/b$ and $\delta(s)$ is the Dirac delta function.

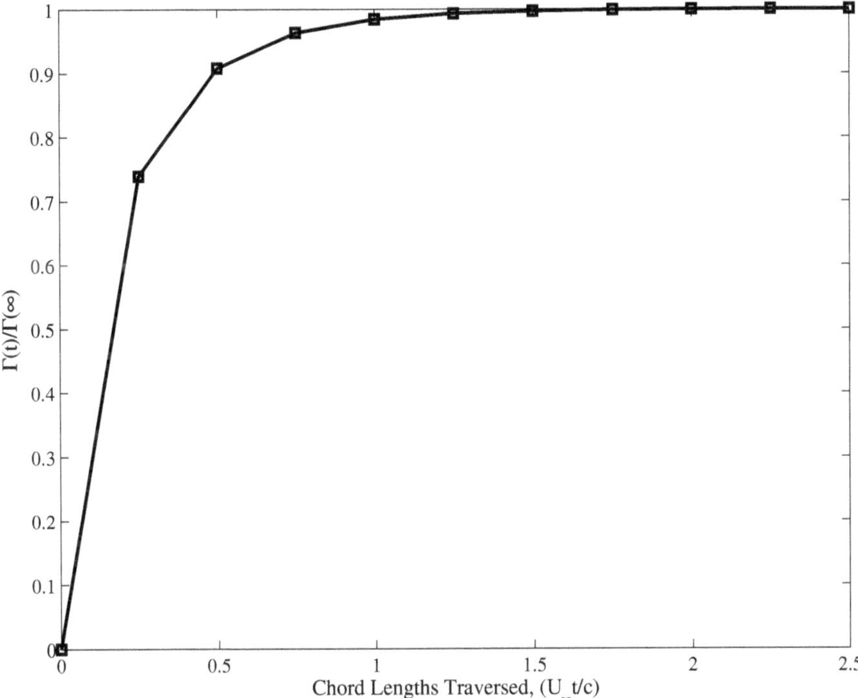

Fig. 3.4 Unsteady circulation, $\Gamma(t)$, of a flat plate suddenly started from rest at $5°$ angle of attack, modeled by the lumped-vortex method with 10 time steps

and the other two vortices are aligned with them. The last row of vortex rings on the panels next to the wing's trailing edge is placed such that the trailing vortex gives rise to the first wake vortex of equal strength, in order to satisfy the Kutta condition. The upwash boundary condition is specified on the 3/4-chord, mid-span location of each panel. Let the upwash produced by a wing panel–identified by the index j–on the 3/4-chord collocation point (x_i, y_i, z_i) of the ith panel be given by the aerodynamic influence coefficient A_{ij}, while that produced by a wake panel identified by index j_w on the same collocation point be denoted B_{ij}. If there are N wing panels and N_w wake panels, then the upwash induced by all the panels on the ith wing panel is given by

$$w_i = \sum_{j=1,\neq i}^{N} A_{ij}\Gamma_j + \sum_{j=1}^{N_w} B_{ij}\Gamma_{wj}, \qquad (i = 1,\ldots,N) \qquad (3.100)$$

which is represented in the following matrix form:

$$\mathbf{w} = \mathbf{A}\boldsymbol{\Gamma} + \mathbf{B}\boldsymbol{\Gamma}_\mathbf{w}, \qquad (3.101)$$

with $\mathbf{A}, \mathbf{B}$ being aerodynamic influence coefficient (AIC) matrices, used to compute the unknown circulation vectors $\boldsymbol{\Gamma}, \boldsymbol{\Gamma}_\mathbf{w}$ from the boundary condition normal to the wing surface,

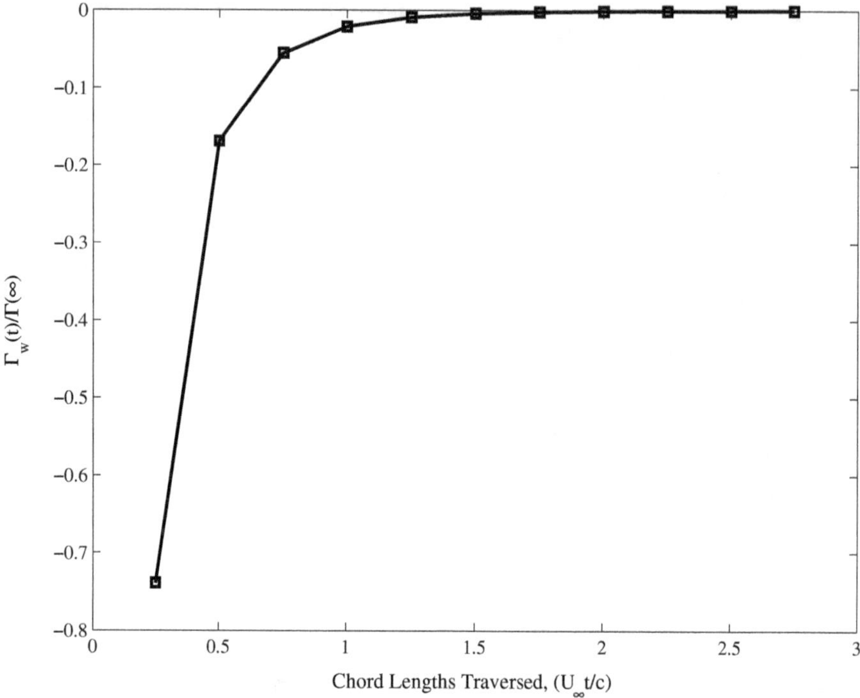

Fig. 3.5 Unsteady wake circulation, $\Gamma_w(t)$, for a flat plate suddenly started from rest at $5°$ angle of attack, modeled by the lumped-vortex method with 10 time steps

$$w_i(t) = -\mathbf{v} \cdot \mathbf{n}(x_i, y_i, z_i, t)$$

satisfied at the collocation points on the wing surface. Here $\mathbf{v}$ denotes the net velocity of the point (x_i, y_i, z_i) due to all kinematic effects (rotation, translation, deformation) of the wing. The net velocity of a wake panel is given by

$$\mathbf{v_w} = \mathbf{U}_\infty + \mathbf{C}\Gamma + \mathbf{D}\Gamma_\mathbf{w}, \tag{3.102}$$

where $\mathbf{U}_\infty$ is the freestream velocity, and $\mathbf{C}$, $\mathbf{D}$ are AIC matrices denoting the effects of wing and wake panels, respectively. The effect of the unsteady wake on the wing requires an iteration with time until a steady state is reached. This procedure includes calculation of the upwash induced by the wake panels on the wing, as well as the velocity induced by the wing and wake on each wake panel, thereby causing the wake to deform. The wake deformation changes the coordinates of the wake panels, and therefore affects $\mathbf{B}$, $\mathbf{C}$, $\mathbf{D}$. Wake circulation is fixed by Helmholtz theorem, which dictates that wake circulation $\Gamma_\mathbf{w}$ is the same for all wake panels coming from the same location on the wing trailing edge (Fig. 3.8), and the Kutta condition, which requires the wing circulation at the trailing edge is the same as that of the first wake panel. The solution procedure is schematically depicted by a flowchart in Fig. 3.9.

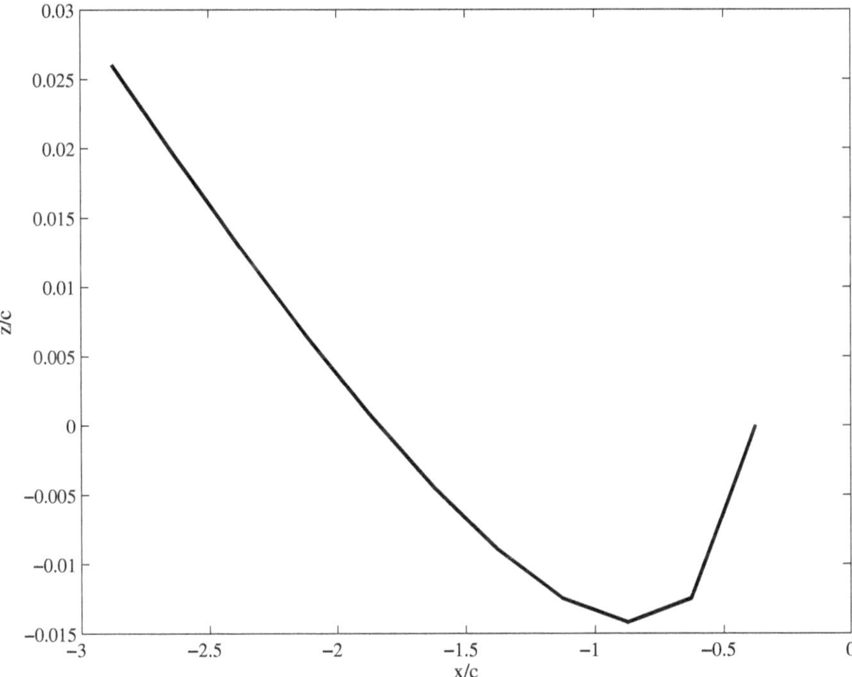

Fig. 3.6 Unsteady wake roll-up for a flat plate suddenly started from rest at 5° angle of attack, after 10 time steps

From earlier discussion of the equivalence between the circulation produced by a vorticity distribution and the constant doublet strength, it is expected that each vortex ring should give rise to exactly the same upwash distribution as that caused by a quadrilateral doublet element of constant strength. Katz and Plotkin [86] provide the closed-form aerodynamic influence coefficients between upwash and doublet (or circulation) strengths (Appendix B) for constant strength doublet elements of quadrilateral shape. However, it is much easier to program the equivalent influence coefficients calculation by the vortex-ring elements using Biot–Savart law. The pressure distribution on the wing is computed by the unsteady Bernoulli equation, expressed as follows:

$$\Delta p = p_\ell - p_u = \rho \left(\frac{v_\Gamma^2}{2}\big|_u - \frac{v_\Gamma^2}{2}\big|_\ell + \frac{\partial \Phi}{\partial t}\big|_u - \frac{\partial \Phi}{\partial t}\big|_\ell \right), \tag{3.103}$$

where v_Γ is the tangential velocity component on the upper (subscript u) and lower (subscript ℓ) surface responsible for circulatory flow, calculated by

$$v_\Gamma = (\mathbf{v} + \mathbf{v_w}) \cdot \boldsymbol{\tau} \frac{\partial \Phi}{\partial \tau}, \tag{3.104}$$

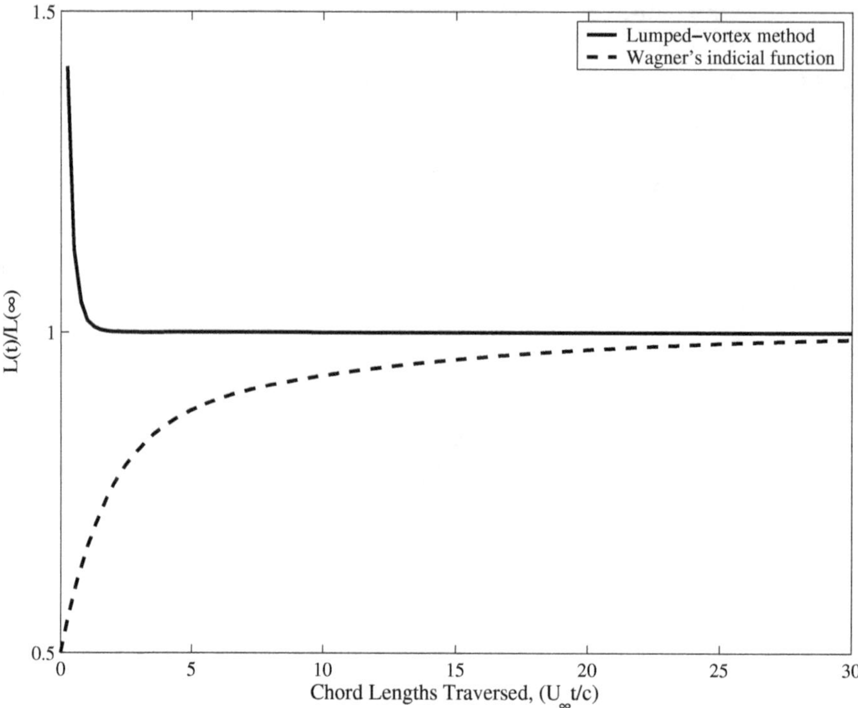

Fig. 3.7 Unsteady lift, $L(t)$, of a flat plate suddenly started from rest at 5 ° angle of attack, compared with Wagner's indicial lift function for 120 time steps

where $\boldsymbol{\tau}$ denotes the local unit vector tangential to the wing surface, which can be resolved in two mutually perpendicular–such as chordwise (**i**) and spanwise (**j**)– directions:

$$v_\Gamma = (\mathbf{v} + \mathbf{v_w}) \cdot \mathbf{i}\frac{\partial \Phi}{\partial x}$$
$$+ (\mathbf{v} + \mathbf{v_w}) \cdot \mathbf{j}\frac{\partial \Phi}{\partial y}. \tag{3.105}$$

In order to carry out the tangential and time derivatives of the velocity potential, Φ, finite-difference approximations as well as the relationship between local vortex-ring circulation and velocity potential difference between upper and lower surfaces, $\Gamma = \Delta\Phi$, are employed [4]. This ring-vortex lattice procedure is coded in a MATLAB code, which is applied to the examples considered here.

For an example application of the ring-vortex lattice method, consider a straight wing of aspect ratio 13.71 and mean aerodynamic chord, $\bar{c} = 0.5833$ m shown in

---

[4] Unfortunately, there is a mistake in Eq. (13.134) of Katz and Plotkin [86], which has been corrected here.

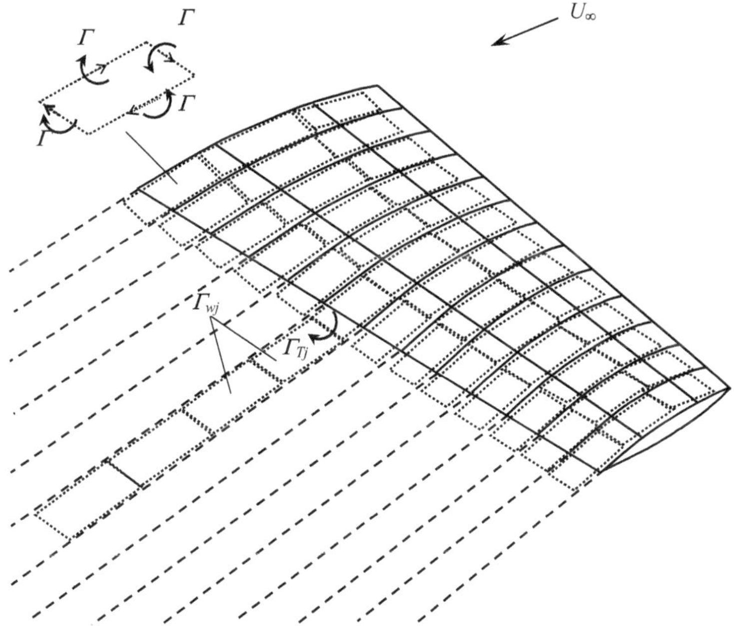

Fig. 3.8 Schematic representation of the vortex-ring model of an aircraft wing

Fig. 3.10, with a NACA 240 series [1] camber line as its thin airfoil section. The steady-state pressure distribution achieved after 0.1 semichord lengths ($b = \bar{c}/2$) of travel for speed $U_\infty = 30$ m/s, $\rho = 1.225$ kg/m^3, 5° angle of attack, is plotted in Fig. 3.11 for 20 chordwise and 10 spanwise divisions. The wing is now set into a pitching oscillation of amplitude 5° with frequency 200 rad/s, with angle of attack given by

$$\alpha(t) = 5\cos(200\Delta t)\ \text{deg.},$$

and the lift and induced drag coefficients are plotted in Fig. 3.12 for 5 semichord lengths of travel. The oscillation in the circulation about the wake for 20 time steps is evident in Fig. 3.13 at the various spanwise stations, and the corresponding unsteady pressure distribution is compared for three selected time instants in Fig. 3.14.

Now consider the wing plan form given above with a NLR-7301 supercritical airfoil shape, which has 17 % maximum thickness (Fig. 3.15). Figures 3.16–3.18 are the results of a combined pitching and heaving oscillation with different forcing frequencies of 100 and 50 rad/s, respectively, plotted for 50 time steps. The variation of lift and induced drag coefficients, as well as the wake circulation is seen to be nonharmonic, and the pressure distribution shows a similar unsteady behavior. Next, a spanwise bending mode of frequency 100 rad/s is considered with the following angle of attack variation:

$$\alpha(y) = \frac{0.5}{U_\infty}(y/s)^2$$

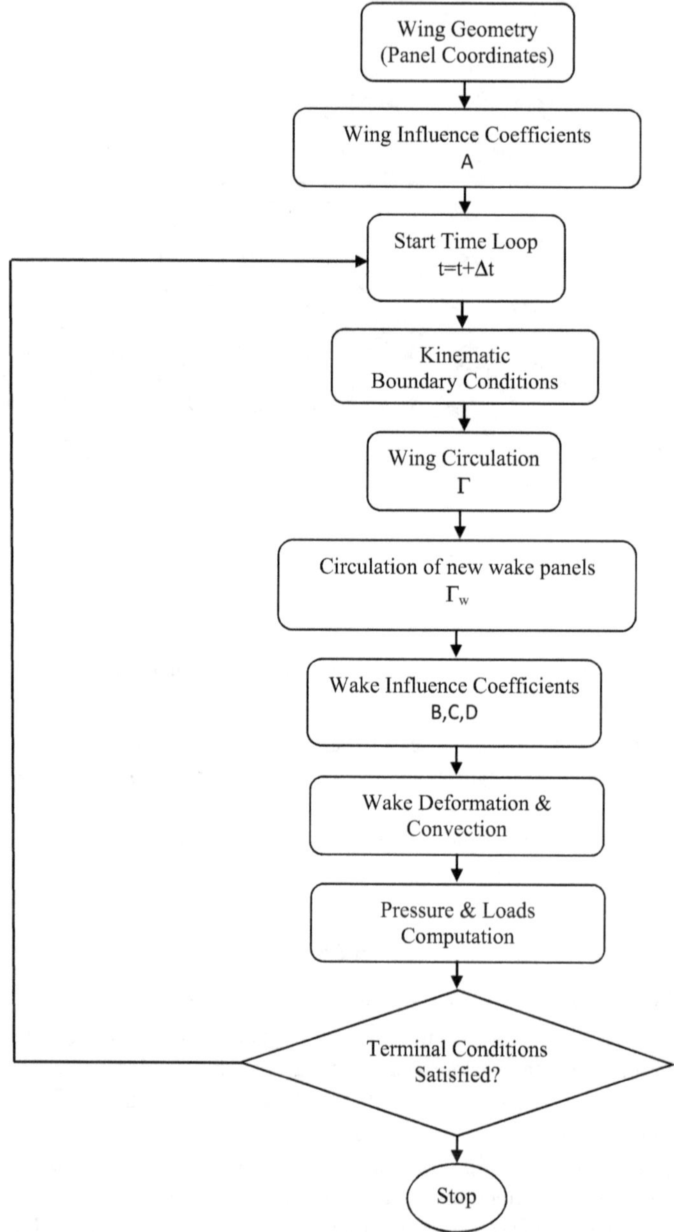

Fig. 3.9 Flowchart for unsteady Vortex-Lattice method

where s denotes the semispan. The resulting pressure distribution on the upper and lower surfaces is plotted in Figs. 3.19 and 3.20, respectively, with the unsteady wake circulation shown in Fig. 3.21. For illustration, only 15 chordwise and 10

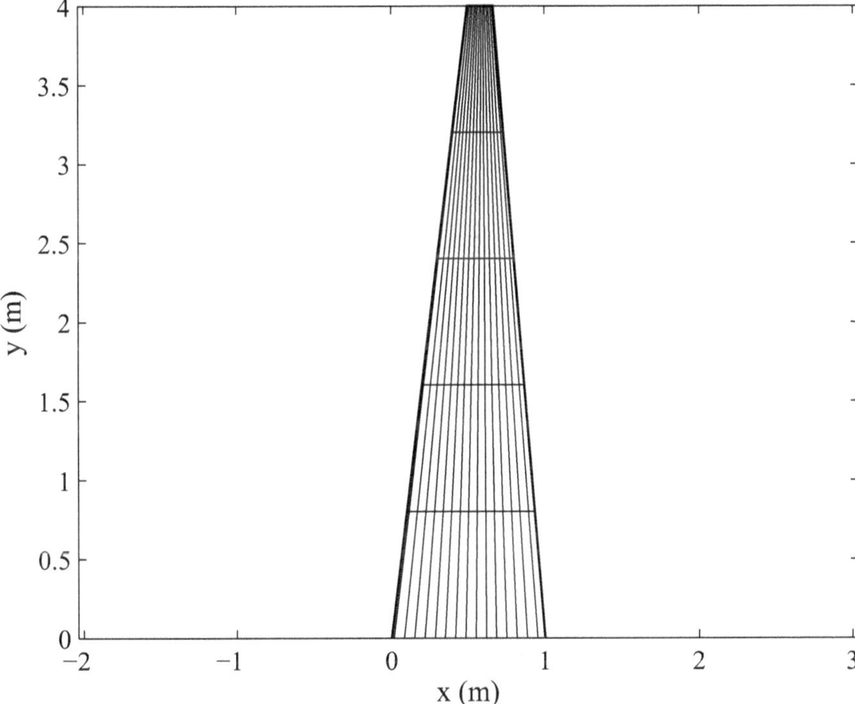

Fig. 3.10 Trapezoidal, straight wing of aspect ratio 13.71 shown with 15 chordwise and 5 spanwise divisions on each side of the semispan, and with equal number of panels on the upper and lower surfaces

spanwise divisions are taken here, but it is expected that a larger number of divisions is required for capturing the initial indicial response (not plotted) in the first 2–3 time steps. The supercritical airfoil section expectedly has a flat pressure distribution in the mid-chord range, with sharper gradients near the leading edge (acceleration) and the trailing edge (deceleration). The smaller extent of adverse (positive) pressure gradient on the upper surface of the supercritical airfoil is mainly responsible for its larger region of laminar flow (thus smaller viscous drag), as well as an increase in the critical Mach number in the high subsonic regime, when compared to a conventional airfoil in which the largest camber falls in the forward- to mid-chord region.

3.4.2 Classical Analytical Solution

Unsteady aerodynamic modeling as practiced today, would not be possible without the seminal contributions of early workers in the closed-form analysis of unsteady, two-dimensional, potential flows past thin airfoils. Birnbaum's [20] pioneering solution based on vortex distribution was further developed by Küssner [90], while

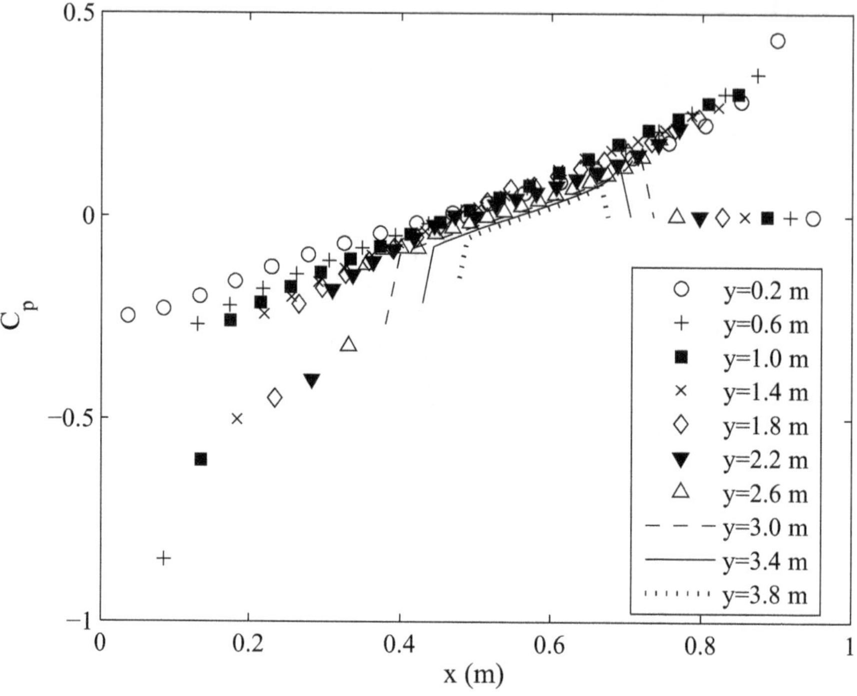

Fig. 3.11 Steady-state pressure distribution for a thin, high aspect-ratio wing with NACA 240 series camberline at 5 ° angle of attack, after 20 time steps

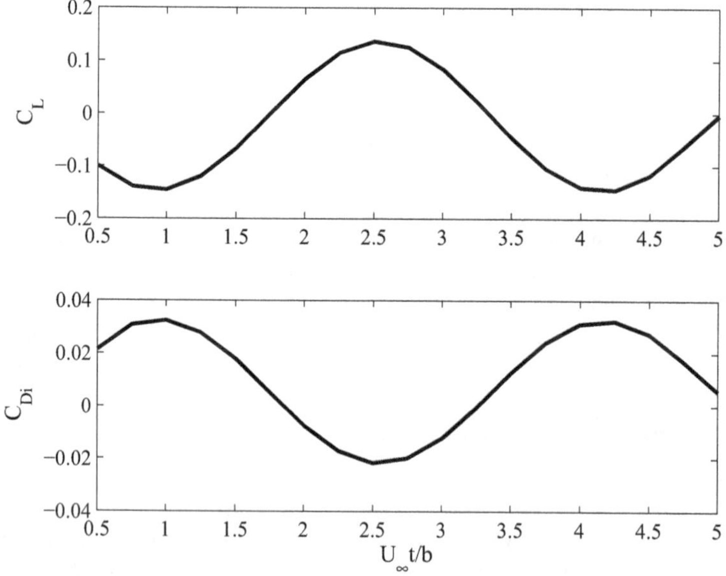

Fig. 3.12 Unsteady lift and induced drag coefficients for a thin, high aspect-ratio wing with NACA 240 series camberline, pitching with 5 ° angle-of-attack amplitude

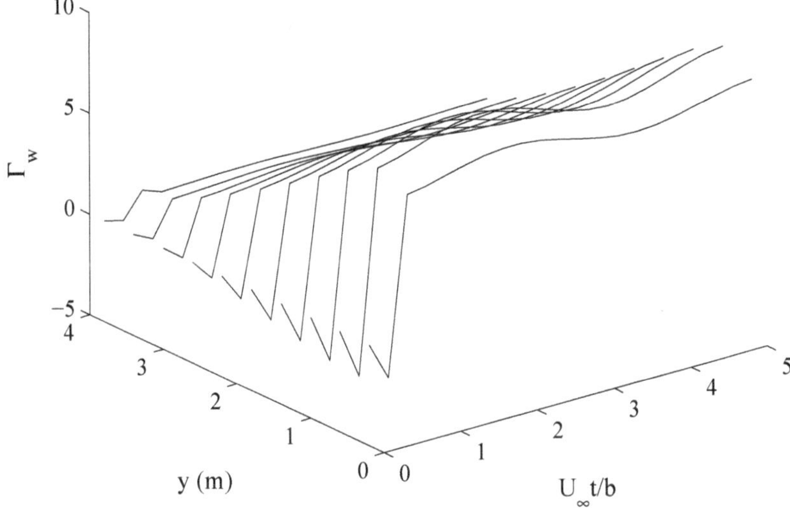

Fig. 3.13 Development of wake circulation for a thin, high aspect-ratio wing with NACA 240 series camberline, pitching with $5°$ angle-of-attack amplitude

the alternative (but equivalent) conformal mapping approach of Wagner [189] was followed by Glauert [63], Theodorsen [173], and von Karman and Sears [187]. The terminology and methods devised by these researchers have been widely practiced in almost all areas of aeroelastic analysis, and have also inspired developments in compressible flows. Here we will briefly review the early analytical work [58].

Theodorsen [173] considered conformal mapping of a pair of source and sink located at diametrically opposite ends of a circle to simulate the noncirculatory (n.c.)[5] flow pattern on a two-dimensional, flat-plate airfoil of chord $2b$ caused by normal velocity disturbance, $w(\xi, t)$, applied on an element $\Delta\xi$ located at point $x = \xi$. Here, x is a nondimensional coordinate measured along the airfoil in the streamwise direction from the mid-chord point, such that the leading edge is at $x = -1$, and the trailing edge corresponds to $x = 1$. When the conformal mapping is applied, the circle is transformed into the chord line, while preserving the velocity potential at the mapped points. The change in the n.c. velocity potential at a point x on the upper surface is derived to be the following:

$$\Delta\phi_{nc}(x,t) = \frac{1}{2\pi}w(\xi,t)b\,\Delta\xi\,L(x,\xi), \tag{3.106}$$

where

$$L(x,\xi) = 2\log\left|\frac{1 - x\xi - \sqrt{1 - x^2}\sqrt{1 - \xi^2}}{x - \xi}\right|. \tag{3.107}$$

---

[5] Not associated with the circulation due to wake.

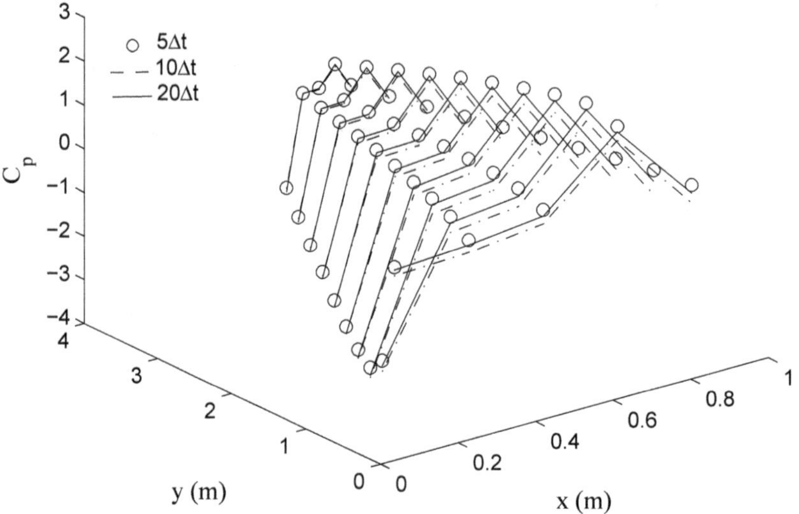

Fig. 3.14 Unsteady pressure distribution for a thin, high aspect-ratio wing with NACA 240 series camberline, pitching with 5 ° angle-of-attack amplitude

This result can be alternatively derived by Green's integral theorem with doublet (or vortex) distribution. The resulting n.c. pressure distribution is then given by the unsteady Bernoulli equation to be the following:

$$
\begin{aligned}
\Delta p_{nc}(x,t) &= -2\rho \left(\frac{\partial}{\partial t} + \frac{U_\infty}{b} \frac{\partial}{\partial x} \right) \Delta\phi_{nc} \\
&= -\frac{2}{\pi}\rho \left(\frac{U_\infty}{b} \frac{\sqrt{1-\xi^2}}{(x-\xi)\sqrt{1-x^2}} \right) bw\Delta\xi \\
&\quad - \frac{\rho}{\pi} b\Delta\xi L(x,\xi)\frac{\partial w}{\partial t}
\end{aligned}
\tag{3.108}
$$

Theodorsen then derived the circulatory potential, $\Delta\phi_c$, to cancel the noncirculatory upwash, $w(\xi,t)$, by an equal and opposite induced upwash, such that there is no pressure singularity, either at an arbitrary point, $x = \xi$, or at the trailing edge, $x = 1$. The circulation on the airfoil $\Delta\Gamma$ required for inducing the necessary upwash is provided by an element Δx_w of the force-free wake (Kelvin's theorem) behind the trailing edge, located at $x = x_w > 1$, with a vorticity distribution, $\gamma_w(x_w,t)$. By either conformal mapping, or Green's integral solution with doublet (or vortex) distribution, the circulatory perturbation potential integrated over all wake elements is obtained to be the following:

$$
\Delta\phi_c(x,t) = \frac{b}{2\pi} \int_1^s \gamma_w(x_w,t) \tan^{-1} \frac{\sqrt{1-x^2}\sqrt{x_w^2-1}}{1-xx_w} dx_w,
\tag{3.109}
$$

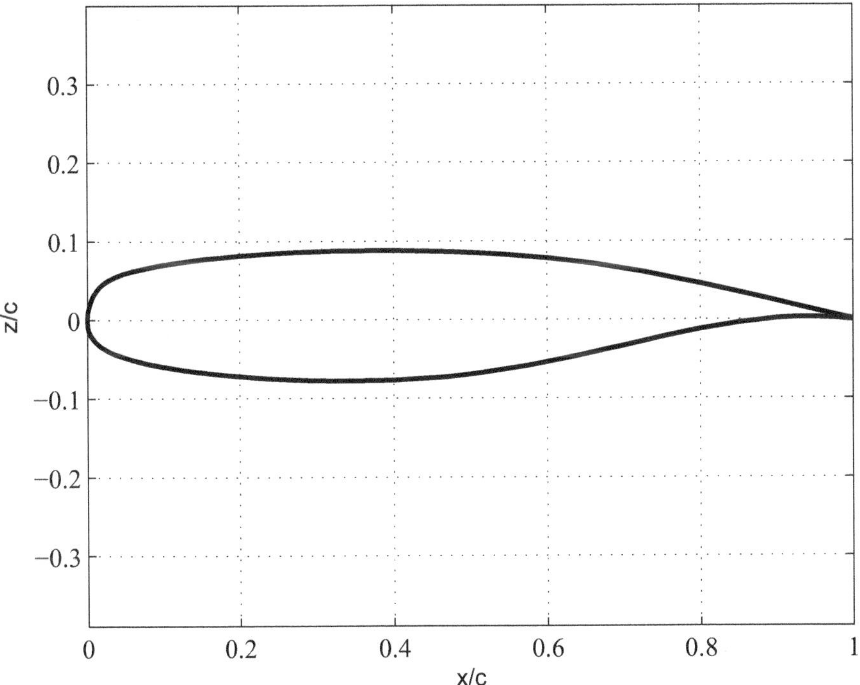

Fig. 3.15 NLR-7301 supercritical airfoil geometry

where the integration limits are for the wake extending from $x_w = 1$ to $x_w = s$, and conforms to the flow started at a time $t = (s - 1)b/U_\infty$ previously. The vorticity generated at the trailing edge convects downstream with the flow, and reaches a position $x = x_w$ after a time $t = (x_w - 1)b/U_\infty$ without any change in its strength. This implies that

$$\gamma_w(x_w, t) = \gamma_w\left[1, (x_w - 1)b/U_\infty\right],$$

which simplifies the solution of the integral in Eq. (3.109).

The circulation around the airfoil (and thus the wake) should be such that the flow leaves smoothly at $x = 1$ (Kutta condition). This condition translates into the requirement of the tangential velocity component at the trailing edge,

$$\frac{\partial}{\partial x}\left(\Delta\phi_{nc} + \Delta\phi_c\right)|_{x=1}. \tag{3.110}$$

being finite, and results in the following integral equation to be soved for $\gamma_w(x_w, t)$, subject to the upwash boundary condition $w(\xi, t)$ on the body surface:

$$\frac{1}{\pi} w(\xi, t)\Delta\xi \sqrt{\frac{1+\xi}{1-\xi}} = -\frac{1}{2\pi}\int_1^s \gamma_w(x_w, t)\sqrt{\frac{x_w + 1}{x_w - 1}}\,\mathrm{d}x_w, \tag{3.111}$$

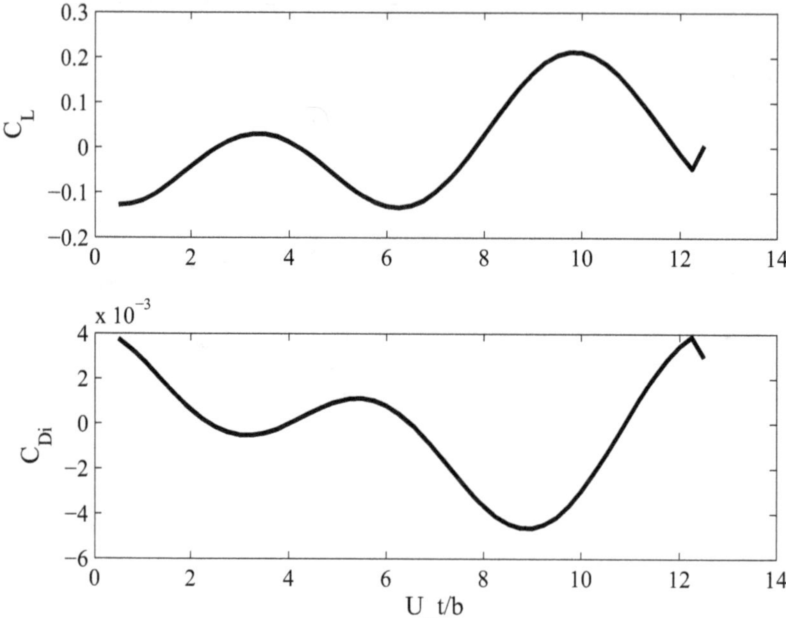

Fig. 3.16 Unsteady lift and induced drag coefficients for a high aspect-ratio wing with thick NLR7301 supercritical airfoil section, in a combined heaving and pitching oscillations of different forcing frequencies

Equation (3.111) parallels the development by Wagner [189] for a step change in $w(\xi, t)$ (such as due to an airfoil impulsively started from rest), as well as of Theodorsen [173] for the oscillatory (harmonic) upwash.

The circulatory pressure difference on the airfoil is derived by accounting for the fact that the unsteadiness in the velocity potential is caused only by the motion of wake $(\partial/\partial t(.) = (U_\infty/b)\partial/\partial x_w(.))$, resulting in the following:

$$\Delta p_c(x,t) = -2\rho \frac{U_\infty}{b} \left(\frac{\partial}{\partial x} + \frac{\partial}{\partial x_w} \right) \Delta\phi_c$$

$$= -\frac{\rho U_\infty}{\pi\sqrt{1-x^2}} \int_1^s \gamma_w(x_w,t) \frac{x+x_w}{\sqrt{x_w^2-1}} dx_w, \tag{3.112}$$

or

$$\Delta p_c(x,t) = \frac{2\rho U_\infty}{\sqrt{1-x^2}} \frac{\int_1^s \gamma_w(x_w,t) \frac{x+x_w}{\sqrt{x_w^2-1}} dx_w}{\int_1^s \gamma_w(x_w,t) \frac{\sqrt{1+x_w}}{\sqrt{x_w-1}} dx_w} \Delta Q(\xi,t), \tag{3.113}$$

where

$$\Delta Q(\xi,t) = \frac{1}{\pi} w(\xi,t)\Delta\xi \sqrt{\frac{1+\xi}{1-\xi}}, \tag{3.114}$$

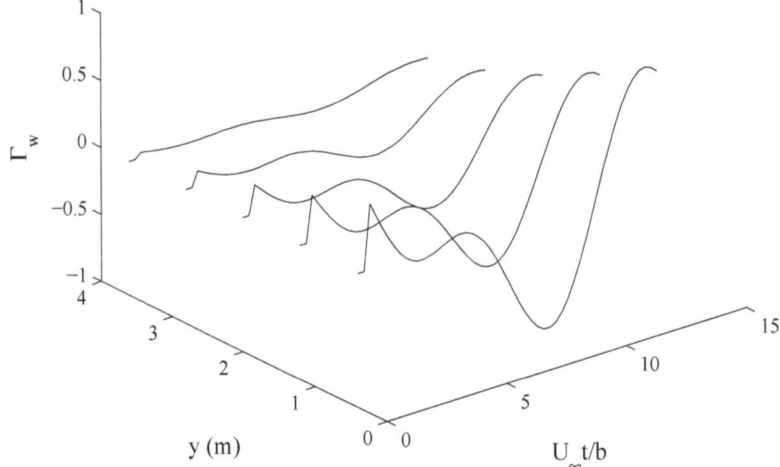

Fig. 3.17 Development of wake circulation for a high aspect-ratio wing with thick NLR7301 supercritical airfoil section, in a combined heaving and pitching oscillations of different forcing frequencies

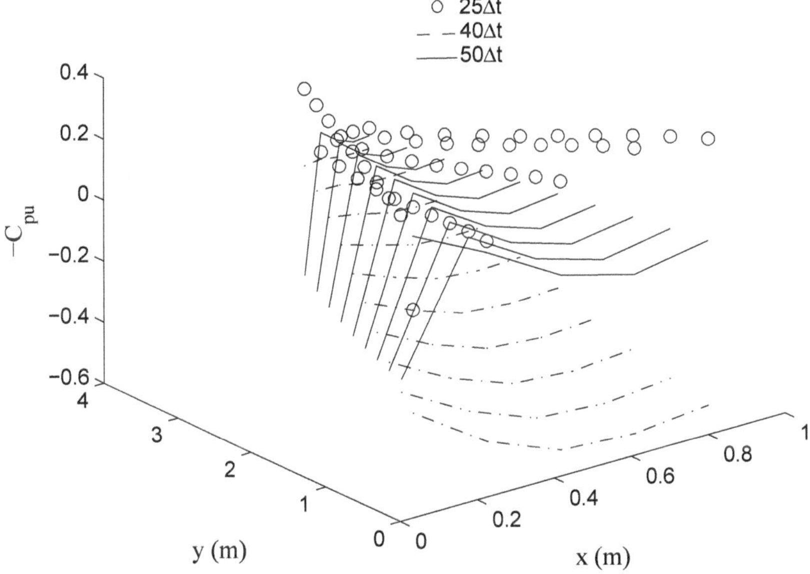

Fig. 3.18 Unsteady pressure distribution on the upper surface of a high aspect-ratio wing with thick NLR7301 supercritical airfoil section, in a combined heaving and pitching oscillations of different forcing frequencies

is the term on the left-hand side of the integral equation due to the noncirculatory upwash prescribed at a point $x = \xi$ on the wing.

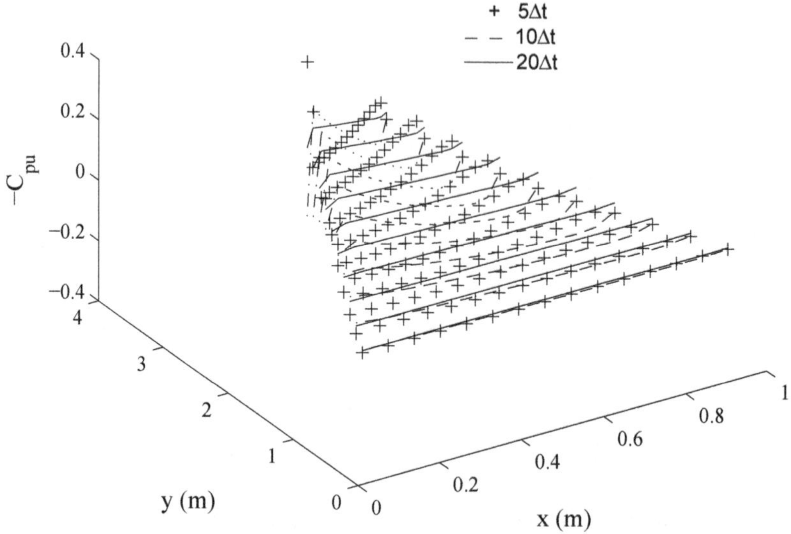

Fig. 3.19 Unsteady pressure distribution on the upper surface of a high aspect-ratio wing with thick NLR7301 supercritical airfoil section in a spanwise bending oscillation mode

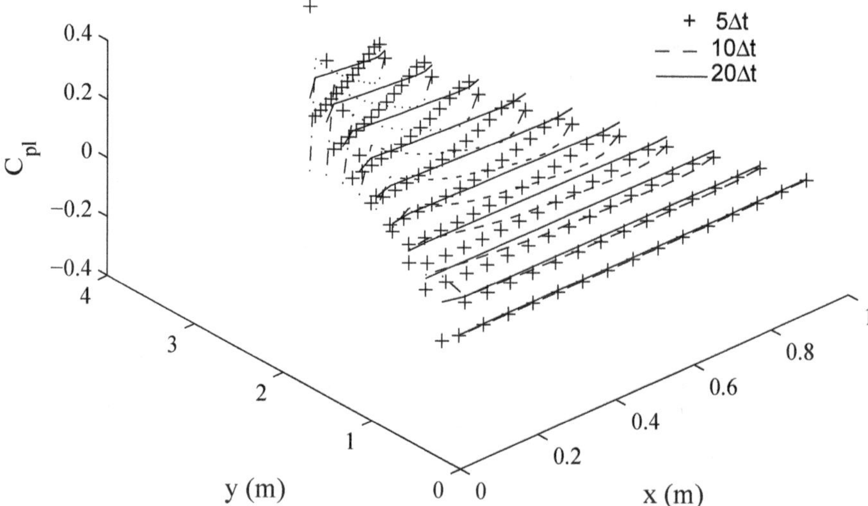

Fig. 3.20 Unsteady pressure distribution on the lower surface of a high aspect-ratio wing with thick NLR7301 supercritical airfoil section in a spanwise bending oscillation mode

While Wagner's [189] analysis of indicial airfoil motion–mentioned in the previous section–will find coverage in Chap. 4 for transient aerodynamics modeling, the remainder of the discussion here is confined to the simple harmonic motion, which is necessary for deriving the unsteady aerodynamic transfer function (Chap. 4). In

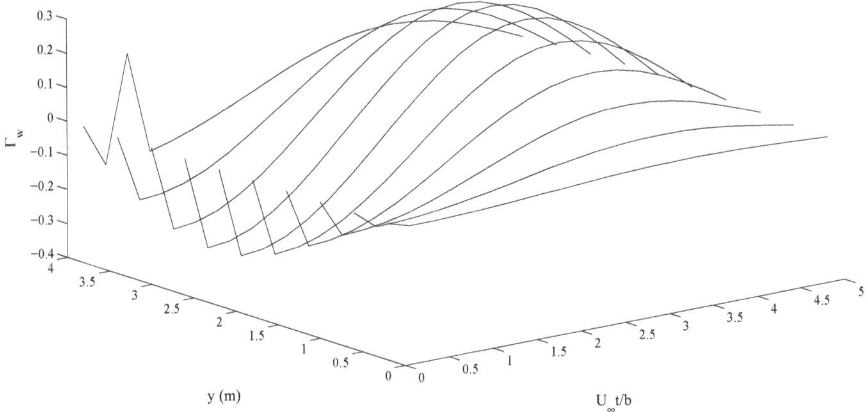

Fig. 3.21 Development of wake circulation for a high aspect-ratio wing with thick NLR7301 supercritical airfoil section in a spanwise bending oscillation mode

the harmonic limit, we have

$$w(x,t) = \bar{w}(x)e^{i\omega t}, \tag{3.115}$$

where ω is the frequency of oscillation and $\bar{w}(x)$ the (complex) upwash amplitude at a given point (prescribed by rigid and elastic motion). Furthermore, the wake response is also harmonic, and is given by the vorticity at a point $x = x_w$:

$$\gamma_w(x_w,t) = \bar{\gamma}e^{i\omega[t-(x_w-1)b/U_\infty]} = \bar{\gamma}e^{i[\omega t - k(x_w-1)]}, \tag{3.116}$$

where $\bar{\gamma}$ is the complex wake vorticity amplitude and

$$k = \omega b/U_\infty \tag{3.117}$$

is the reduced frequency representing number of waves in a wake length of 2π semichords, b. Clearly, k is the governing parameter of circulatory incompressible, irrotational flow. In the harmonic case, the wake is assumed to have developed to its full extent ($s \to \infty$) before small amplitude perturbation, $\bar{\gamma}$, is applied. This is analogous to an infinite sheet of vortices fixed to the wing, and oscillating at the excitation frequency. A change in the vorticity of the wake element at $x = x_w$ affects the circulation around the wing only after time $t = (x_w - 1)b/U_\infty$, therefore a phase-lag is inherent in the circulatory pressure distribution. However, in the limit $s \to \infty$ the exponential term on the right-hand side of Eq. (3.116) vanishes, and Δp_c can be expressed as follows:

$$\Delta p_c(x,t) = \Delta \bar{p}_c(x)e^{i\omega t}, \tag{3.118}$$

where

$$\Delta \bar{p}_c(x) = \frac{2\rho U_\infty}{\sqrt{1-x^2}} \frac{\int_1^s \frac{x+x_w}{\sqrt{x_w^2-1}}e^{-ikx_w}\mathrm{d}x_w}{\int_1^s \frac{1+x_w}{\sqrt{x_w^2-1}}e^{-ikx_w}\mathrm{d}x_w} \Delta \bar{Q}(\xi), \tag{3.119}$$

where

$$\bar{\Delta Q}(\xi) = \frac{1}{\pi}\bar{w}(\xi)\Delta\xi\sqrt{\frac{1+\xi}{1-\xi}}, \tag{3.120}$$

The chief problem with the formulation, Eq. (3.119), is the evaluation of the improper integrals in the limit $s \to \infty$ due to the oscillatory integrands. The difficulty is partly resolved by writing

$$\frac{\int_1^s \frac{x_w}{\sqrt{x_w^2-1}}e^{-ikx_w}\,\mathrm{d}x_w}{\int_1^s \frac{1+x_w}{\sqrt{x_w^2-1}}e^{-ikx_w}\,\mathrm{d}x_w} = 1 - \frac{\int_1^s \frac{e^{-ikx_w}}{\sqrt{x_w^2-1}}\,\mathrm{d}x_w}{\int_1^s \frac{e^{-ikx_w}}{\sqrt{x_w^2-1}}\,\mathrm{d}x_w + \int_1^s \frac{x_w e^{-ikx_w}}{\sqrt{x_w^2-1}}\,\mathrm{d}x_w}, \tag{3.121}$$

which leaves the only problematic integral as the following one:

$$I(k) = \int_1^s \frac{xe^{-ikx}}{\sqrt{x^2-1}}\,\mathrm{d}x \tag{3.122}$$

In the limit $s \to \infty$, this improper integral can be evaluated in two ways:

1. Consider the reduced frequency to be a complex number, such that $ik = a + ib$, $a > 0$, for which the integrand

$$\frac{xe^{-ikx}}{\sqrt{x^2-1}} = \frac{xe^{bx}e^{-iax}}{\sqrt{x^2-1}}$$

   vanishes identically in the limit $x \to \infty$. Thus we write

$$I(k) = \frac{\mathrm{d}}{\mathrm{d}k}\int_1^\infty \frac{ie^{-ikx}}{\sqrt{x^2-1}}\,\mathrm{d}x = \frac{\mathrm{d}}{\mathrm{d}k}\frac{\pi}{2}H_0^{(2)}(k) = -\frac{\pi}{2}H_1^{(2)}(k) \tag{3.123}$$

   where $H_n^{(2)}(.)$ is the Hankel function of second kind and order n [6]. Such a method of evaluating an improper, harmonic integral by converting the frequency to a complex number is called *analytic continuation*, and is equivalent to extending the simple harmonic motion to a more general one of either a growing ($ik = a+ib, a > 0$), or decaying oscillation ($ik = a+ib, a < 0$) at the given frequency k. Analytic continuation is a useful device, and will be further explored in Chap. 4.

2. Karman and Sears [187] proposed adding and subtracting a quasi-steady term from the integrand, which corresponds to the unsteady upwash assumed as being equal to its steady-state value. The additional quasi-steady term vanishes in the limit $s \to \infty$.

The integral $I(k)$ substituted into Eqs. (3.121) results in the following:

$$C(k) = \frac{\int_1^s \frac{x_w}{\sqrt{x_w^2 - 1}} e^{-ikx_w} \, \mathrm{d}x_w}{\int_1^s \frac{1+x_w}{\sqrt{x_w^2 - 1}} e^{-ikx_w} \, \mathrm{d}x_w}$$

$$= 1 - \frac{H_0^{(2)}(k)}{H_0^{(2)}(k) - i H_1^{(2)}(k)}$$

$$= \frac{H_1^{(2)}(k)}{H_1^{(2)}(k) + i H_0^{(2)}(k)}, \tag{3.124}$$

where $C(k)$ is called the *Theodorsen function*. This completes the derivation of analytical lifting pressure distribution, now expressed in a closed form as follows:

$$\Delta \bar{p}_c(x) = \frac{2\rho U_\infty}{\sqrt{1-x^2}} \left\{ C(k) + x\left[1 - C(k)\right]\right\} \Delta \bar{Q}(\xi)$$

$$= \frac{2\rho U_\infty}{\pi} \bar{w}(\xi)\Delta\xi \sqrt{\frac{1+\xi}{1-\xi}} \sqrt{\frac{1-x}{1+x}} \left[C(k) - 1 + \frac{1}{1-x}\right]. \tag{3.125}$$

The main utility of the Theodorsen function lies in its closed-form expression, which does not require iteration for the wake vorticity. This allows a direct use of $C(k)$ in a flutter analysis of low-speed aircraft. The work of Theodorsen was built upon the integral formulation for bound vorticity by Birnbaum [20] approximated by power series in reduced frequency, which was further improved by Küssner [90], and Glauert's [63] application of conformal mapping to the problem and deriving a Fourier series representation for circulation, followed by von Karman and Sears [187] demonstration that the infinite integrals occurring in Glauert's series coefficients can be approximated by Bessel functions. In an alternative approach to Theodorsen's, Schwarz [150] presented a solution to the two-dimensional, incompressible, oscillatory problem by applying Söhngen's [160] inversion to the integral equation of a vortex sheet representing the airfoil and wake, with Kutta condition applied at the trailing edge. This approach was further developed by Söhngen [161] and Küssner and Schwarz [92] using Cauchy principal value and Fourier series, respectively. Garrick [58] later demonstrated a solution to the velocity potential integral equation using a separation of variables. Smilg and Wasserman [159] tabulated the values of the Theodorsen function for prescribed cases, and proposed a strip-wise application for flutter calculation of unswept, finite-span wings.

Analytical model of the Theodorsen function is a valuable mathematical tool, and has also served as an example for similar models for the compressible flow regimes. However, Theodorsen's model is only a first-order approximation of wake development, and hence cannot account for the wake shape (roll-up) as a vortex-lattice method can (see above). Furthermore, it does not include thickness and camber effects. Thus, low-speed flutter stability analysis of aircraft wings and helicopter blade dynamics are its only practical applications.

3.5   Integral Equation for Linear Compressible Flow

The integral equation governing lift distribution and upwash for compressible flows can be derived in two alternative ways: (a) Green's integral solution with velocity potential sources (or doublets), and (b) small-perturbation acceleration potential formulation by unsteady Bernoulli equation. While the acceleration potential method has been covered above, the Green's integral formulation of the previous section was confined to incompressible flows. We therefore begin with the extension of the latter for the linearized compressible flow.

3.5.1   Velocity Potential Formulation by Green's Theorem

Green's divergence theorem [88] leads to the following integral relationship between any two continuously differentiable functions f and ϕ, which along with their first order spatial derivatives are finite and single valued in a domain $\mathcal{V}$ bounded by a surface $\mathcal{S}$ with outward normal $\mathbf{n}$:

$$\int_{\mathcal{V}} \left(f \nabla^2 \phi - \phi \nabla^2 G \right) d\mathcal{V} = \int_{\mathcal{S}} \left(G \frac{\partial \phi}{\partial n} - \phi \frac{\partial G}{\partial n} \right) d\mathcal{S} \qquad (3.126)$$

We have seen above the special case in which ϕ was the solution to the Laplace equation, and G was a singularity (Green's) function that also satisfied the Laplace equation. For the following linearized, compressible wave equation expressed in terms of the small perturbation velocity potential, ϕ, in a frame moving with the freestream:

$$\nabla^2 \phi = \frac{1}{a_\infty^2} \frac{\partial^2 \phi}{\partial t^2}, \qquad (3.127)$$

the field potential analogous to Eq. (3.69) is given by the following *Kirchoff's formula*, which is also a result of classical acoustics (*Huygens' principle*):

$$\phi(x,y,z,t) = \frac{1}{4\pi} \int_{\mathcal{S}} \left[\frac{1}{r} \left(\frac{\partial \phi}{\partial n} \right) \Big|_{t-r/a_\infty} - \frac{\partial}{\partial n} \frac{(t-r/a_\infty)\phi}{r} \right] d\mathcal{S}, \qquad (3.128)$$

where $r = \sqrt{(x-\xi)^2 + (y-\eta)^2 + (z-\zeta)^2}$ is the distance between an infinitesimal disturbance produced at (ξ, η, ζ) and the field point (x, y, z). An elegant representation of Eq. (3.128) is the following:

$$\phi(x,y,z,t) = \frac{1}{4\pi} \int_{\mathcal{S}} f \left(t - \frac{r}{a_\infty} \right) d\mathcal{S}, \qquad (3.129)$$

where

$$f(t) = \frac{1}{r} \frac{\partial \phi}{\partial n} + \frac{1}{a_\infty r} \frac{\partial \phi}{\partial t} \frac{\partial r}{\partial n} \qquad (3.130)$$

Clearly, an infinitesimal disturbance produced at (ξ, η, ζ) is felt only after a delay (or time lag) at field point (x, y, z) by r/a_∞. Thus time appears everywhere in the compressible flow case as the term $t - r/a_\infty$, and the resulting perturbation potential ϕ is called the *retarded potential*. Furthermore, the time lag limits a disturbance from a point source to remain within a sphere which continuously expands with time, and hence the unsteady behavior (coupling of space and time variables) is inherent in the compressible case. In the incompressible flow limit ($a_\infty \to \infty$), the dependence on time vanishes from the governing equation, which is transformed from the wave equation to the Laplace equation.

The fundamental solutions of the wave equation appearing in Green's integral are the disturbance due to a stationary point source of unit strength, given by

$$\phi_s(r,t) = -\frac{1}{4\pi r} f\left(t - \frac{r}{a_\infty}\right), \tag{3.131}$$

and that due to a stationary point doublet of unit strength,

$$\phi_d(r,t) = -\frac{1}{4\pi}\frac{\partial}{\partial n}\left[\frac{1}{r} f\left(t - \frac{r}{a_\infty}\right)\right], \tag{3.132}$$

where the function $f(t)$ represents the fluctuation of strength with time at a distance r from the source. In the harmonic limit, the source and doublet strengths pulsate with frequency ω, resulting in

$$\phi_s(r,t) = -\frac{1}{4\pi r} e^{i\left(t - \frac{r}{a_\infty}\right)} = -\frac{1}{4\pi r} e^{-i\kappa r} e^{i\omega t} \tag{3.133}$$

$$\phi_d(r,t) = -\frac{1}{4\pi}\frac{\partial}{\partial n}\left(\frac{1}{r} e^{-i\kappa r}\right) e^{i\omega t}, \tag{3.134}$$

where $\kappa = \omega/a_\infty$ is an important flow parameter called the *wave number*. These forms of unit source and doublet solutions automatically satisfy the far-field Sommerfeld condition, $\phi(\infty, t) = 0$, and thus a distribution of pulsating sources or doublets of varying strength on the wing surface can be used to model its net perturbation to the flowfield. However, since the wing is a moving object, the effect of moving sources (or doublets) in an otherwise undisturbed medium must be analyzed.

A further discussion of the Green's solution can be carried out alternatively by (a) the classical acoustic principles (Helmholtz formulation and Doppler effect), (b) the invariance principle of the wave equation by Galilean and Lorentz coordinate transformation [93], or (c) linear superposition of sources of a strength that is constant with time [135]. Since the last method promises to offer a better insight into the physics of unsteady compressible flows, we shall briefly consider it here along the lines of Garrick [58].

Consider a moving point source of unit strength creating an isolated impulsive disturbance in the flowfield at time $t = \tau$, given by

$$f(t) = f(\tau)\delta(t - \tau), \tag{3.135}$$

where $\delta(.)$ is the Dirac delta function. The cumulative effect of the moving source, $\xi(\tau), \eta(\tau), \zeta(\tau)$, is given by the linear superposition (convolution) integral

$$\phi_s(r,t) = -\frac{1}{4\pi} \int_{-\infty}^{t-r/a_\infty} \frac{f(\tau)}{r} \delta\left(t - \tau - \frac{r}{a_\infty}\right) d\tau, \qquad (3.136)$$

When we model a uniform wing motion with speed U_∞ in the negative x direction by a distribution of sources arranged along the ξ axis, we have

$$r = \sqrt{(x + U_\infty \tau)^2 + y^2 + z^2}$$

and the sampling property of the convolution integral results in

$$\phi_s(r,t) = -\frac{1}{4\pi} \left[\frac{f(\tau)}{r} \frac{d\tau}{d\theta} \right]_{\tau = t - r/a_\infty}, \qquad t > \tau \qquad (3.137)$$

where

$$\theta = \tau - \left(t - \frac{r}{a_\infty}\right)$$

Solving the quadratic equation for τ, we have

$$\tau = \frac{1}{\beta^2}\left(t + \frac{U_\infty x}{a_\infty^2} - \frac{R}{a_\infty}\right)$$

where

$$R = \sqrt{(x + U_\infty \tau)^2 + \beta^2(y^2 + z^2)}$$

$$\beta^2 = 1 - M^2$$

and $M = U_\infty/a_\infty$ is the freestream Mach number.

The velocity potential due to the moving unit source can be now expressed as follows:

$$\phi_s(r,t) = -\frac{1}{4\pi R} f\left[\frac{1}{\beta^2}\left(t + \frac{Mx}{a_\infty} - \frac{R}{a_\infty}\right)\right] \qquad (3.138)$$

The following Galilean transformation to the flowfield coordinates moving with the freestream:

$$x + U_\infty t \Rightarrow x - \xi; \quad y \Rightarrow y - \eta; \quad z \Rightarrow z - \zeta,$$

where the source coordinates are (ξ, η, ζ), results in the separate subsonic and supersonic formulations.

Subsonic Flow

$$\phi_s(r,t) = -\frac{1}{4\pi R} f(t - \tau_1)$$
(3.139)

where

$$R = \sqrt{(x - \xi)^2 + \beta^2 \left[(y - \eta)^2 + (z - \zeta)^2 \right]}$$

$$\tau_1 = \frac{-M(x - \xi) + R}{a_\infty \beta^2}$$

$$\beta^2 = 1 - M^2$$

This result implies that the solution for a moving source of unit strength can be derived from that of the stationary source merely by replacing r by R in the amplitude, and by $a_\infty \tau_1$ in the phase (or time lag). For a moving doublet, the velocity potential is derived by taking a derivative of Eq. (3.139) in the spatial coordinate n along the doublet axis.

For a harmonically pulsating, moving source, the result is the following:

$$\phi_s(r,t) = -\frac{1}{4\pi R} e^{-i\kappa[R + M(x-\xi)]} e^{i\omega t}$$
(3.140)

where

$$\kappa = \frac{\omega}{a_\infty \beta^2}$$

is the wave number.

Now we consider the mean surface (midplane) of the wing as a flat surface S, on the upper part ($z = 0+$) of which a distribution of velocity potential sources of strength per unit area, $A(\xi, \eta)$, is placed. On the lower surface of the midplane ($z = 0-$), we have a distribution of sinks of the same strength. The wing is moving in an initially undisturbed medium in the negative x direction. The cumulative perturbation velocity potential produced by the wing at a field point (x, y, z) is thus given by the sum of individual perturbations (Eq. (3.139)) of all the sources and sinks as the following integral:

$$\phi(x, y, z, t) = -\frac{1}{4\pi} \iint_S \frac{A(\xi, \eta)}{R} f(t - \tau_1) \, d\eta d\xi,$$
(3.141)

The integral equation is to be solved for the unsteady source-sink strength, $A(\xi, \eta) f(t - \tau_1)$, given a distribution of the velocity potential $\phi(x, y, z, t)$. As remarked earlier, such a solution requires numerical inversion of the integral equation by specifying ϕ at a large number of grid points in the entire flowfield. This is always a cumbersome process, and can be avoided in the following manner.

If one takes the derivative in the z direction of Eq. (3.141), the result is a distribution of sources and sinks of equal strengths separated by an infinitesimal distance, thereby implying a distribution of velocity potential doublets pointing in the negative z-direction into the wing's midplane. The cumulative effect of such doublets results in the following subsonic integral equation :

$$w(x, y, z, t) = -\frac{1}{4\pi} \iint_S A(\xi, \eta) \frac{\partial}{\partial z} \left(\frac{1}{R} \right) f(t - \tau_1) \, d\eta d\xi, \tag{3.142}$$

where $w(x, y, z, t) = \partial\phi/\partial z$ denotes the unsteady upwash at a field point (x, y, z). This integral equation must be solved for the time-dependent doublet strength distribution on the wing midplane,

$$A(\xi, \eta) f(t - \tau_1),$$

given an upwash distribution $w(x, y, z, t)$ on the wing. This requires taking the limit $z \to 0$, for the wing's midplane:

$$w(x, y, 0, t) = -\lim_{z \to 0} \frac{1}{4\pi} \iint_S A(\xi, \eta) \frac{\partial}{\partial z} \left(\frac{1}{R} \right) f(t - \tau_1) \, d\eta d\xi$$

Since the upwash is continuous across the wing midplane, $z = 0$, the integral equation is uniquely solved if one prescribes the unsteady upwash distribution on the wing due to its vertical deformation, $h(x, y, t)$:

$$w(x, y, 0, t) = \frac{dh}{dt} = \frac{\partial h}{\partial t} + U_\infty \frac{\partial h}{\partial x}$$

By substituting the velocity potential solution for the harmonic case,

$$w(x, y, 0, t) = \bar{w} e^{i\omega t} = \left(i\omega \bar{h} + U_\infty \frac{\partial \bar{h}}{\partial x} \right) e^{i\omega t},$$

we have the following relationship between the complex amplitudes of the upwash and the source strength on the wing:

$$\bar{w}(x, y) = -e^{-i\kappa M x} \lim_{z \to 0} \frac{\partial}{\partial z} \frac{1}{4\pi} \iint_S \bar{A}(\xi, \eta) \frac{1}{R} e^{-i\kappa(R - M\xi)} d\eta d\xi. \tag{3.143}$$

Supersonic Flow

The velocity potential due to a source of unit strength moving in the negative direction with supersonic speed is given as follows:

$$\phi_s(r, t) = -\frac{1}{4\pi R} [f(t - \tau_1) + f(t - \tau_2)] \tag{3.144}$$

where

$$R = \sqrt{(x - \xi)^2 - \beta^2 \left[(y - \eta)^2 + (z - \zeta)^2 \right]}$$

$$\tau_1 = \frac{M(x - \xi) - R}{a_\infty \beta^2}$$

$$\tau_2 = \frac{M(x - \xi) + R}{a_\infty \beta^2}$$

and

$$\beta^2 = M^2 - 1$$

A physical interpretation of this result is obtained by considering the fact that a supersonic source moving in the negative x direction creates a *Mach cone* of disturbance described by

$$(x - \xi)^2 = \beta^2 \left[(y - \eta)^2 + (z - \zeta)^2\right]$$

Hence, $R = 0$ at a point (x, y, z) on the Mach cone, and R is a real (positive) number at a point inside it. Outside the Mach cone, R is imaginary, and thus undefined, thereby implying that a point located outside the cone is unaffected by the pressure disturbance created by the moving source. Clearly, the Mach cone presents a discontinuity in the velocity potential flowfield. Now, consider a field point (x, y, z) located inside the Mach cone at a time instant t. Such a point experiences the perturbation of an advancing (moving in positive x direction) spherical wave front represented by $f(t - \tau_1)$ (as in the subsonic case), and also that due to a retreating spherical wave front $f(t - \tau_2)$. The sum of the two perturbations is the total velocity potential at the given point in space and time. The distances $a_\infty \tau_1$ and $a_\infty \tau_2$ are the respective radii of the advancing and retreating wave fronts at the given instant, and give the respective locations of the source at the past time instants τ_1 and τ_2 from the concerned field point.

For a harmonically pulsating, moving source, the result is the following:

$$\phi_s(r, t) = -\frac{1}{4\pi R} \left(e^{-i\omega\tau_1} + e^{-i\omega\tau_2}\right) e^{i\omega t} \tag{3.145}$$

or

$$\phi_s(r, t) = -\frac{1}{2\pi R} e^{-i\kappa M(x - \xi)} \cos(\kappa R) e^{i\omega t} \tag{3.146}$$

where

$$\kappa = \frac{\omega}{a_\infty(M^2 - 1)} = \frac{\omega}{a_\infty \beta^2}$$

is the wave number.

For a superposition of sources of strength per unit area $A(\xi, \eta)$ on the upper surface of the thin wing, $z = 0+$, and an equal strength sink distribution on the lower surface, $z = 0-$, the following integral relationship can be derived between the velocity potential at a field point (x, y, z) and the unknown strength of the source-sink distribution, $A(\xi, \eta) f(t - \tau_1) + f(t - \tau_2)$, in a manner similar to that given above for the subsonic case:

$$\phi(x, y, z, t) = -\frac{1}{4\pi} \iint_{S_R} \frac{A(\xi, \eta)}{R} \left[f(t - \tau_1) + f(t - \tau_2) \right] d\eta d\xi, \qquad (3.147)$$

where the area of integration S_R is now only that part of the wing which lies inside the forward-facing Mach cone emanating from the field point (point O in Fig. 3.22). This is due to the fact (noted above) that at any given time t, only the sources lying inside the forward-facing Mach cone can affect the field point. Consequently, the integration limits can be defined as follows:

$$\phi(x, y, z, t) = -\frac{1}{4\pi} \int_0^{x_m} \int_{y_1(\xi)}^{y_2(\xi)} \frac{A(\xi, \eta)}{R} \left[f(t - \tau_1) + f(t - \tau_2) \right] d\eta d\xi, \qquad (3.148)$$

where

$$x_m = x \pm z\beta \qquad (3.149)$$

and

$$y_1(\xi) = y - \sqrt{\frac{1}{\beta^2}(x - \xi)^2 - z^2}$$

$$y_2(\xi) = y + \sqrt{\frac{1}{\beta^2}(x - \xi)^2 - z^2}. \qquad (3.150)$$

When the field point is located on the midplane, we have $z = 0$, thus $x = x_m$. In other cases, the region of integration is bounded by the Mach hyperbola (shown in Fig. 3.22), which is the intersection of the Mach cone and the midplane. The correct sign for the maximum upstream extent (or apex) of the Mach hyperbola, x_m, in Eq. (3.149) depends upon whether the field point lies above or below the midplane of the wing. If $z > 0$, the negative sign is taken, while the positive sign is used for $z < 0$. The values of spanwise extent of the Mach cone is given by Eq. (3.150). If the wing lies completely inside the Mach cone, we have $S_R = S$. Here we note the important fact that unlike the subsonic case, a source located on the upper surface of the wing cannot affect the flow below the wing (and vice versa) due to the zone of influence being confined to the Mach cone. Thus the two flow regions lying on the either side of the midplane are mutually independent in the supersonic flow. In summary, the physical characteristics unique to the supersonic case are the following:

1. Flow perturbations caused by a given point are confined to the Mach cone emanating in the downstream direction from that point.
2. The flow at a point in the supersonic flowfield can be affected only by the points lying inside the upstream facing Mach cone emanating from the given point.
3. The points above the wing do not affect the flow below the wing, and vice versa.

An important consequence of the last feature is whereas the subsonic flow requires that both upper and lower surface source and sink pairs (doublets) contribute to the

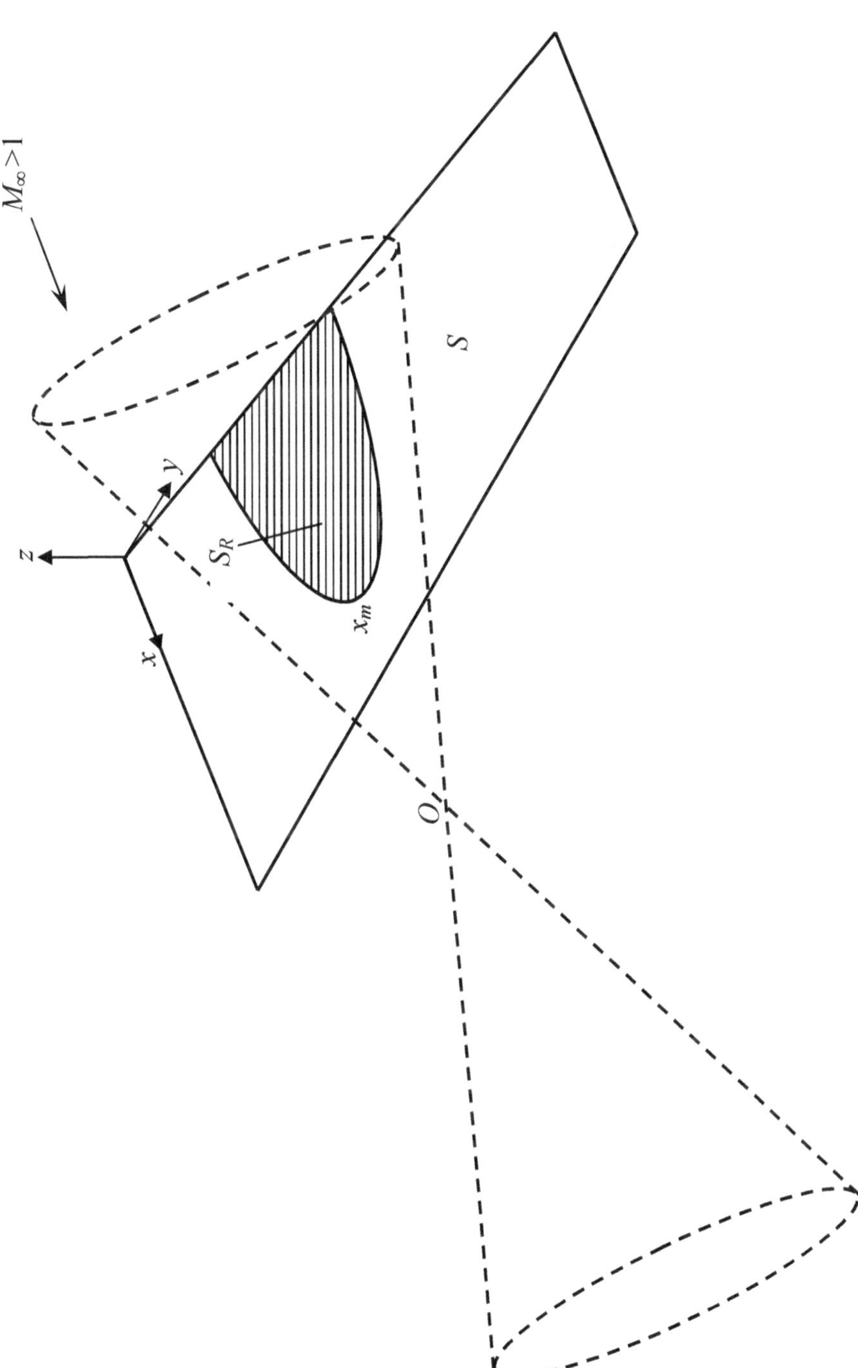

Fig. 3.22 Mach cones emanating from a point O in a supersonic flow, and the region of integration for wing's influence

net velocity potential at a field point, the supersonic flow at a given point is only affected by the sources (or sinks) on the relevant side of the midplane. Therefore, while a velocity potential doublet is the natural singularity solution of the subsonic flow, the velocity potential source is the characteristic solution for the supersonic case.

For a distribution of sources and sinks on the upper and lower surfaces of the midplane of a moving, thin wing, the following integral relationship can be derived between the upwash at a field point (x, y, z) and the unknown strength of the source-sink distribution, $A(\xi, \eta)$, by taking a derivative in the z direction on the wing:

$$w(x, y, 0, t) = -\lim_{z \to 0} \frac{1}{4\pi} \frac{\partial}{\partial z} \int_0^{x_m} \int_{y_1(\xi)}^{y_2(\xi)} \frac{A(\xi, \eta)}{R} \left[f(t - \tau_1) + f(t - \tau_2) \right] d\eta d\xi,$$

$$(3.151)$$

However, taking the derivative is more difficult in the supersonic case due to the dependence of the integration limits on z. The matter is resolved by using the following transformation of the spanwise integration variable's variation on the Mach hyperbola:

$$\eta(\xi) = \frac{1}{2} \left[(y_2 - y_1) \cos \theta + (y_2 + y_1) \right], \qquad (3.152)$$

or

$$d\eta = -\frac{1}{2} (y_2 - y_1) \sin \theta d\theta, \qquad (3.153)$$

a substitution of which into Eq. (3.151) produces

$$w(x, y, 0, t) = -\sqrt{M^2 - 1} \lim_{z \to 0} \frac{1}{4\pi} \frac{\partial}{\partial z} \int_0^{x_m} \int_0^{\pi} A(\xi, \theta) (f_1 + f_2) d\theta d\xi, \quad (3.154)$$

where

$$f_1 = f \left[t - \frac{M(x - \xi)}{a_\infty \beta^2} + \frac{y_0 \sin \theta}{a_\infty \beta} \right]$$

$$f_2 = f \left[t - \frac{M(x - \xi)}{a_\infty \beta^2} - \frac{y_0 \sin \theta}{a_\infty \beta} \right], \qquad (3.155)$$

and

$$y_0 = \frac{1}{2} (y_2 - y_1). \qquad (3.156)$$

After expanding the integral in Eq. (3.154), and using integration by parts, the following result is finally derived [57]:

$$w(x, y, 0, t) = -\sqrt{M^2 - 1} \lim_{z \to 0} \frac{1}{2} \frac{\partial x_m}{\partial z} A(x_m, y) f \left(t - \frac{Mz}{a_\infty \beta} \right), \qquad (3.157)$$

or,

$$w(x, y, 0+, t) = \frac{1}{2}\beta^2 A(x, y) f(t)$$

$$w(x, y, 0-, t) = -\frac{1}{2}\beta^2 A(x, y) f(t) \tag{3.158}$$

which directly relates the upwash to the source and sink strengths, respectively, on the upper and lower surfaces.

In the case of harmonically pulsating sources (i.e., the wing undergoing simple harmonic motion), the velocity potential is expressed as

$$\phi(x, y, z, t) = \bar{\phi}(x, y, z)e^{i\omega t},$$

and its complex amplitude is related to that of the upwash in the midplane by [6]

$$\bar{\phi}(x, y, 0\pm) = \mp\frac{1}{2\pi\beta} \int_0^x \int_{y_1(\xi)}^{y_2(\xi)} \bar{w}(\xi, \eta, 0)\frac{\left(e^{-i\omega\tau_1} + e^{-i\omega\tau_2}\right)}{\sqrt{(\eta - y_1)(y_2 - \eta)}}d\eta d\xi, \tag{3.159}$$

which is simplified to the following:

$$\bar{\phi}(x, y, 0\pm) = \mp\frac{1}{\pi}e^{-i\kappa M x} \int_0^x \int_{y_1(\xi)}^{y_2(\xi)} \bar{w}(\xi, \eta, 0)\frac{e^{i\kappa M\xi}}{\hat{R}}\cos(\kappa \hat{R})d\eta d\xi, \tag{3.160}$$

where

$$\hat{R} = \sqrt{(x - \xi)^2 - \beta^2(y - \eta)^2}$$

and

$$\kappa = \frac{\omega}{a_\infty \beta^2}$$

is the wave number.

After the velocity potential is computed from the time-dependent upwash prescribed on the wing, the differential pressure distribution on the wing,

$$\Delta p(x, y, t) = \bar{\Delta p}(x, y)e^{i\omega t},$$

is calculated by the unsteady Bernoulli equation, Eq. (3.53):

$$\Delta p(x, y, t) = p(x, y, 0+, t) - p(x, y, 0-, t) = 2p(x, y, 0, t)$$

$$= -2\rho_\infty \left[\left(\frac{\partial\phi}{\partial t} + U_\infty\frac{\partial\phi}{\partial x}\right)\right]_{z=0}. \tag{3.161}$$

---

[6] Note the upper limit of the streamwise integral is $x_m = x$ in the limit $z \to 0$.

Hence, the pressure-upwash integral equation is the following:

$$\frac{\bar{\Delta p}(x,y)}{\rho_\infty} = \frac{2}{\pi}\left(i\omega + U_\infty\frac{\partial}{\partial x}\right)\int_0^x \int_{y_1(\xi)}^{y_2(\xi)} \bar{w}(\xi,\eta,0)\frac{e^{-i\kappa M(x-\xi)}}{\hat{R}}\cos(\kappa\hat{R})d\eta d\xi$$

$$(3.162)$$

When numerical solution based upon the velocity potential integral equation is attempted, the problem of subsonic wing edges is inevitably encountered. If the freestream Mach number M is such that a wing edge makes a smaller angle from the freestream direction (x-axis) than the angle of the downstream facing Mach cone, $\sin^{-1}(1/M)$, then the integration region includes a part of the (x,y) plane which is *not* on the wing, but lies between the Mach cone and the wing edge. Since the upwash $w(x,y,0,t)$ cannot be geometrically prescribed at an off-wing point, this ambiguity must be resolved by other considerations (which, unfortunately, have little physical basis). Such a problem is avoided entirely by basing the solution procedure on the acceleration potential, rather than the velocity potential.

3.5.2 Acceleration Potential Formulation

The basic integral relationship between upwash at a field point due to a pressure distribution on the wing for the unsteady linearized (subsonic or supersonic) flow can be derived from Eq. (3.54) by taking the z-derivative of both the sides and substituting the expression for upwash,

$$w = \frac{\partial \phi}{\partial z},\qquad(3.163)$$

resulting in

$$w(x,y,z,t) = -\frac{1}{\rho_\infty U_\infty}\frac{\partial}{\partial z}\int_{-\infty}^x p\left(\xi,y,z,t-\frac{x-\xi}{U_\infty}\right)d\xi.\qquad(3.164)$$

Since pressure disturbance is related to perturbation acceleration potential by Eq. (3.55), the integral equation Eq. (3.164) is referred to as the acceleration potential formulation. This integral equation must be solved for the pressure difference across the mean surface, $z = 0$, given a prescribed upwash distribution, Eq. (3.51), on it:

$$w(x,y,0,t) = -\lim_{z\to 0}\frac{1}{\rho_\infty U_\infty}\frac{\partial}{\partial z}\int_{-\infty}^x p\left(\xi,y,z,t-\frac{x-\xi}{U_\infty}\right)d\xi.\qquad(3.165)$$

The upper limit of integration, x, is handled differently for subsonic and supersonic cases. In the supersonic case, the pressure disturbance cannot travel upstream of the Mach cone, therefore the upper limit of integration is defined by the upstream extent of the Mach cone, x_m (Fig. 3.22), whereas no such restriction exists for the subsonic case. For the harmonically oscillating wing, we have

$$w(x, y, 0, t) = \bar{w}(x, y)e^{i\omega t}; \quad (x, y) \in S, \tag{3.166}$$

which results in the following integral expression for the complex upwash amplitude (magnitude and phase) in the frequency domain :

$$\bar{w}(x, y) = -\lim_{z \to 0} \frac{1}{\rho_\infty U_\infty} e^{-i\omega x/U_\infty} \frac{\partial}{\partial z} \int_{-\infty}^{x} \left[\bar{p}\,(\xi, y, z)\, e^{i\omega \xi/U_\infty} \right] d\xi, \tag{3.167}$$

where

$$p(x, y, z, t) - p_\infty = \bar{p}(x, y, z)e^{i\omega t}, \tag{3.168}$$

is the harmonic pressure difference with a complex amplitude, $\bar{p}$.

It is interesting to note the equivalence of the integral equation, Eq. (3.167), with the Green's identity resulting from the superposition of harmonically pulsating, acceleration potential doublets. Such a formulation is employed in the subsonic doublet-lattice method [7], and the subsonic/supersonic doublet-point method [180, 181], all of which are discussed below. The disturbation acceleration potential ψ is directly related to the pressure difference (hence the disturbance velocity potential) in the flowfield by

$$\psi = -\frac{p - p_\infty}{\rho_\infty} = \frac{\partial \phi}{\partial t} + U_\infty \frac{\partial \phi}{\partial x},$$

and satisfies the following wave equation (identical to that for ϕ) in a frame moving with the freestream:

$$\nabla^2 \psi = \frac{1}{a_\infty^2} \frac{\partial^2 \psi}{\partial t^2} \tag{3.169}$$

Now, since both ψ and ϕ satisfy the same governing equation, they can be used to fulfill the Green's integral equation, Eq. (3.126), wherein a velocity potential source solution, ϕ_s, is equivalent to the acceleration potential doublet solution, ψ_d, in creating the same pressure disturbance at a given field point. Since we have already derived the velocity potential solution for source, it is now straightforward to obtain the equivalent acceleration doublet solution.

In the further discussion, we will confine our attention to the harmonic case, for which the complex amplitude $\bar{\psi}$ satisfies the Helmholtz equation,

$$\nabla^2 \bar{\psi} = \frac{1}{a_\infty^2} \left(U_\infty \frac{\partial}{\partial x} + i\omega \right)^2 \bar{\psi}. \tag{3.170}$$

The upwash boundary condition on the vibrating wing planform S, with complex vertical deflection amplitude, $\bar{h}(x, y)$, is given by

$$\bar{w}(x, y) = \left(U_\infty \frac{\partial}{\partial x} + i\omega \right) \bar{h}(x, y), \quad (x, y) \in S \tag{3.171}$$

Subsonic Flow

Noting the equivalence between ψ and ϕ in Green's integral, and utilizing the source velocity potential solution for the subsonic case (Eq. (3.140)), we have the following for a moving acceleration potential doublet of unit harmonic strength:

$$\psi(\xi, \eta, \zeta, t) = -\frac{p - p_\infty}{\rho_\infty} = -\frac{1}{4\pi} \frac{\partial}{\partial z} \left(\frac{1}{R} e^{-i\kappa[R + M(x - \xi)]} \right) e^{i\omega t}, \qquad (3.172)$$

where

$$r = \sqrt{(x - \xi)^2 + (y - \eta)^2 + (z - \zeta)^2}$$

$$R = \sqrt{(x - \xi)^2 + \beta^2(y - \eta)^2 + (z - \zeta)^2}$$

$$\kappa = \frac{\omega}{a_\infty \beta^2}$$

and $\beta^2 = 1 - M^2$. Here, the field point is at (x, y, z), whereas the doublet is located at (ξ, η, ζ) (which is also the convention in the velocity potential formulation).

By substituting Eq. (3.172) into Eq. (3.167), the upwash amplitude induced by a jump in the strength, $\Delta\bar{\psi} = -4\pi$, of a harmonically pulsating acceleration potential doublet at (ξ, η), when crossing from lower to upper surface (i.e., in positive z direction) of the mean wing plane S, where

$$\Delta\bar{\psi} = -\frac{\bar{p}_\ell - \bar{p}_u}{\rho_\infty}, \qquad (3.173)$$

is given by the following *kernel function*[7]:

$$K(x_0, y_0, \omega, M) = -\lim_{z \to 0} e^{-i\omega x_0 / U_\infty} \frac{\partial^2}{\partial z^2} \int_{-\infty}^{x_0} \frac{e^{i \frac{\omega}{U_\infty \beta^2} \left[\lambda - M\sqrt{\lambda^2 + \beta^2(y_0^2 + z^2)} \right]}}{\sqrt{\lambda^2 + \beta^2(y_0^2 + z^2)}} d\lambda,$$

$$(3.174)$$

where $x_0 = x - \xi$, $y_0 = y - \eta$. Therefore, the upwash amplitude $\bar{w}$ induced by the pressure difference (or lift per unit area) distribution,

$$\Delta p = \bar{p}_\ell - \bar{p}_u = -\Delta\bar{\psi}\rho_\infty,$$

---

[7] The strength 4π of the doublet for defining the kernel is the convention adopted by the workers in this area, and relates to a sphere of unit radius centered at the doublet as the elemental control volume in Green's integral.

on the wing is given by the following integral:

$$\bar{w}(x, y) = \frac{1}{4\pi \rho_\infty U_\infty} \iint_S K(x_0, y_0, \omega, M)\bar{\Delta}p(\xi, \eta)\mathrm{d}\eta\mathrm{d}\xi \qquad (3.175)$$

The simple form of this integral equation is to be contrasted with the velocity potential formulation (Eq. (3.143)). The numerical scheme for inversion of this integral requires an evaluation of the improper kernel function (to be discussed later).

Supersonic Flow

The moving acceleration potential doublet of unit strength in supersonic flow yields the following result by virtue of Eq. (3.146):

$$\frac{p(\xi, \eta, \zeta) - p_\infty}{\rho_\infty} = \frac{1}{4\pi} \frac{\partial}{\partial z} \left(\frac{e^{-i\omega\tau_1} + e^{-i\omega\tau_2}}{R} \right) e^{i\omega t}$$

$$= \frac{1}{2\pi} \frac{\partial}{\partial z} \left(\frac{e^{-i\kappa M(x-\xi)} \cos(\kappa R)}{R} \right) e^{i\omega t} \qquad (3.176)$$

where

$$R = \sqrt{x_0^2 - \beta^2 r_0^2}; \quad r_0 = \sqrt{y_0^2 + z_0^2}$$

$$\tau_1 = \frac{Mx_0 - R}{a_\infty \beta^2}$$

$$\tau_2 = \frac{Mx_0 + R}{a_\infty \beta^2}$$

$$\beta^2 = M^2 - 1$$

$$\kappa = \frac{\omega}{a_\infty \beta^2}$$

and $x_0 = x - \xi$, $y_0 = y - \eta$. Since the influence of the doublet is confined within the Mach cone, the perturbation exists only if $x_0 > \beta r_0$. This fact can be indicated in the following manner, by using the unit step function, $u_s(.)$:

$$\frac{p - p_\infty}{\rho} = \frac{1}{2\pi} \frac{\partial}{\partial z} \left(\frac{e^{-i\kappa Mx_0} \cos(\kappa R)}{R} \right) u_s(x_0 - \beta r_0)e^{i\omega t}$$

By substituting Eq. (3.176) into Eq. (3.167), the upwash amplitude caused by a unit harmonic strength acceleration potential doublet is the following supersonic kernel function :

$$K(x_0, y_0, M, \omega) = -\lim_{z \to 0} e^{-i\omega x_0/U_\infty} \frac{\partial^2}{\partial z^2} \int_{\beta r_0}^{x_0}$$

$$\left\{ \frac{e^{-i\frac{\omega}{U_\infty \beta^2}\left[\lambda - M\sqrt{\lambda^2 - \beta^2 r_0^2}\right]} + e^{-i\frac{\omega}{U_\infty \beta^2}\left[\lambda + M\sqrt{\lambda^2 - \beta^2 r_0^2}\right]}}{\sqrt{\lambda^2 - \beta^2 r_0^2}} \right\} d\lambda, \qquad (3.177)$$

or

$$K(x_0, y_0, M, \omega) = -\lim_{z \to 0} 2e^{-i\omega x_0/U_\infty} \frac{\partial^2}{\partial z^2} \int_{\beta r_0}^{x_0} \frac{e^{-i\frac{\omega\lambda}{U_\infty \beta^2}} \cos\left(\frac{\omega M}{U_\infty \beta^2}\sqrt{\lambda^2 - \beta^2 r_0^2}\right)}{\sqrt{\lambda^2 - \beta^2 r_0^2}} d\lambda.$$

$$(3.178)$$

The supersonic kernel has singularities on the Mach cone ($\lambda = \beta r_0$), which is the lower limit of the integral. These are to be treated carefully, as will be discussed below.

3.6 Subsonic Kernel Function and the Doublet-Lattice Method

The subsonic integral equation is expressed differently for planar and nonplanar cases. This is necessary due to singularities in the kernel function of the integral equation. The planar case solely considers the effect of pressure distribution on an essentially flat mean surface ($z = 0$), whereas in a nonplanar case the mean lifting surface configuration could consist of several flat panels, which need not be all in the same plane. In such cases, the vertical separation, z, between the surface creating a pressure disturbance, and the point on which the upwash is calculated must be accounted for. Our discussion here will be confined to the planar case for simplicity, but the reader can refer to publications [7, 96] where the nonplanar kernel is addressed.

We recapitulate that the upwash induced by an oscillatory subsonic flow past a planar lifting surface S (Fig. 3.1)—regarded as a sheet of harmonically pulsating acceleration potential doublets—is expressed in a nondimensional form as follows [192]:

$$\frac{\bar{w}}{U_\infty}(x, y) = -\frac{1}{4\pi U_\infty^2} \iint_S K(x_0, y_0, k, M)\, \Delta\bar{\psi}(\xi, \eta) d\xi d\eta, \qquad (3.179)$$

where (x, y) is the nondimensional location of the point on the wing at which the upwash amplitude, $\bar{w}$, is prescribed, and $K(x_0, y_0, k, M_\infty)$ is the subsonic kernel function (derived above) relating the upwash to the unknown amplitude, $\Delta\bar{\psi}$, of jump in the strength of a harmonically pulsating acceleration potential doublet at (ξ, η), when crossing from the lower to the upper surface of the mean wing plane denoted by S, $x_0 = x - \xi$, $y_0 = y - \eta$, and

$$k = \frac{\omega b}{U_\infty}, \qquad (3.180)$$

is the *reduced frequency* of harmonic motion with b as the characteristic length. The doublet strength can be expressed as the nondimensional pressure difference,

$$\frac{\Delta \bar{p}}{q_\infty} = \frac{\bar{p}_\ell - \bar{p}_u}{q_\infty} = -2\frac{\Delta \bar{\psi}}{U_\infty^2}, \tag{3.181}$$

Thus the following alternative nondimensional expression is found in the literature [7, 22]:

$$\frac{\bar{w}}{U_\infty}(x, y) = \frac{1}{8\pi} \iint_S K\,(x_0, y_0, k, M_\infty)\,\frac{\Delta \bar{p}}{q_\infty}(\xi, \eta)\mathrm{d}\xi\,\mathrm{d}\eta, \tag{3.182}$$

and will be utilized here.

The integral equation, Eq. (3.182), is to be solved for the unknown pressure amplitude distribution, $\Delta \bar{p}$, given a prescription of the upwash amplitude, $\bar{w}$, at selected points on the wing. Since the kernel function, $K(x_0, y_0, k, M)$, which physically relates the influence of the pressure distribution on the upwash, is crucial to the solution of the integral equation, the former should be in a suitable form for numerical evaluation.

For a planar lifting surface, the subsonic kernel function can be expressed as follows:

$$K\,(x_0, y_0, k, M) = -\lim_{z\to 0}\frac{\partial^2}{\partial z^2}e^{-ikx_0}\int_{-\infty}^{x_0}\frac{1}{R}e^{ik(\lambda-MR)/\beta^2}\,\mathrm{d}\lambda, \tag{3.183}$$

with $\beta = \sqrt{1 - M^2}$, and $R = \sqrt{\lambda^2 + \beta^2 y_0^2 + \beta^2 z^2}$.

The integral of Eq. (3.183) for the kernel function is both improper and singular, therefore its evaluation requires special treatment, especially at the lower limit of the integral and for the singularity in the limit $y_0 \to 0$. Such an evaluation offers the main computational difficulty in estimating the harmonic subsonic air loads. Watkins et al. [192] have shown that the kernel can be expanded as follows:

$$
\begin{aligned}
-K\,(x_0, y_0, k, M) = k^2 e^{-ikx_0} &\left\{ \frac{1}{k\mid y_0\mid}K_1\,(k\mid y_0\mid) + \frac{i\pi}{2k\mid y_0\mid}\left[I_1\,(k\mid y_0\mid)\right.\right.\\
&\left. - L_1\,(k\mid y_0\mid)\right] - \frac{ikM\mid y_0\mid + \beta}{M\beta(ky_0)^2}e^{-(ikM\mid y_0\mid)/\beta}\\
&+ \int_0^{M/\beta}\sqrt{1+\chi^2}e^{-ik\mid y_0\mid\chi}\mathrm{d}\chi\\
&- \frac{i}{M(ky_0)^2}\int_0^{kx_0}e^{i\left[\lambda - M\sqrt{\lambda^2+\beta^2(ky_0)^2}\right]/\beta^2}\mathrm{d}\lambda\\
&+ \left[\frac{Mkx_0 + \sqrt{(kx_0)^2 + \beta^2(ky_0)^2}}{M(ky_0)^2\sqrt{(kx_0)^2 + \beta^2(ky_0)^2}}\right]\\
&\left. e^{i\left[kx_0 - M\sqrt{(kx_0)^2+\beta^2(ky_0)^2}\right]/\beta^2}\right\},
\end{aligned}
\tag{3.184}
$$

where $I_1(.)$ is the modified Bessel function of first kind and first order, $L_1(.)$ the modified Struve function of first order, and $K_1(.)$ the modified Bessel function of second kind and second order, all of which can be found in a handbook on mathematics [6]. For treating the kernel function's singularities for the planar case, Watkins et al. [192], following the earlier method of Schwarz [150], separate the kernel function into a singular part, K', and a nonsingular part, $(K - K')$:

$$K = K' + (K - K'),$$

and suggest Mangler's [109] approach to derive the finite part (or principal value) of the following singular kernel:

$$K'(x_0, y_0, k, M) = -e^{-ikx_0} \left[-\frac{x_0 + \bar{R}}{y_0^2 \bar{R}} + ik\frac{1}{\bar{R}} \right.$$
$$\left. -\frac{k^2}{2\beta^2} \frac{x_0 - M\bar{R}}{\bar{R}} - \frac{k^2}{2} \log \frac{k(\bar{R} - x_0)}{2(1 - M)} \right], \tag{3.185}$$

where

$$\bar{R} = \sqrt{x_0^2 + \beta^2 y_0^2}.$$

The nonsingular kernel is continuous for all values of x_0, y_0, k, M in the subsonic range, and is expressed in the limit $y_0 \to 0$ as follows [192]:

$$\lim_{y_0 \to 0} (K - K') = -e^{-ikx_0} \left\{ \left(\frac{\beta^2}{2x_0^2} + ik\frac{1 + M}{2x_0} \right) e^{ikx_0/(1+M)} \right.$$
$$-\frac{\beta^2}{2x_0^2} - \frac{ik}{x_0} + \frac{k^2}{2} \left[\frac{M + 2}{M + 1} - 2\gamma - \log\left(\frac{kx_0}{M + 1} \right) \right.$$
$$\left. \left. + \text{Ci}\left(\frac{kx_0}{M + 1} \right) + i\,\text{Si}\left(\frac{kx_0}{M + 1} \right) - i\frac{\pi}{2} \right] \right\}, \tag{3.186}$$

where $\gamma = 0.577216$ is Euler's constant, and $\text{Si}(x), \text{Ci}(x)$ are sine-integral and cosine-integral functions of x, respectively:

$$\text{Si}(x) = \frac{\pi}{2} - \int_x^\infty \frac{\sin t}{t} dt$$

$$\text{Ci}(x) = -\int_x^\infty \frac{\cos t}{t} dt,$$

which are numerically evaluated [6].

The formulation of subsonic integral equation as a solution to the boundary value problem for unsteady pressure distribution on harmonically oscillating lifting surface was first introduced for the two-dimensional case by Possio [134] and followed by

Küssner [93]. The form of the kernel given in Eq. (3.184) was derived by Watkins, Runyan, and Woolston [192] by extending the two-dimensional airfoil formulation of Possio to the three-dimensional lifting surface. Landahl [96] offered an alternative formulation of the kernel, which could also be applied to nonplanar surfaces, and is even simpler than Eq. (3.184) for the planar case. This form of the kernel lends itself to practical numerical computation, without the recourse to Bessel and Struve functions.

The solution of the integral equation for pressure distribution—given a prescribed upwash distribution generated by structural vibration—can be carried out by two alternative methods. The first of these, called the *kernel function collocation method* assumes a form of pressure distribution by a finite series whose coefficients are determined algebraically from the upwash prescribed at discrete points on the surface (generally the wing edges and control surface hinge lines). The kernel function's singularities directly downstream of the pressure (acceleration potential) doublet ($y_0 \to 0$) are accommodated by using high-order quadrature in the spanwise direction. The kernel function collocation methods of Watkins et al. [194], Laschka [98], Landahl [97], Rowe [145], and Lottati and Nissim [103] are noteworthy.

The second solution approach, called the *doublet-lattice method* (DLM), approximately solves the integral equation by discretizing the pressure-upwash relationship into a matrix of *aerodynamic influence coefficients* (AIC). Here, the integral equation is discretized by assuming that the lifting surface can be approximated by a finite number trapezoidal panels, each of which is treated as a separate wing. The panels are arranged in columns parallel to the freestream, and the quarter-chord line of each panel is taken as a line doublet of acceleration potential (pressure) of unknown strength. The doublet strengths (therefore AICs) are evaluated by satisfying the upwash boundary condition at the three-quarter chord location (called *control point*) at the mid-span of each panel.[8] Since the control points are not located on the line doublets, the kernel function singularities are avoided by using interpolating functions approximation of the kernel while integrating within each panel. The AIC matrix can then be directly employed for the computation of unsteady generalized aerodynamic loads vector by the principle of virtual work. The doublet-lattice method is therefore of practical utility in aeroelastic computations where chordwise rigid sections can be also treated as panels for aerodynamic computation, thereby employing the same indexing for the structural degrees of freedom as well as panel control points. The upwash distribution is then directly derived from the vibration modes shapes. For this reason, doublet-lattice method has found wide application in the aircraft industry for flutter analysis, although it is not very well founded in aerodynamic theory. Due to its direct applicability to ASE modeling, the doublet-lattice method is detailed here (rather than the competing kernel function collocation method).

---

[8] Such a choice of line doublet and control point is borrowed from the *vortex-lattice method* [86], where a quarter-chord line vortex and a three-quarter chord control point for enforcing the upwash boundary condition automatically satisfy the Kutta condition for steady incompressible flow past a flat plate.

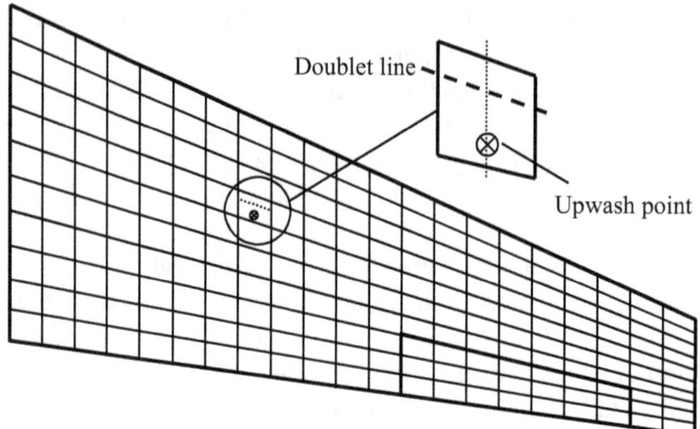

Fig. 3.23 Paneling of a wing for the doublet-lattice method

The doublet-lattice method is a systematic procedure evolved by Albano and Rodden [7] for solving the subsonic integral equation. It is motivated by the incompressible vortex-lattice, quasi-steady model of Hedman [72] for carrying out pressure calculations on lifting surfaces, but utilizes the acceleration potential doublet, rather than the velocity potential vortex, as the elementary solution of the Helmholtz (or wave) equation. The resulting integral equation associated with the linearized, unsteady subsonic flow (see above) represents the solution of the boundary-value problem by Green's identity [114] driven by unsteady motion of well-defined geometries. The alternative solution procedure [97, 98, 146, 192] of prescribing upwash for kernel function collocation is less than systematic, because of a discontinuous upwash distribution at control surface hinge lines and wing edges. In contrast, Albano and Rodden [7] use the aerodynamic influence matrix, $\mathbf{A}$, to approximate the integral equation by a discrete relationship between upwash distribution vector, $\bar{\mathbf{w}}$, and pressure difference distribution, $\Delta\bar{\mathbf{p}}$, evaluated over a number, N, of elemental boxes on the lifting surface. The discretized integral equation is expressed in a nondimensional form as follows:

$$\frac{\bar{w}}{U_\infty}(x_i, y_i) = \sum_{j=1, \neq i}^{N} A_{ij} \frac{\Delta\bar{\mathbf{p}}}{q_\infty}(\xi_j, \eta_j), \qquad (i = 1, \ldots, N) \qquad (3.187)$$

which is collected in the following matrix form:

$$\frac{\bar{\mathbf{w}}}{U_\infty} = \mathbf{A}\frac{\Delta\bar{\mathbf{p}}}{q_\infty}. \qquad (3.188)$$

Such a discretization procedure by trapezoidal boxes is depicted in Fig. 3.23, and is borrowed from the preexisting Mach-box method [129] for the supersonic case. The boxes are arranged in strips parallel to the freestream such that the boxes' sides, fold lines (dihedrals) and hinge lines fall on the box boundaries. The unknown doublet strength, $\Delta\bar{p}/q_\infty$, in each box is assumed to be smeared along the quarter-chord line. Once the upwash distribution is specified at the three-quarters chord, mid-span collocation point of the box by a normalized structural mode shape, $h(x, y)$, for a reduced frequency, $k = \omega b/U_\infty$,

$$\frac{\bar{w}}{U_\infty} = \frac{\partial h}{\partial x} + ikh, \tag{3.189}$$

it is substituted into Eq. (3.188), and the unknown pressure distribution is computed by inverting the aerodynamic influence matrix :

$$\frac{\Delta \bar{\mathbf{p}}}{q_\infty} = \mathbf{A}^{-1}\bar{\mathbf{w}}U_\infty, \tag{3.190}$$

The doublet-lattice method primarily involves the calculation of the aerodynamic influence matrix, $\mathbf{A}$, by integrating the kernel function relating the pressure (or sending) box with an upwash (or receiving) box. Such an integration, however, must be carried out in the spanwise direction in Mangler's sense [109], due to the singularities of the kernel function as $z_0 \to 0$ and $y_0 \to 0$. Before such an integration is carried out numerically, the kernel function of the subsonic integral equation, Eq. (3.182), is separated into two parts as follows [7]:

$$K = -e^{-ikx_0} \left(K_1 T_1 + K_2 T_2 \right)/r_1^2, \tag{3.191}$$

where

$$T_1 = \cos\left(\gamma(s) - \gamma(\sigma)\right), \tag{3.192}$$

$$T_2 = \left\{\frac{z_0}{r_1}\cos\gamma(s) - \frac{y_0}{r_1}\sin\gamma(s)\right\}\left\{\frac{z_0}{r_1}\cos\gamma(\sigma) - \frac{y_0}{r_1}\sin\gamma(\sigma)\right\}, \tag{3.193}$$

$$K_1 = I_1 + \frac{Mr_1}{R}e^{-ik_1u_1}\frac{1}{\sqrt{1 + u_1^2}}, \tag{3.194}$$

$$K_2 = -3I_2 - ik_1\frac{M^2 r_1^2}{R^2}e^{-ik_1u_1}\frac{1}{\sqrt{1 + u_1^2}}$$

$$- \frac{Mr_1}{R}\left[2 + (1 + u_1^2)\frac{\beta^2 r_1^2}{R^2} + \frac{Mr_1u_1}{R}\right]e^{-ik_1u_1}\frac{1}{\left(1 + u_1^2\right)^{3/2}}, \tag{3.195}$$

$$I_1 = \int_{u_1}^{\infty} e^{-ik_1u}\frac{1}{\left(1 + u^2\right)^{3/2}}\,du, \tag{3.196}$$

$$I_2 = \int_{u_1}^{\infty} e^{-ik_1u}\frac{1}{\left(1 + u^2\right)^{5/2}}\,du, \tag{3.197}$$

$$u_1 = \frac{MR - x_0}{\beta^2 r_1}, \tag{3.198}$$

$$k_1 = kr_1, \tag{3.199}$$

$$\beta = \sqrt{1 - M^2}, \tag{3.200}$$

$$R = \sqrt{x_0^2 + \beta^2 r_1^2}, \tag{3.201}$$

$$r_1 = \sqrt{y_0^2 + z_0^2}, \tag{3.202}$$

and the differences between the nondimensional Cartesian coordinates of the upwash point, (x, y, z), and load point, (ξ, η, ζ), are

$$x_0 = x - \xi, \qquad y_0 = y - \eta, \qquad z_0 = z - \zeta,$$

while the dihedral angle, $\gamma(.)$, of the nonplanar surface is evaluated at the upwash point and load points, respectively, by the curvilinear coordinates, s and σ, as shown in Fig. 3.24. In the planar case, we have $\gamma(s) = \gamma(\sigma) = 0$, which implies

$$T_1 = 1 \tag{3.203}$$

$$T_2 = \frac{z_0^2}{r_1^2} \tag{3.204}$$

It is to be noted that in the planar case, $z_0 = 0$, implying $T_2 = 0$, *except* for the singular case of

$$\lim_{y_0 \to 0} T_2 = 1.$$

However, while K_1 is singular for the limit as $y_0 \to 0$, K_2 is nonsingular and hence its contribution, $T_2 K_2$, identically vanishes for the planar case $z_0 = 0$, even in the limit, $y_0 \to 0$. Thus, for simplicity our discussion will be confined to the planar case,

$$K = -\lim_{z_0 \to 0} e^{-ikx_0} K_1 / r_1^2. \tag{3.205}$$

The influence of a line doublet of the sending box on the upwash collocation point is evaluated for a swept doublet line at the 1/4-chord location in the sending box. This is depicted in Fig. 3.25, using the notation of Blair [23]. The nondimensional integral equation for the planar case is expressed as follows:

$$\frac{\bar{w}}{U_\infty}(x, y) = \frac{1}{8\pi} \iint_S K(x - \xi, y - \eta, k, M) \frac{\Delta \bar{p}}{q_\infty}(\xi, \eta) d\xi d\eta, \tag{3.206}$$

where

$$K(x_0, y_0) = -\lim_{z_0 \to 0} \frac{K_1}{y_0^2 + z_0^2} e^{-ikx_0}, \tag{3.207}$$

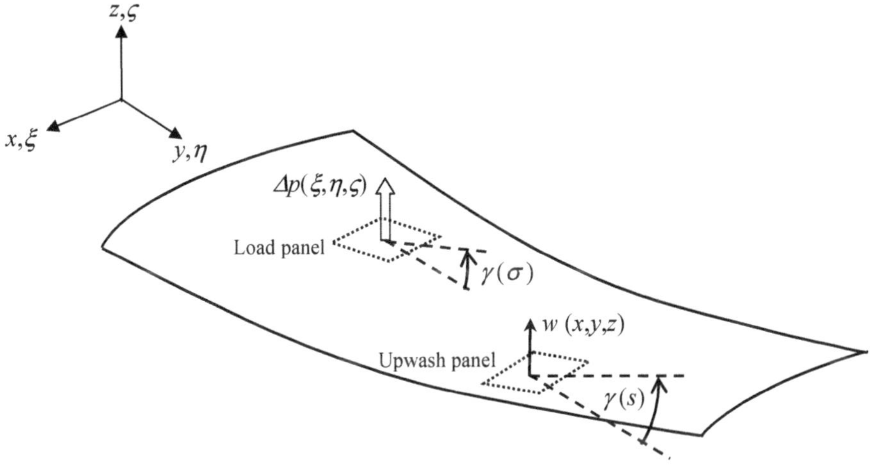

Fig. 3.24 Load point, (ξ, η, ζ), upwash point, (x, y, z), and the respective curvilinear coordinates, σ, and s, for the dihedral angle of associated panels

$$K_1 = I_1 + \frac{M \mid y_0 \mid}{\sqrt{x_0^2 + \beta^2 y_0^2}} e^{-ik_1 u_1} \frac{1}{\sqrt{1 + u_1^2}}, \tag{3.208}$$

the integral I_1 is given by Eq. (3.196) with

$$k_1 = k \mid y_0 \mid, \tag{3.209}$$

and

$$u_1 = \frac{M\sqrt{x_0^2 + \beta^2 y_0^2} - x_0}{\beta^2 \mid y_0 \mid}. \tag{3.210}$$

In order to treat the kernel function's second-order singularity as $y_0 \to 0$, Mangler's [109] principal value is taken by evaluating the limit $z_0 \to 0$ in Eq. (3.207) as the final step. This requires approximating the factor,

$$\bar{K}(x_0, y_0) = K_1(x_0, y_0)e^{-ikx_0}, \tag{3.211}$$

by the following parabolic function for the swept doublet line (Fig. 3.25) in the sending box :

$$\bar{K}(x_0, y_0) = A_0 + A_1\ell + A_2\ell^2, \tag{3.212}$$

where ℓ is the nondimensional coordinate running along the doublet line, and A_0, A_1, A_2 are the unknown complex coefficients to be determined from evaluations of $\bar{K}(x_0, y_0)$ at the mid-point, (x_c, y_c), and the left and right ends of the doublet

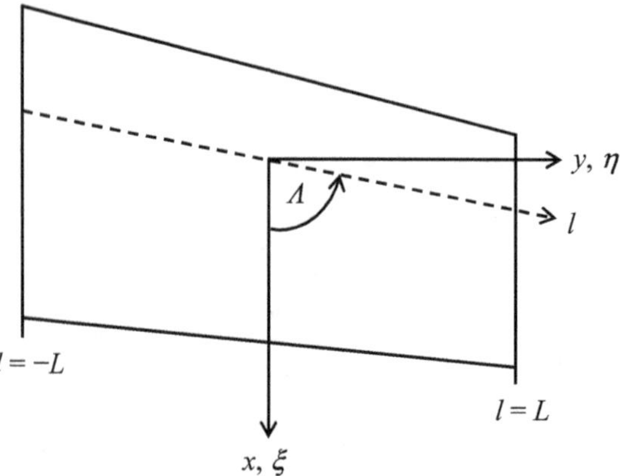

Fig. 3.25 Geometry of swept doublet line in the sending box of the doublet-lattice method

line, (x_L, y_L), and (x_R, y_R), respectively. These points correspond to $\ell = 0$, $\ell = -L$, and $\ell = L$, respectively, which identifies the coefficients to be the following:

$$A_0 = \bar{K}(x_c, y_c) \tag{3.213}$$

$$A_1 = \frac{\bar{K}(x_R, y_R) - \bar{K}(x_L, y_L)}{2L} \tag{3.214}$$

$$A_2 = \frac{\bar{K}(x_R, y_R) - 2\bar{K}(x_c, y_c) + \bar{K}(x_L, y_L)}{2L^2} \tag{3.215}$$

Upon substituting Eqs. (3.213)–(3.215) into the integral equation, Eq. (3.206), putting $\eta = \ell \sin \Lambda$, and assuming the doublet strength to be uniformly spread over the average box chord $\Delta\xi$, we have the following approximation:

$$\frac{\bar{w}}{U_\infty}(x, y) = \frac{\Delta\xi}{8\pi} \frac{\Delta\bar{p}}{q_\infty} \lim_{\epsilon \to 0} \int_{-L}^{L} \frac{A_0 + A_1\ell + A_2\ell^2}{(y - \ell \sin \Lambda)^2 + \epsilon^2} \mathrm{d}\ell, \tag{3.216}$$

which is abbreviated as follows [23]:

$$\frac{\bar{w}}{U_\infty}(x, y) = \frac{\Delta\xi}{8\pi} \frac{\Delta\bar{p}}{q_\infty}(B_0 + B_1 + B_2) \tag{3.217}$$

The coefficients B_0, B_1, B_2 are computed by taking the limit $\epsilon \to 0$ as the final step in the following manner:

$$B_0 = \lim_{\epsilon \to 0} \int_{-L}^{L} \frac{A_0}{(y - \ell \sin \Lambda)^2 + \epsilon^2} \mathrm{d}\ell$$

$$= \lim_{\epsilon \to 0} \frac{A_0}{\epsilon \sin \Lambda} \tan^{-1} \left[\frac{\ell \sin \Lambda - y}{\epsilon} \right]_{-L}^{L}$$

$$= \lim_{\epsilon \to 0} \frac{A_0}{\epsilon \sin \Lambda} \tan^{-1} \left(\frac{2L \sin \Lambda}{\epsilon^2 + y^2 - L^2 \sin^2 \Lambda} \right)$$

$$= \frac{2LA_0}{y^2 - L^2 \sin^2 \Lambda} \tag{3.218}$$

Similarly,

$$B_1 = \lim_{\epsilon \to 0} \int_{-L}^{L} \frac{A_1 \ell}{(y - \ell \sin \Lambda)^2 + \epsilon^2} \, d\ell$$

$$= \frac{A_1}{\sin^2 \Lambda} \log \left(\frac{y - L \sin \Lambda}{y + L \sin \Lambda} \right) + \frac{2LyA_1}{(y^2 - L^2 \sin^2 \Lambda) \sin \Lambda} \tag{3.219}$$

$$B_2 = \lim_{\epsilon \to 0} \int_{-L}^{L} \frac{A_2 \ell^2}{(y - \ell \sin \Lambda)^2 + \epsilon^2} \, d\ell$$

$$= \frac{2LA_2}{\sin^2 \Lambda} + \frac{4yA_2}{\sin \Lambda} \log \left(\frac{y - L \sin \Lambda}{y + L \sin \Lambda} \right) + \frac{2Ly^2 A_2}{(y^2 - L^2 \sin^2 \Lambda) \sin^2 \Lambda} \tag{3.220}$$

The coordinate y in these equations represents the distance of the upwash point of the receiving box from the center of the swept doublet line in the sending box (Fig. 3.25).

The kernel factor, $\bar{K}$, is to be evaluated carefully, as it has a singularity in the limit $| y_0 | \to 0$, for which the term u_1 tends to either ∞ or $-\infty$, depending upon whether $x_0 < 0$, or $x_0 \geq 0$. This singularity occurs for the receiving box lying directly downstream of the sending box, and results in the integral I_1 becoming improper for the planar case. The evaluation of the integral is numerically carried out as follows.

1. $| y_0 | \neq 0$, $u_1 \geq 0$: Integration by parts yields

$$I_1 = \int_{u_1}^{\infty} e^{-ik_1 u} \frac{1}{\left(1 + u^2\right)^{3/2}} \, du$$

$$= \left[e^{-ik_1 u} \frac{u}{\sqrt{1 + u^2}} \right]_{u_1}^{\infty} + ik_1 \int_{u_1}^{\infty} e^{-ik_1 u} \frac{u}{\sqrt{1 + u^2}} \, du$$

$$= -u_1 e^{-ik_1 u_1} \frac{1}{\sqrt{1 + u_1^2}} + e^{-ik_1 u_1} - ik_1 \int_{u_1}^{\infty} e^{-ik_1 u} \, du$$

$$\quad + ik_1 \int_{u_1}^{\infty} e^{-ik_1 u} \frac{u}{\sqrt{1 + u^2}} \, du$$

$$= e^{-ik_1 u_1} \left(1 - \frac{u_1}{\sqrt{1 + u_1^2}} \right) - ik_1 \int_{u_1}^{\infty} e^{-ik_1 u} \left(1 - \frac{u}{\sqrt{1 + u^2}} \right) du \tag{3.221}$$

The series approximation of Laschka [98] is employed next for numerical integration as follows:

$$1 - \frac{u}{\sqrt{1 + u^2}} \simeq \sum_{n=1}^{11} a_n e^{-ncu}, \tag{3.222}$$

Table 3.1 Series coefficients of Laschka [98]

	$c = 0.372$
n	a_n
1	0.24186198
2	−2.7918027
3	24.991079
4	−111.59196
5	271.43549
6	−305.75288
7	−41.18363
8	545.98537
9	−644.78155
10	328.72755
11	−64.279511

the coefficients of which are listed in Table 3.1. This results in the following expression for the integral:

$$I_1 = e^{-ik_1 u_1} \left(1 - \frac{u_1}{\sqrt{1 + u_1^2}} - ik_1 J_1 \right), \tag{3.223}$$

where

$$J_1 = e^{ik_1 u_1} \int_{u_1}^{\infty} e^{-ik_1 u} \left(1 - \frac{u}{\sqrt{1 + u^2}} \right) du \simeq \sum_{n=1}^{11} \frac{a_n e^{-ncu}}{nc + ik_1}. \tag{3.224}$$

2. $| y_0 | \neq 0, \ u_1 < 0$:

$$I_1 = \int_{u_1}^{\infty} e^{-ik_1 u} \frac{1}{\left(1 + u^2 \right)^{3/2}} du$$

$$= \int_{u_1}^{0} e^{-ik_1 u} \frac{1}{\left(1 + u^2 \right)^{3/2}} du + \int_{0}^{\infty} e^{-ik_1 u} \frac{1}{\left(1 + u^2 \right)^{3/2}} du$$

$$= \int_{u_1}^{0} e^{-ik_1 u} \frac{1}{\left(1 + u^2 \right)^{3/2}} du + 1 - ik_1 J_0, \tag{3.225}$$

where

$$J_0 = \int_{0}^{\infty} e^{-ik_1 u} \left(1 - \frac{u}{\sqrt{1 + u^2}} \right) du \simeq \sum_{n=1}^{11} \frac{a_n}{nc + ik_1}, \tag{3.226}$$

and the integral term is evaluated by taking advantage of the integrand's symmetry about $u = 0$, using $v = -u$ and $v_1 = -u_1 > 0$:

$$\int_{u_1}^{0} e^{-ik_1 u} \frac{1}{\left(1 + u^2\right)^{3/2}} du = \int_{0}^{v_1} e^{ik_1 v} \frac{1}{\left(1 + v^2\right)^{3/2}} dv$$

$$= \left[e^{ik_1 v} \frac{v}{\sqrt{1 + v^2}} \right]_{0}^{z_1} - ik_1 \int_{0}^{v_1} e^{ik_1 v} \frac{v}{\sqrt{1 + v^2}} dv$$

$$= e^{ik_1 v_1} \frac{v_1}{\sqrt{1 + v_1^2}} - ik_1 \int_{0}^{v_1} e^{ik_1 v} \left(1 - \sum_{n=1}^{11} a_n e^{-ncv} \right) dv$$

$$= 1 + e^{ik_1 v_1} \left(\frac{v_1}{\sqrt{1 + v_1^2}} - 1 + ik_1 \sum_{n=1}^{11} \frac{a_n \, e^{-ncv_1}}{-nc + ik_1} \right)$$

$$- ik_1 \sum_{n=1}^{11} \frac{a_n}{-nc + ik_1} \tag{3.227}$$

3. $\mid y_0 \mid = 0,\ x_0 < 0$:

$$I_1 = 0 \tag{3.228}$$

4. $\mid y_0 \mid = 0,\ x_0 \geq 0$:

$$I_1 = 2 \tag{3.229}$$

In order to calculate the influence of the line doublets lying on the other side of the plane of symmetry (if it exists), signs of the coefficients B_1, B_2 are reversed, and the spanwise locations of the sending box coorners are also reversed in sign. After the influence coefficient matrix is computed for all collocation points, it is inverted in order to yield the pressure distribution by Eq. (3.190).

It must be emphasized that there is no rigorous theoretical basis for the doublet-lattice method, other than linear superposition of elementary solutions in the manner of the vortex-lattice method. The placement of the doublet line at 1/4-chord and the upwash collocation point at 3/4-chord of each box is a purely numerical device, and is simply borrowed from the vortex-lattice method, which uses it in order to satisfy the Kutta condition at the trailing edge in the steady, incompressible limit. Consequently, there is no guarantee that the doublet-lattice method would produce physically accurate steady-state results, especially for higher Mach numbers. However, the oscillatory data predicted by the doublet-lattice method is quite well established by agreement with experimental data, and is thus historically employed in aeroelastic analyses. Caution must be exercised in the selection of the number of boxes in both chordwise and spanwise directions, because the pressure distribution produced by the discrete paneling can often be discontinuous. This could be at problem at high reduced frequencies where the rapid pressure fluctuations could be realistically captured only by a much finer paneling than that required at lower frequencies. The required box density can also increase with a higher mode shape, which involves an increasingly wavy upwash distribution. Furthermore, in cases of small aspect ratio planforms, the aspect ratio of the boxes can also be important.

A variation of the doublet-lattice method is the *doublet-point method* [180] in which the line doublet in each panel is replaced by a point doublet whose strength is directly proportional to the concentrated lift force at the panel's centroid. The singularity of the kernel function is treated by separating the kernel into singular and nonsingular parts, and evaluating the singular integral in the sense of Mangler [109]. The doublet-point method is an attractive alternative to the doublet-lattice method, because it not only gives a realistic pressure distribution in the steady limit ($k \to 0$)— which the DLM surprisingly does not—but can also be extended to the supersonic regime ($M > 1$) for which the DLM is inapplicable.

Formulations alternative to the acceleration potential integral equation, Eq. (3.182), include the *velocity potential method* [80, 59] wherein the classical Helmholtz wave equation is solved over the wing and the wake, and the *Green's function method* [114]. Both of these methods require a model of the unsteady wake. Although such methods have the advantages of higher accuracy as well as applicability to thick wings and wing–body combinations undergoing large amplitude oscillations, their computational complexity could be an order of magnitude higher than that of acceleration potential based methods, thereby proving less suitable for ASE modeling.

The doublet-lattice method for planar surfaces detailed here is coded into a MATLAB computer program, and is used to generate subsonic unsteady aerodynamic data throughout this book. Some numerical results of the code's application are discussed below.

3.6.1 Numerical Results

The first validation of the doublet-lattice code is on a rectangular wing of aspect ratio 2.0 with a trailing-edge, full-span flap of 40 % chord. Figures 3.26 and 3.27 show the hinge-moment coefficient magnitude and phase, respectively, of the flap oscillating at various reduced frequencies, and with various chordwise and spanwise divisions of equal lengths in each direction. The results are compared with the experimental data of Beals and Targoff [16] as reported in Albano and Rodden [7]. Good agreement of the magnitude and phase trends with the experimental data at lower reduced frequencies can be observed. The doublet-lattice method overpredicts the phase at higher frequencies, which is somewhat improved by increasing the number of boxes. The discrepancy in the magnitude when compared to experimental results increases at higher frequencies. This indicates that the high-frequency pressure waves due to flap oscillation are only approximately captured by the doublet-lattice method. The real and imaginary parts of the unsteady pressure distribution are plotted in Figs. 3.28

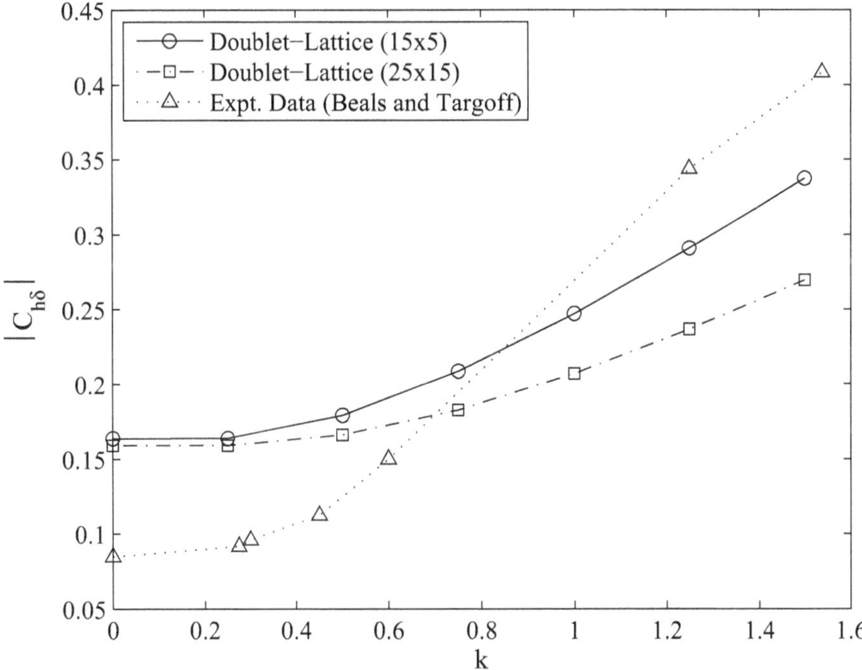

Fig. 3.26 Magnitude of hinge-moment coefficient of a rectangular wing of aspect ratio 2.0 with an oscillating flap of 40 % chord at $M = 0.24$

and 3.29 for $k = 1.5$, respectively, clearly showing the pressure discontinuity at the flap hinge line.

The next study is for the AGARD wing of aspect ratio 1.4509, taper ratio 0.709, and leading-edge sweep angle 39°, whose plan form shape is depicted in Fig. 3.30, with coordinates rendered nondimensional with the semispan s. The computed pressure distributions for reduced frequencies, $k = 0.4, 0.8, 1.2, 1.6$ are plotted in Figs. 3.31–3.34. The unsteady lift due to the rigid plunge mode at various reduced frequencies is compared is Fig. 3.35 with that of the commercial doublet-lattice code H7WC and the time-domain panel method, both as reported by Blair and Williams [24]. The convergence of the results with the increased number of boxes is evident in this figure. However, while the comparison between the three schemes is good at lower reduced frequencies, the higher frequency results show a slight disagreement. This is due to the fact that higher frequency pressure waves are differently captured by the different integration methods.

Figure 3.36 shows the upwash distribution on the AGARD wing produced by the combination of a chordwise and a spanwise bending mode, and Figs. 3.37–3.39 are the corresponding real and imaginary parts of the unsteady pressure distribution computed by the doublet-lattice method for $k = 0.6$.

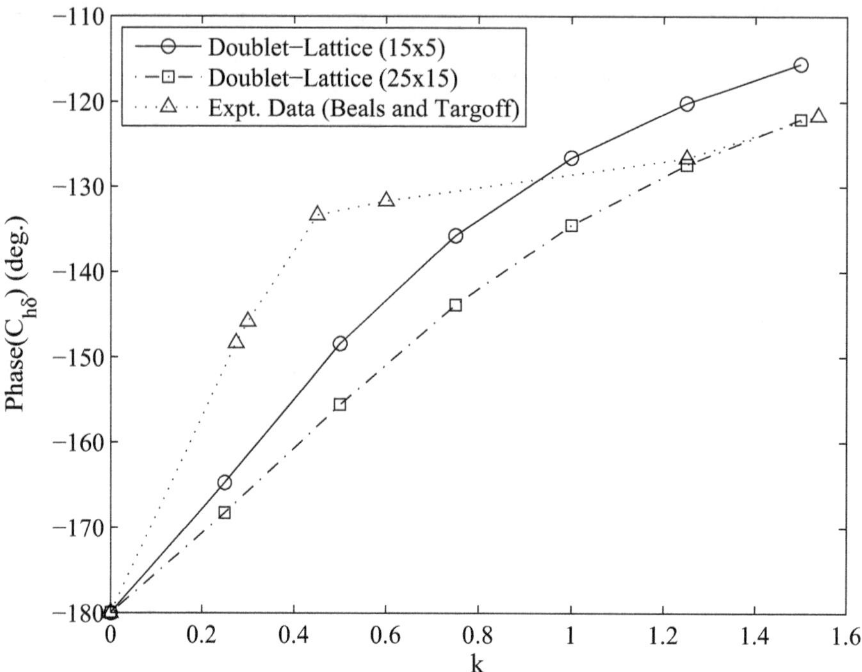

Fig. 3.27 Phase of hinge-moment coefficient of a rectangular wing of aspect ratio 2.0 with an oscillating flap of 40 % chord at $M = 0.24$

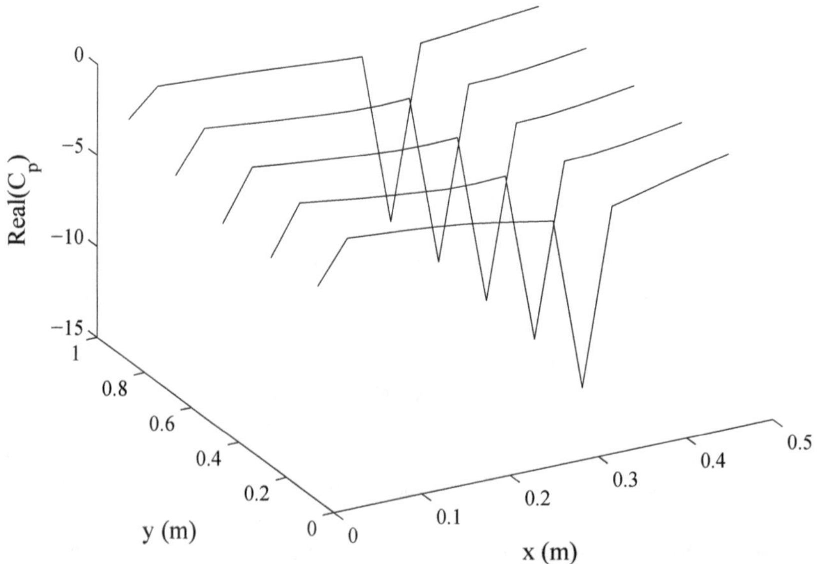

Fig. 3.28 Real part of unsteady pressure distribution on a rectangular wing of aspect ratio 2.0 with an oscillating flap of 40 % chord at $M = 0.24$ and $k = 1.5$ using a doublet lattice of 15 chordwise and 5 spanwise boxes

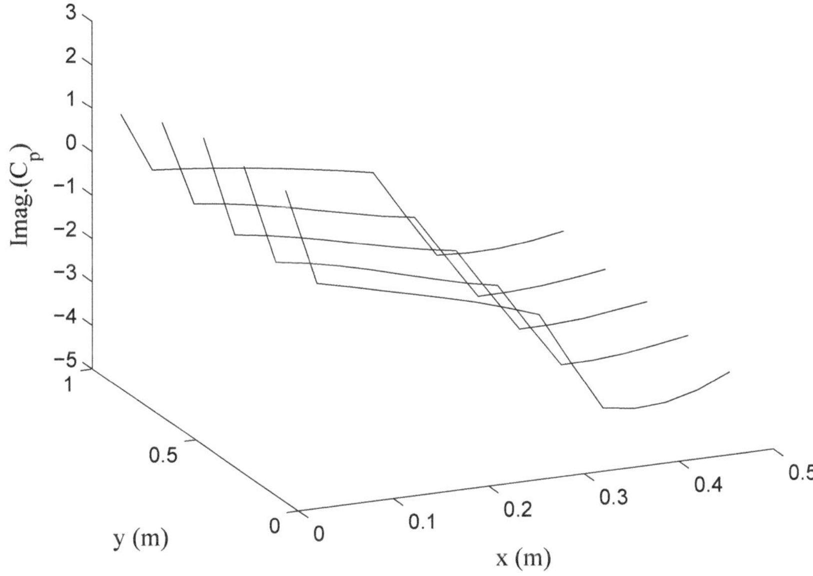

Fig. 3.29 Imaginary part of unsteady pressure distribution on a rectangular wing of aspect ratio 2.0 with an oscillating flap of 40 % chord at $M = 0.24$ and $k = 1.5$ using a doublet lattice of 15 chordwise and 5 spanwise boxes

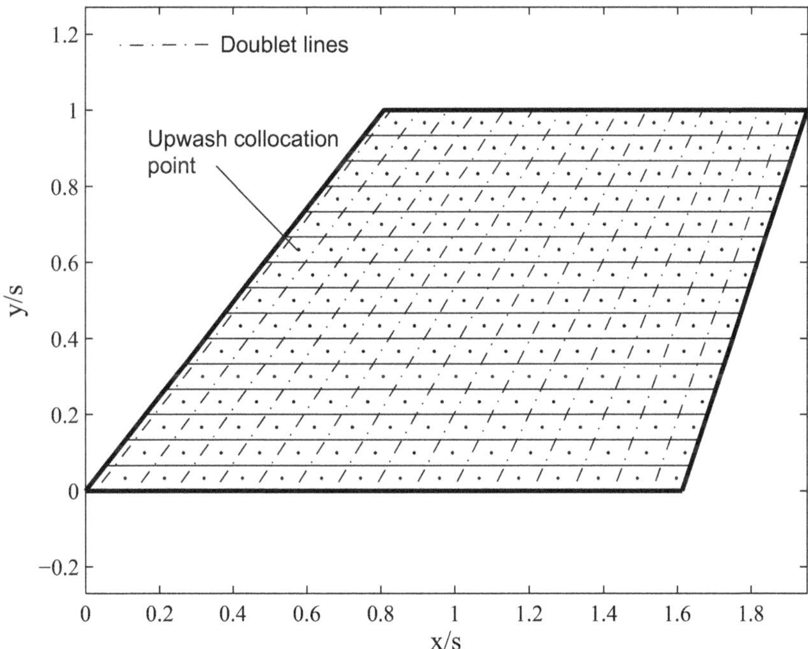

Fig. 3.30 AGARD wing of aspect ratio 1.49 discretized with 15 spanwise and 15 chordwise divisions for doublet lattice calculation

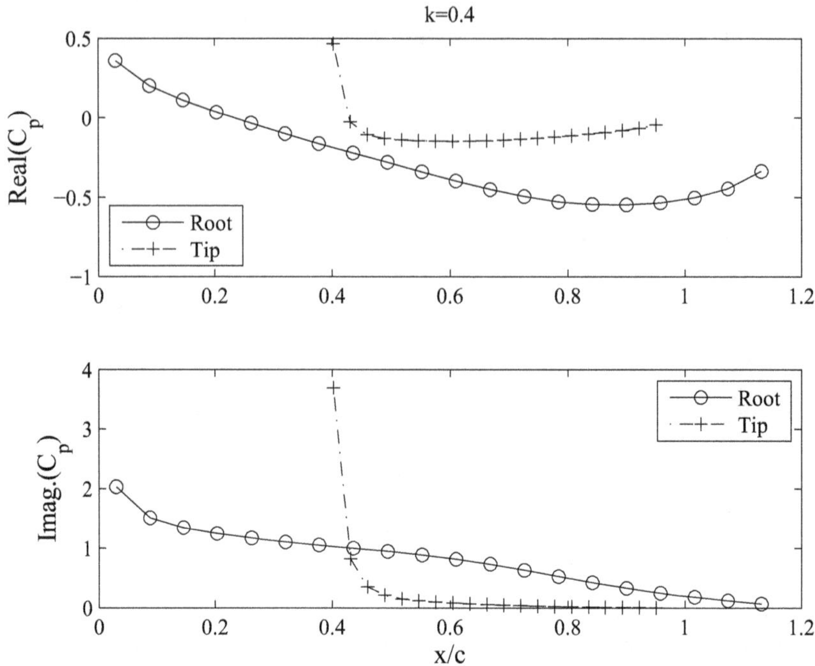

Fig. 3.31 Unsteady pressure distribution on AGARD wing due to a rigid plunge mode at $M = 0.8$ and $k = 0.4$

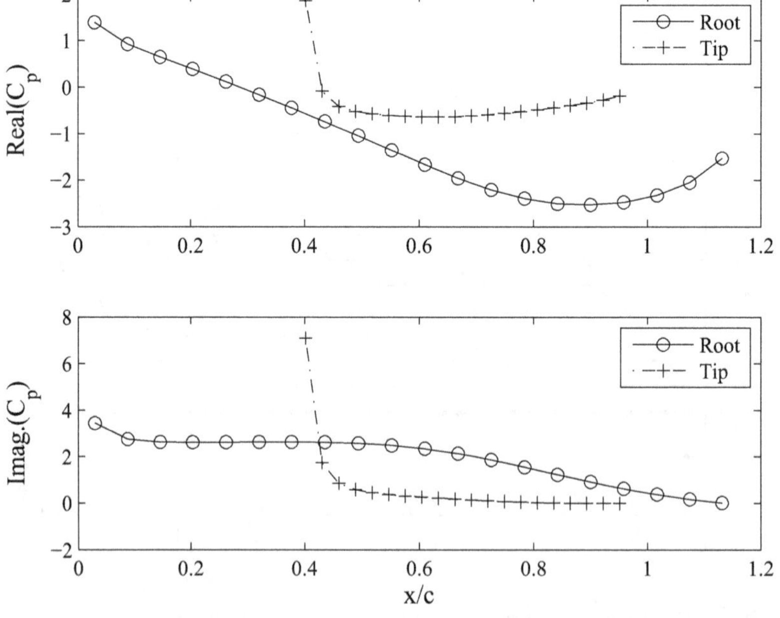

Fig. 3.32 Unsteady pressure distribution on AGARD wing due to a rigid plunge mode at $M = 0.8$ and $k = 0.8$

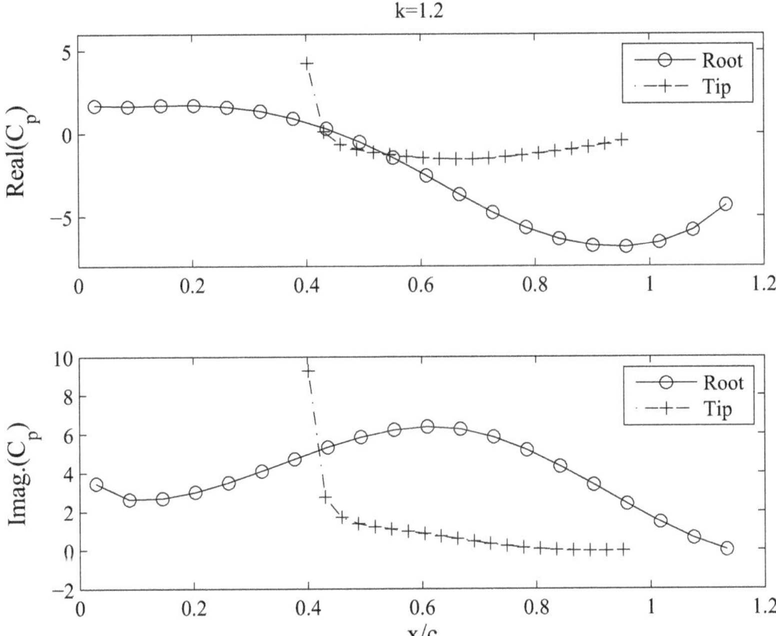

Fig. 3.33 Unsteady pressure distribution on AGARD wing due to a rigid plunge mode at $M = 0.8$ and $k = 1.2$

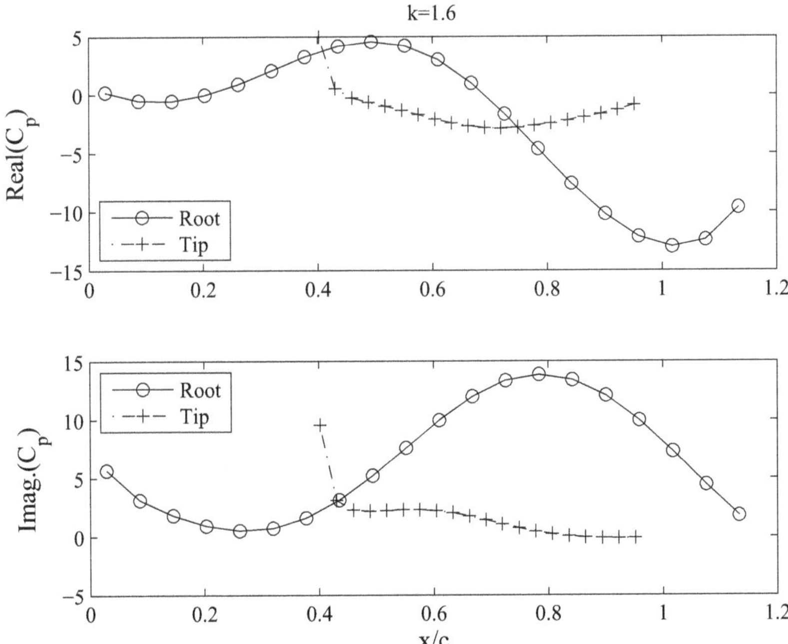

Fig. 3.34 Unsteady pressure distribution on AGARD wing due to a rigid plunge mode at $M = 0.8$ and $k = 1.6$

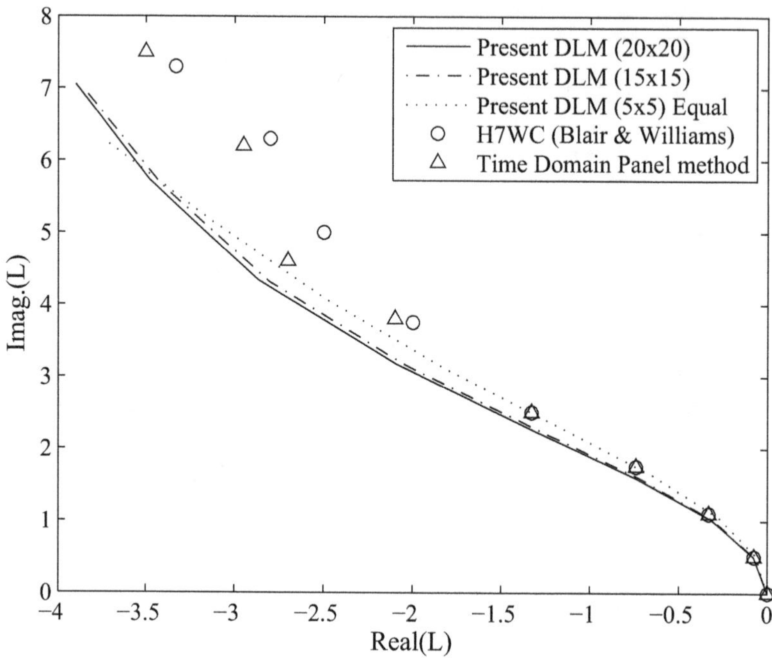

Fig. 3.35 Unsteady lift on AGARD wing due to a rigid plunge mode at $M = 0.8$ compared with the results of H7WC and time-domain panel codes [24]

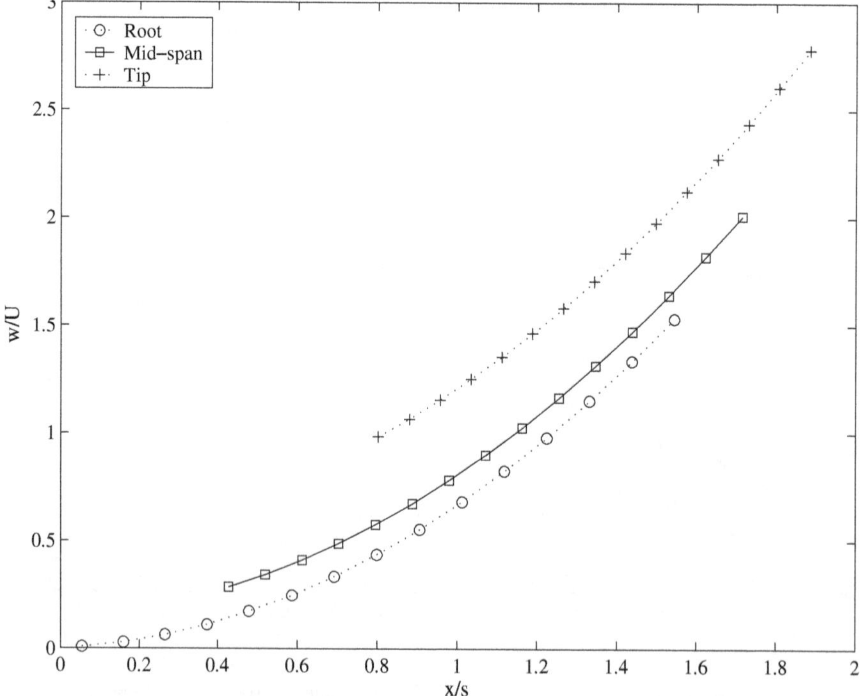

Fig. 3.36 Unsteady upwash distribution on AGARD wing due to chordwise and spanwise bending modes at $M = 0.8$ and $k = 0.6$ based upon semispan, s

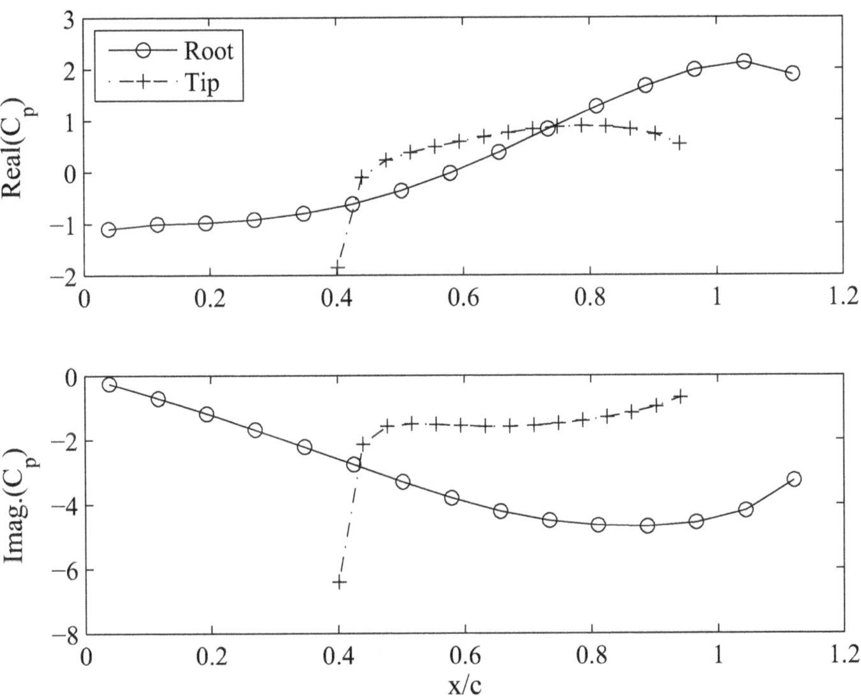

Fig. 3.37 Unsteady pressure distribution on AGARD wing at the root and tip due to chordwise and spanwise bending modes at $M = 0.8$ and $k = 0.6$

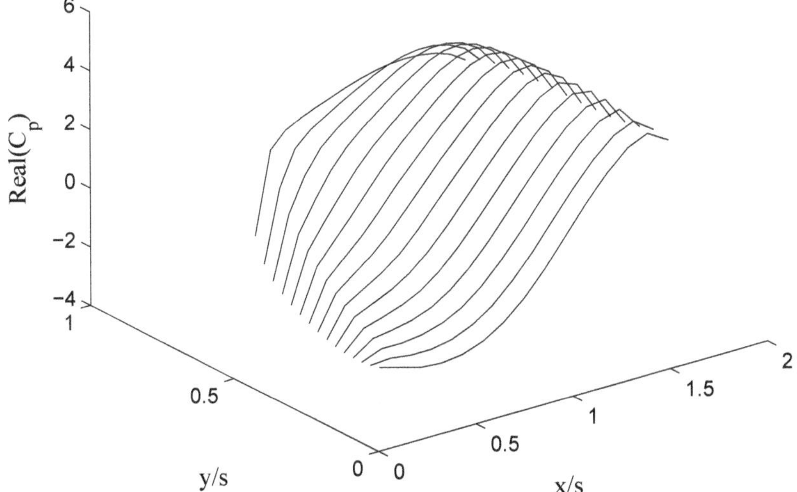

Fig. 3.38 Real part of unsteady pressure distribution on AGARD wing due to chordwise and spanwise bending modes at $M = 0.8$ and $k = 0.6$

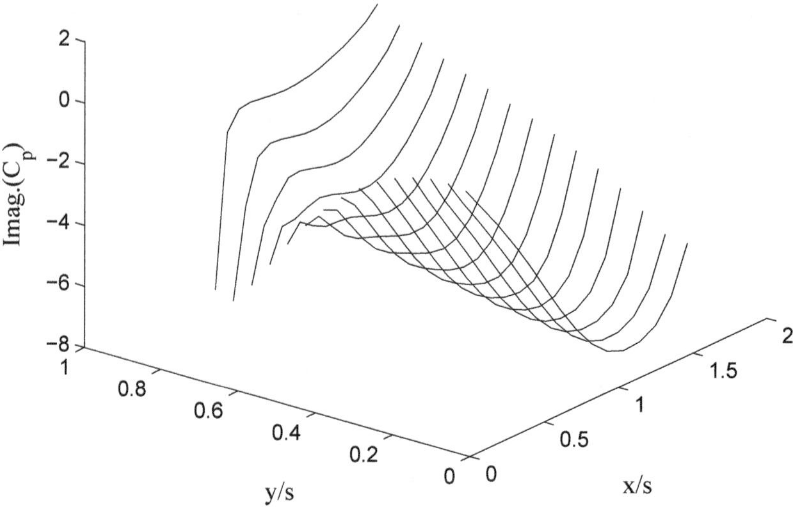

Fig. 3.39 Imaginary part of unsteady pressure distribution on AGARD wing due to chordwise and spanwise bending modes at $M = 0.8$ and $k = 0.6$

The final numerical study is based on NASA-Langley's *drone for aeroelastic testing*, (DAST-ARW1). This wing is modified for the present by a trapezoidal plan form depicted in Fig. 3.40 for a sample grid consisting of 20 equal spanwise and 10 chordwise boxes, which does not include the leading-edge extension of the original wing. The first computation is for a linear combination of three structural modes, whose mode shapes are plotted in Figs. 3.41–3.43. Here, $h(x, y)$ is the vertical deflection and $dh/dx(x, y)$ is the chordwise slope of the vertical deflection required for satisfying the upwash boundary condition according to Eq. (3.189). The normalization of the deflections and slopes is carried out by dividing by the magnitudes $|h|$ and $|dh/dx|$, respectively. The computed real and imaginary parts of the resulting pressure distribution for 30 spanwise and 10 chordwise divisions are shown in Fig. 3.44. Where the pressure waviness and increased magnitudes near the tip due to bending and torsion modes are evident.

3.7 Supersonic Lifting Surface Methods

The supersonic integral equations derived by the alternative approaches of the velocity and acceleration potentials (see above), can be solved by employing a numerical procedure based upon approximate discretization. There are many such methods available in the literature, and have differing formulations based upon the way the upwash–pressure relationship is inverted. Our focus here will be on the three-dimensional lifting surface procedures.

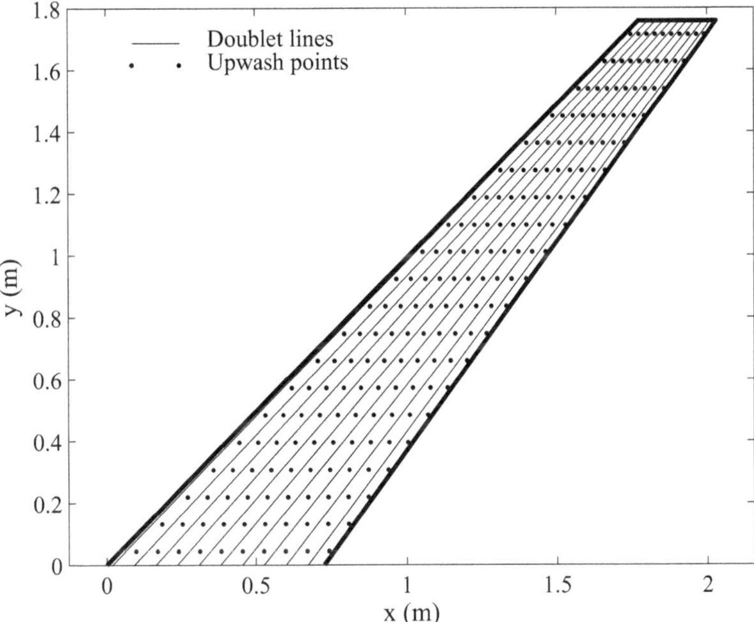

Fig. 3.40 DAST-ARW1 wing discretized for doublet lattice calculation

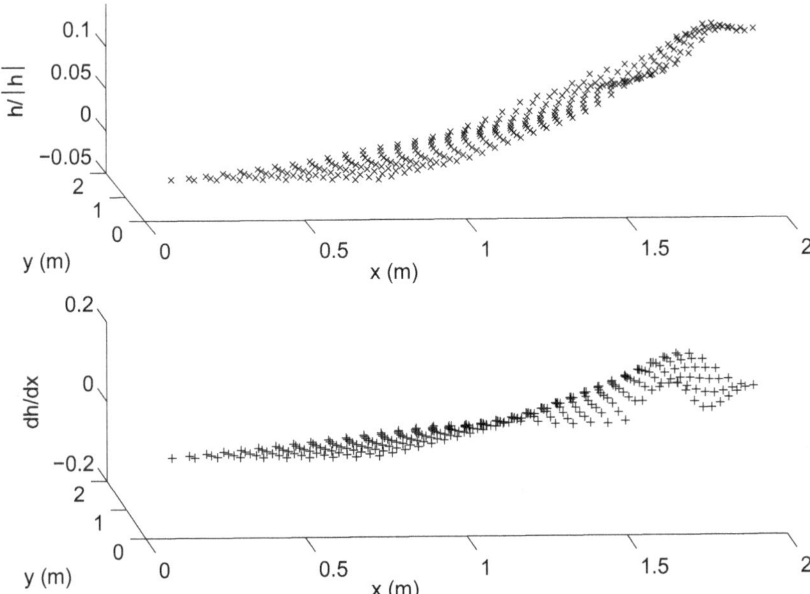

Fig. 3.41 Vertical deflection and chordwise slope for the first structural mode of DAST-ARW1 wing with natural frequency 9.3 Hz

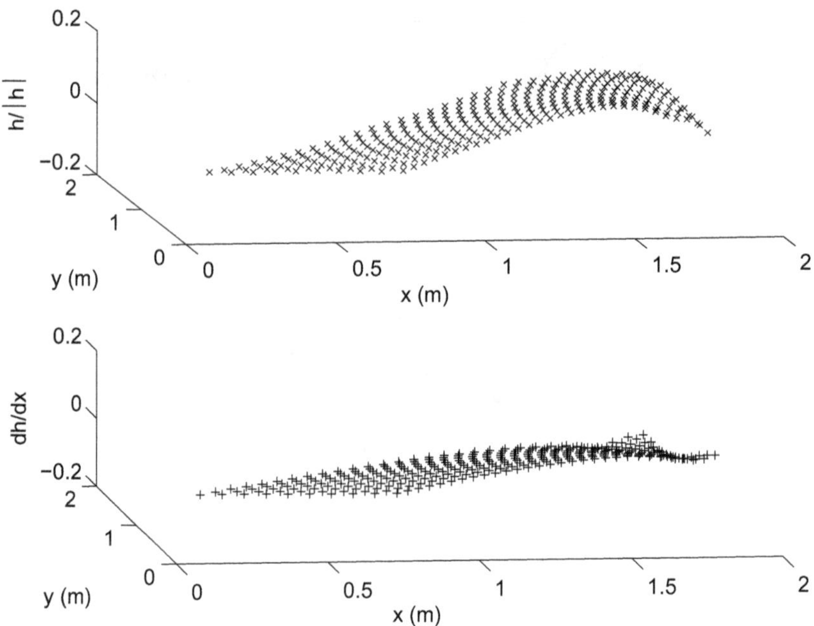

Fig. 3.42 Vertical deflection and chordwise slope for the second structural mode of DAST-ARW1 wing with natural frequency 32.72 Hz

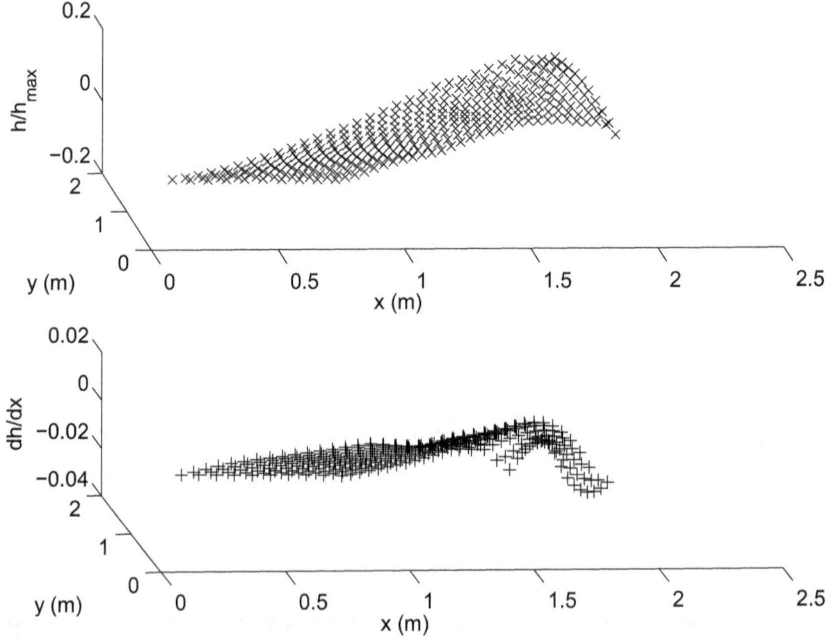

Fig. 3.43 Vertical deflection and chordwise slope for the third structural mode of DAST-ARW1 wing with natural frequency 48.91 Hz

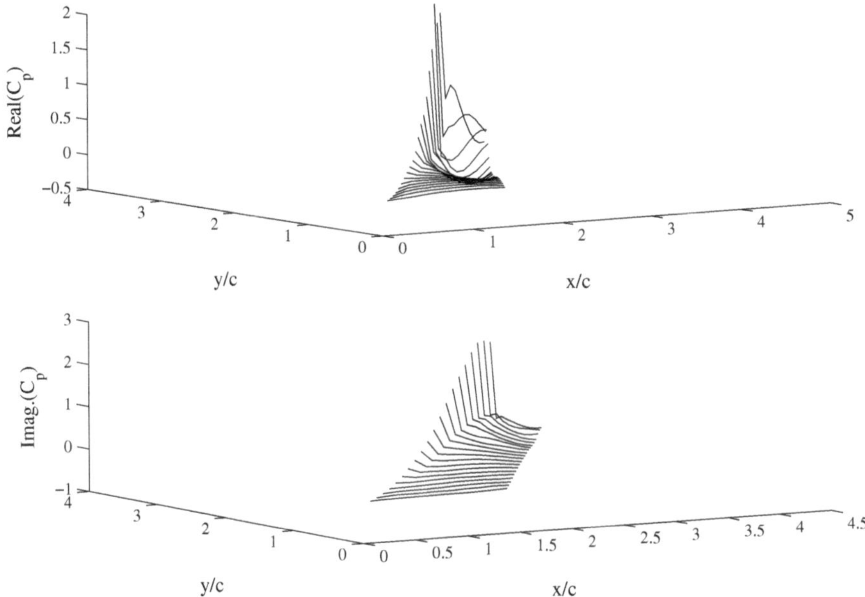

Fig. 3.44 Unsteady pressure distribution on DAST-ARW1 wing due to three bending/torsion modes at $M = 0.7$ and $k = 0.6$

3.7.1   Mach-Box Method

While Possio [133] devised an acceleration potential based formulation for the two-dimensional supersonic case quite early, the first practical supersonic lifting surface method finding a common application in aircraft industry of 1950s–1980s is based upon the velocity potential approach. This is the *Mach-box method* [129], which employs a grid of rectangular boxes for approximating the lifting surface. Each box width is chosen such that the diagonal was a Mach line, which facilitates integrations over boxes lying in the forward-facing Mach cone (Fig. 3.22), because such integrals would be either over complete or half boxes. However, this also implies that the grid is not exactly aligned with the wing plan form shape (like it is in the doublet-lattice method), but instead has jagged leading and trailing edges. Each box is assumed to have a separate vertical deflection (h) at the center (x_c, y_c) as well as pitch (α) and roll (θ) rotations about it. Hence, the net upwash amplitude at a point (x, y) of the jth box is given by

$$\bar{w}(x_j, y_j) = i\omega \left[h_j + (x - x_c)\alpha_j + (y - y_c)\theta_j \right], \qquad (j = 1, \ldots, N) \qquad (3.230)$$

The pressure loading amplitude at a given point (x, y) on the wing caused by the jth box of plan form area S_j is calculated by the velocity potential integral equation as follows:

$$\frac{\bar{\Delta p}(x, y)}{\rho_\infty} = \frac{2}{\pi} \left(i\omega + U_\infty \frac{\partial}{\partial x} \right) \iint_{S_j} \bar{w}(\xi, \eta) \frac{e^{-i\kappa M(x-\xi)}}{\hat{R}} \cos{(\kappa \hat{R})} \mathrm{d}\eta \mathrm{d}\xi \qquad (3.231)$$

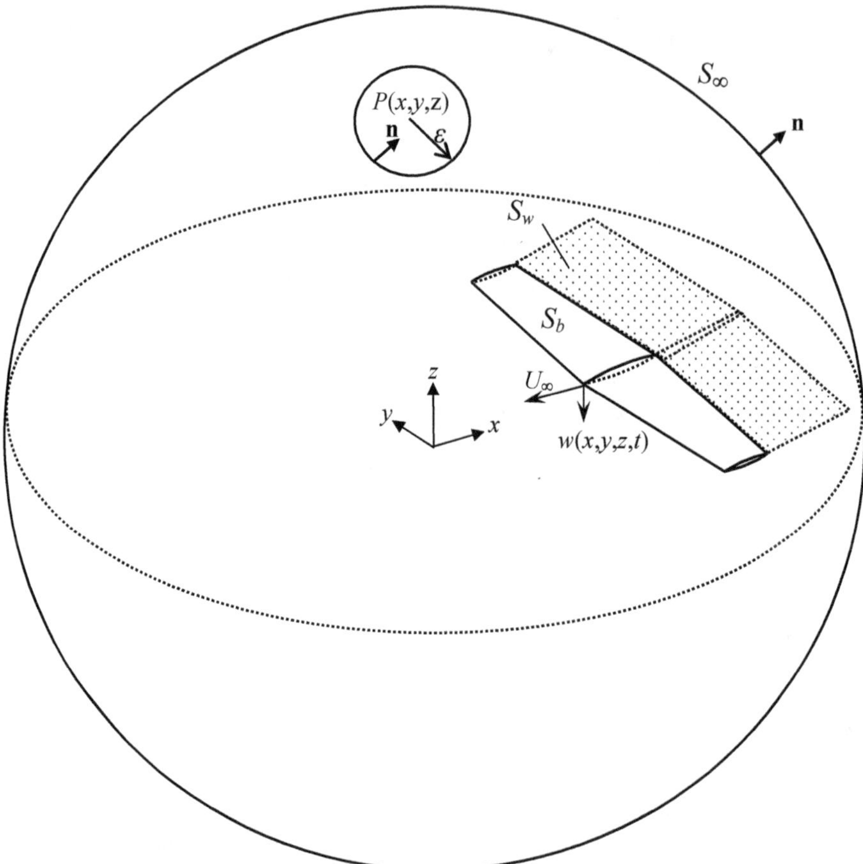

Fig. 3.45 Geometry of the flowfield around a wing for Green's integral theorem formulation

where $\beta = \sqrt{M^2 - 1}$

$$\hat{R} = \sqrt{(x - \xi)^2 - \beta^2(y - \eta)^2}$$

and

$$\kappa = \frac{\omega}{a_\infty \beta^2}.$$

A simple approximation is to take an average (constant) upwash evaluated at the receiving box centroid, which means taking the term $\bar{w}$ out of the integral in Eq. (3.231). While this can be accurate for boxes of very small dimensions, in most cases it could cause unrealistic pressure distributions. The influence of a box upon itself is constant and equal to the pressure at the centroid. The individual influences on a receiving (pressure) box due to all the sending (upwash) boxes in the forward-facing Mach cone are assembled into the following matrix form:

$$\frac{\Delta \bar{\mathbf{p}}}{q_\infty} = \mathbf{C}\frac{\bar{\mathbf{w}}}{U_\infty}, \tag{3.232}$$

where $\mathbf{C}$ is the matrix of complex aerodynamic influence coefficients (AICs). The generalized aerodynamic forces for each structural mode can be individually computed by the principle of virtual work.

The presence of subsonic wing edges requires placing boxes in the off-wing region between the concerned edge and the Mach cone. These are called the *diaphragm* boxes, and require special treatment. This is carried out by partitioning Eq. (3.232) into regular wing boxes and diaphragm boxes, and by applying zero pressure difference in the diaphragm boxes.

The Mach-box method is a systematic and reasonably accurate method for generalized forces and flutter calculations, but can give spurious pressure distributions due to subsonic and jagged edges. In order to improve its numerical accuracy, more sophisticated integration methods can be adopted than taking a constant pressure in each box. One such approach is the use of Gaussian quadrature [31]. Another shortcoming of the Mach-box method is the dependence of the grid on the Mach number, which makes it difficult to be used in a matched flutter analysis (i.e., matching the flight velocity and Mach number at which flutter occurs).

3.7.2 *Doublet-Point Method*

In order to avoid the problems associated with the velocity potential Mach-box method, several alternative strategies based upon either velocity potential, or the acceleration potential have been adopted. An alternative velocity potential formulation to the Mach-box method is that of Evvard [28, 51], where the requirement of considering the diaphragm region for subsonic edges is completely eliminated. The velocity potential integral equation has also led to the potential-gradient method [82], [74], which introduced upwash integration along receiving box boundary in order to remove the Mach number grid dependence of the Mach-box method. However, wake elements are necessary in this boundary-element type scheme, which cause modeling difficulties. The constant pressure panel method [9] is an improvement of the potential gradient method, where the wake elements between two planar surfaces are eliminated.

While the velocity potential methods for supersonic oscillatory aerodynamics are useful, they require a different scheme and computational grid (such as the doublet-lattice method) for subsonic flutter calculations. In contrast, a method based upon the acceleration potential integral equation would require little change in the grid geometry, from the doublet-lattice method for the subsonic flow. As in the subsonic case, the supersonic acceleration potential methods fall into two categories: (a) *Kernel function collocation* methods, where the supersonic kernel is expanded by a polynomial, whose coefficients are computed by satisfying the upwash boundary condition at selected points. (b) *Integral equation solution* of the pressure distribution by inversion on a discretized geometry, like the subsonic doublet-lattice method. The kernel function techniques [35, 104] require a priori knowledge of all pressure discontinuities such as control-surface hinge lines and intersecting surfaces, for which the collocation points must be suitably selected. This makes the method application specific. Under the second category fall the harmonic-gradient method [29], and the doublet-point method [181]. Of these, we shall consider the latter, because it requires no change at all in the grid geometry when applied to subsonic and supersonic flows. Furthermore, our discussion of it here will be brief, since it has many features in common with the subsonic doublet-lattice method (which we have already covered in some detail). Our treatment will focus only on the planar part for simplicity. The extension for the nonplanar kernel [67] is given by the nonplanar supersonic doublet-point method [167].

The doublet-point method is based upon a discretization of the lift (or pressure) distribution on the wing into a number n of load points, (ξ_j, η_j):

$$\frac{\Delta \bar{p}}{q_\infty}(\xi, \eta) = \sum_{j=1}^{n} F_j \delta(\xi - \xi_j)\delta(\eta - \eta_j), \tag{3.233}$$

where $\delta(.)$ is the Dirac delta function. Then the discretized integral equation is given by

$$\frac{\bar{w}}{U_\infty}(x_i, y_i) = \frac{1}{8\pi} \sum_{j=1}^{n} F_j K(x_i - \xi_j, y_i - \eta_j, k, M), \tag{3.234}$$

where the upwash amplitude $\bar{w}$ is averaged over n trapezoidal boxes into which the wing plan form is discretized, and $K(x_0, y_0, k, M)$ is the nondimensional kernel function, k as the reduced frequency and M the Mach number. For the subsonic case, the load points are taken at the 1/4-chord, mid-span locations of the boxes, while the upwash boundary condition is applied at the 3/4-chord point (like the DLM). Application of Eq. (3.234) involves a simple evaluation of the kernel for each pair of load and upwash points, while treating the kernel singularity for upwash points located directly downstream of the load point in Mangler's sense [109] through an averaging procedure.

For the supersonic case, both load and upwash points are collocated at a box's mid-chord, mid-span point. Here, the kernel function has additional singularities on the Mach cones emanating downstream from the load point, which must be accounted for in the upwash averaging process. This process involves separating the supersonic kernel function into steady and unsteady factors. the supersonic kernel (derived earlier) is expressed in a nondimensional form as follows:

$$K(x_0, y_0, k, M) = \lim_{z \to 0} e^{ikx_0} \left[\frac{M^2}{R} \left(\frac{e^{-ikX_1}}{x_0 + X_1} + \frac{e^{-ikX_2}}{x_0 + X_2} \right) + \int_{X_1}^{X_2} \frac{e^{-ikv}}{\left(r_0^2 + v^2\right)^{3/2}} dv \right]$$
$$\tag{3.235}$$

where

$$R = \sqrt{x_0^2 - \beta^2 r_0^2}; \qquad r_0 = y_0^2 + z^2$$

and

$$X_1 = \frac{x_0 - MR}{\beta^2}; \qquad X_2 = \frac{x_0 + MR}{\beta^2}$$

In the steady case ($k = 0$), the planar kernel is simply the following:

$$K_s = \lim_{z \to 0} \frac{2x_0}{Rr^2} \tag{3.236}$$

Since the steady factor has singularities at directly downstream points and on the Mach cone, it varies rapidly over each box. Therefore, it needs to be averaged over a rectangular area A with the same width as that of the receiving box, and with leading edge at the box mid-chord point. Depending upon the ten different ways [181] in which the averaging area can intersect the Mach cone emanating from the load point, the integral is performed to obtain the average steady part:

$$K_s = \frac{1}{A} \lim_{z \to 0} \iint_S \frac{2x_0}{Rr^2} \mathrm{d}x_0 \mathrm{d}y_0 \tag{3.237}$$

Thus we write

$$K = K_s \bar{K} = K_s \lim_{z \to 0} e^{ikx_0} \left[\frac{M^2 r_0^2}{2x_0} \left(\frac{e^{-ikX_1}}{x_0 + X_1} + \frac{e^{-ikX_2}}{x_0 + X_2} \right) \right.$$
$$\left. + \frac{Rr^2}{2x_0} \int_{X_1}^{X_2} \frac{e^{-ikv}}{\left(r_0^2 + v^2 \right)^{3/2}} \mathrm{d}v \right] \tag{3.238}$$

In contrast with the steady kernel, the unsteady factor $\bar{K}$ has a singularity only at the apex of the Mach cone $x_0 = 0$, and is therefore only varying slowly over the receiving box, and its value can be approximated as that at the box center. The evaluation of the unsteady factor involves the expansion of the integral term as follows:

$$\int_{X_1}^{X_2} \frac{e^{-ikv}}{\left(r_0^2 + v^2 \right)^{3/2}} \mathrm{d}v = \frac{X_2 e^{-ikX_2}}{r_0^2 \sqrt{r_0^2 + X_2^2}} - \frac{X_1 e^{-ikX_1}}{r_0^2 \sqrt{r_0^2 + X_1^2}} + ik \int_{X_1}^{X_2} \frac{v e^{-ikv}}{r_0^2 \sqrt{r_0^2 + v^2}} \mathrm{d}v \tag{3.239}$$

The integral on the right-hand side of Eq. (3.239) is numerically evaluated by using the same series approximation [98] as that employed for Eq. (3.222), with the coefficients listed in Table 3.1.

The supersonic doublet-point method is encoded in the N5KM program of McDonnell Aircraft Company (now a part of Boeing), and has been used for many decades of aeroelastic modeling with good results. Its competitor, the harmonic-gradient code [29] ZONA51 of Arizona State University, is also in continuous practical use since the 1990s, especially by researchers at NASA-Langley. Other panel methods have also evolved, such as the SPNLRI, GUL, and CAR of Nationaal Lucht- en Ruimtevaartlaboratorium (NLR)-Amsterdam developed by the team of M.H.L. Hounjet, and the CHB code of Svenska Aeroplan AB (SAAB)-Scania by Valter Stark's team.

3.8 Transonic Small-Disturbance Solution by Green's Function Method

It is generally difficult to solve the unsteady aerodynamics problem associated with oscillating lifting surfaces in the transonic flowfield. The difficulty is due to the presence of nonlinear terms in the governing partial differential equations, which are necessary for capturing the phenomena of nearly normal shock waves. While an accurate steady-state solution of the transonic flow requires either Euler, or full-potential methods, for thin surfaces oscillating with small amplitudes, the approximate solution by transonic small-disturbance (TSD) model can be sufficiently accurate in aeroelastic applications. However, even TSD solutions could be cumbersome, requiring iterative, time-dependent evaluations over a large computational grid, such as computational elasticity program (CAP)-TSD [14, 15]. The memory and CPU time of such a computation could approach that of the Euler and full-potential methods [79], thereby negating the advantages of the much simpler governing equation.

Due to the success enjoyed by the panel methods (doublet-lattice, doublet-point, etc.) for linearized flows, there has been some effort to devise a similarly efficient scheme for the TSD equation. The field-panel method [188], which is based upon Morino and Kuo's [114] boundary-value solution by Green's theorem, is such a method. Other methods which could be considered in an efficient application are based upon corrections to the doublet-lattice method for the transonic case, such as the method of Pitt and Goodman [130]. However, since the doublet-lattice-like methods have an uncertain theoretical foundation, their transonic correction is fraught with similar numerical experimentation, and thus cannot be expected to be valid in a general application. Hence, one has to look for a simple but theoretically rigorous method. A TSD solution by the Green's function approach, which incorporates acceleration potential doublets on the wing mean surface and velocity potential sources for the wake and field points, is such a candidate. For illustration of this approach, let us consider the transonic doublet-lattice method (TDLM) as formulated by Lu and Voss [105] for planar mean surface configurations. However, before discussing the TDLM formulation, it is necessary to study the extension of Green's function method for aircraft wings in unsteady transonic flow.

3.8.1   *Transonic Green's Integral Equation*

The Green's integral formulation presented here is along the lines of Morino et al. [115], and can be considered an extension of the approach given earlier in this chapter for incompressible flows. Some of the concepts that were briefly touched upon earlier, will be detailed here for clarity. Consider a frame attached to the freestream and convecting downstream with it at velocity $(U_\infty, 0, 0)^T$. In such a frame, the substantial (or Eulerian) derivative is given by

$$\left(\frac{D}{Dt}\right)_\infty (.) = \frac{\partial}{\partial t}(.) + U_\infty \frac{\partial}{\partial x}(.) \tag{3.240}$$

The Laplacian of the perturbation velocity potential, Φ, in this frame is the following:

$$\nabla^2 \Phi = \frac{1}{a^2}\left(\frac{D^2\Phi}{Dt^2}\right)_\infty, \tag{3.241}$$

where a is the local speed of sound. By substituting the unsteady Bernoulli equation for the isentropic case,

$$\frac{\partial \Phi}{\partial t} + \frac{q^2}{2} + \frac{a^2}{\gamma - 1} = \frac{a_\infty^2}{\gamma - 1}, \tag{3.242}$$

into Eq. (3.241), where

$$\nabla \phi = \mathbf{q}, \tag{3.243}$$

is the perturbation velocity vector, the speed of sound is eliminated and the following FPE is obtained:

$$\nabla^2 \Phi - \frac{1}{a_\infty^2}\frac{\partial^2 \Phi}{\partial t^2} = \chi, \tag{3.244}$$

where

$$\chi = \frac{1}{a_\infty^2}\left[(a^2 - a_\infty^2)\nabla^2\Phi + 2\mathbf{q}\cdot\frac{\partial \mathbf{q}}{\partial t} + \frac{1}{2}\mathbf{q}\cdot\nabla q^2\right] \tag{3.245}$$

represents the nonlinear terms required to describe the transonic flowfield. The left-hand side of Eq. (3.244) equated to zero ($\chi = 0$) is the linear wave equation utilized for subsonic and supersonic flows.

The boundary conditions consist of the impermeable (solid) wall condition on the body, $(x, y, z) \in S_b$,

$$\frac{\partial \Phi}{\partial n} = \mathbf{q}_b \cdot \mathbf{n}, \tag{3.246}$$

where $\mathbf{q} = \mathbf{q}_b$ for $(x, y, z) \in S_b$, and $\mathbf{n}$ is the local outward normal to the surface, the condition in the far field,

$$\Phi = 0, \qquad x^2 + y^2 + z^2 \to \infty, \tag{3.247}$$

as well as the discontinuities presented by the wake and shock waves. The wake, $(x, y, z) \in S_w$, is the infinitesimally thin region trailing behind the wing, where the vorticity generated by the wing is transported downstream. Being a region of rotational flow, Kelvin's theorem is inapplicable normal to the wake, therefore the velocity potential on the wake is undefined. This causes a mathematical problem of an arbitrary (or ambiguous) value of Φ on the wake. In the two-dimensional case, this problem is handled by taking a cut along the wake surface, and applying a different value of Φ at its either side, which leads to a discontinuous velocity potential across the wake. For the three-dimensional case, the momentum conservation leads to a zero-pressure jump across the wake, $\Delta p = 0$, where $\Delta(.)$ is the operator representing the change as one crosses the wake region from the lower to the upper surface. Additionally, mass conservation requires the velocity should be continuous normal to the wake. These conditions, substituted into the unsteady, isentropic Bernoulli equation, Eq. (3.242), result in the following:

$$\left(\frac{D \Delta \Phi}{Dt} \right)_w = \frac{\partial \Delta \Phi}{\partial t}(.) + \mathbf{q}_w \cdot \nabla (\Delta \Phi), \tag{3.248}$$

where

$$\mathbf{q}_w = \frac{1}{2} \left(\mathbf{q}_1 + \mathbf{q}_2 \right), \tag{3.249}$$

and

$$\Delta \Phi = \Phi_2 - \Phi_1. \tag{3.250}$$

The mathematical ambiguity (or nonunicity) of the value of the jump in the velocity potential (or doublet strength) across the wake, $\Delta \Phi$, is physically resolved by following fluid particles as they slide off the wing's trailing edge into the wake. As stated earlier, Kelvin's theorem cannot be applied to a closed curve comprising such particles due to shear stress at the solid surface, which causes them to rotate. However, conforming with the experimental observation that the flow leaves the trailing edge of an airfoil nearly tangentially to the mid-plane, Kutta hypothesis (or condition) can be applied to fix the correct circulation at the trailing edge. Kutta condition translates into the following equation for the value of $\Delta \Phi$ on the wake at the trailing edge:

$$\Delta \Phi_w = \lim_{x_w \to x_{te}} \Delta \Phi = \lim_{x_u \to x_{te}} \Phi(x_u) - \lim_{x_\ell \to x_{te}} \Phi(x_\ell). \tag{3.251}$$

Here, the subscripts u, ℓ stand for the points on the upper and lower surfaces, respectively, and x_{te} represents the streamwise location of the trailing edge. Thus, the fluid particles coming from the upper and lower surfaces meet smoothly at the trailing-edge, thereby producing a continuity of $\Delta \Phi$ at that point. This implies the presence of a concentrated vortex at the trailing edge. The value of $\Delta \Phi_w$ remains constant as a fluid particle convects along the wake, by virtue of Eq. (3.248). Hence, a problem that could not be closed mathematically is resolved by physical considerations. However, the evolution of the wake geometry with time presents another problem, which requires a numerical approximation. The initial condition can be taken as follows:

$$\Phi(x, y, z, 0) = 0, \tag{3.252}$$

which implies the flow everywhere is undisturbed at $t = 0$. This includes the case of a wing suddenly started from rest, $\mathbf{q}(t), t > 0$. In the limiting case of $t = 0+$, the time immediately after $t = 0$, while the wing has a nonzero velocity, the wake has not had a sufficient time to develop and $\Delta\Phi_w(0+) = 0$ is a valid assumption. The wake circulation develops continuously with time as the fluid particles convect downstream from the trailing edge.

We now proceed to the integral formulation, which involves solution to the partial equation Eq. (3.244) subject to the boundary conditions specified by Eqs. (3.246)–(3.248). Since the solution is sought outside a solid surface, S_b, and a wake surface, S_w, on both of which normal (or gradient) boundary conditions are specified, such a boundary-value problem is termed an *exterior Neumann problem*. The region interior to the wing and wake boundary is assumed to have an infinitesimally thin extent lying between the upper and lower surfaces of the wing and wake. Consider a flowfield volume V, bounded by the body and wake region on the inner side and the far field boundary, S_∞, on the outer side, where

$$(x, y, z) \in S_\infty : \sqrt{x^2 + y^2 + z^2} \to \infty,$$

as shown in Fig. 3.45. The boundary of V can be denoted by $S = S_b \bigcup S_w \bigcup S_\infty$, and its outward normal by $\mathbf{n}$. Green's integral theorem [88] for the exterior Neumann problem is expressed as follows:

$$\int\int\int_V \left(f\nabla^2 g - g\nabla^2 f\right) dV = \int\int_S \left(f\frac{\partial g}{\partial n} - g\frac{\partial f}{\partial n}\right) dS$$
$$+ \int\int_{S_\infty} \left(f\frac{\partial g}{\partial n} - g\frac{\partial f}{\partial n}\right) dS \tag{3.253}$$

where f, g are piecewise smooth functions. Let us now consider a field point P with coordinates (ξ, η, ζ), at which the solution to the governing partial differential equation, Eq. (3.244) for a specific time τ is sought, given by

$$g(\xi, \eta, \zeta, \tau) = \Phi(\xi, \eta, \zeta, \tau).$$

Furthermore, let $f(x - \xi, y - \eta, z - \zeta)$ be the following *Green's function*,

$$f(x - \xi, y - \eta, z - \zeta, t - \tau) = -\frac{1}{4\pi r}\delta(t - \tau + r/a_\infty), \tag{3.254}$$

which satisfies the linear part of Eq. (3.244), rewritten as the following wave equation:

$$\nabla^2 f - \frac{1}{a_\infty^2}\frac{\partial^2 f}{\partial t^2} = \delta(x - \xi, y - \eta, z - \zeta)\delta(t - \tau), \tag{3.255}$$

with the terminal boundary condition

$$\lim_{t \to \infty} f(x - \xi, y - \eta, z - \zeta, t - \tau) = 0. \tag{3.256}$$

Here

$$r = \sqrt{(x - \xi)^2 + (y - \eta)^2 + (z - \zeta)^2},$$

and $\delta(.)$ is the Dirac delta function. Note that the wave equation solution has the properties $\lim_{r \to \infty} f = 0$, $\lim_{r \to \infty} \partial f / \partial n = 0$, and $f = 0$ for $t > \tau$.

Substituted into Eq. (3.253), the selected solutions, f, g cause the second integral on the right-hand side to vanish due to the far-field boundary condition, Eq. (3.247), resulting in

$$\iiint_V \left(f \nabla^2 \Phi - \Phi \nabla^2 f \right) dV = \iint_S \left(f \frac{\partial \Phi}{\partial n} - \Phi \frac{\partial f}{\partial n} \right) dS. \tag{3.257}$$

Now consider a small sphere, S_ϵ, of radius ϵ surrounding the point P, which is located inside the far-field boundary, S_∞, and outside the boundary of the potential flow region, $S = S_b \bigcup S_w$, as shown in Fig. 3.45. This small volume must be excluded from the volume integral of Eq. (3.257) as the point P lies outside the flowfield. The result of spatial integration is then the following:

$$\iiint_V \left(f \nabla^2 \Phi - \Phi \nabla^2 f \right) dV = \iint_{S \bigcup S_\epsilon} \left(f \frac{\partial \Phi}{\partial n} - \Phi \frac{\partial f}{\partial n} \right) dS = 0, \tag{3.258}$$

or, since the direction of $\mathbf{n}$ on the small sphere is radially inward,

$$- \iint_{S_\epsilon} \left(f \frac{\partial \Phi}{\partial n} - \Phi \frac{\partial f}{\partial n} \right) dS_\epsilon + \iint_S \left(f \frac{\partial \Phi}{\partial n} - \Phi \frac{\partial f}{\partial n} \right) dS = 0. \tag{3.259}$$

When the wave solution expression given by Eq. (3.254) is substituted into Eq. (3.259)[9], and the result is integrated with time, we have the following in the limit of the radius of the spherical volume vanishing, ($\epsilon \to 0$):

$$\Phi(\xi, \eta, \zeta, t) = \int_{-\infty}^{t} \iint_S \left(f \frac{\partial \Phi}{\partial n} - \Phi \frac{\partial f}{\partial n} \right) dS \, d\tau. \tag{3.260}$$

If the point P is located either outside the far-field surface S_∞, or inside the the boundary of the potential flow, S, Green's integral theorem results in

$$0 = \int_{-\infty}^{t} \iint_S \left(f \frac{\partial \Phi}{\partial n} - \Phi \frac{\partial f}{\partial n} \right) dS \, d\tau. \tag{3.261}$$

---

[9] The spatial integration over the sphere yields the following:

$$- \iint_{S_\epsilon} \left(f \frac{\partial \Phi}{\partial n} - \Phi \frac{\partial f}{\partial n} \right) dS_\epsilon = \left(\epsilon \frac{\partial \Phi}{\partial n} - \Phi \right) \delta(t - \tau + \epsilon/a_\infty)$$

Therefore, a convenient representation of Green's theorem inside the flowfield is the following:

$$\hat{\Phi} = E(x, y, z, t)\Phi(x, y, z, t) = \int_{-\infty}^{t}\iint_{S}\left(f\frac{\partial\Phi}{\partial n} - \Phi\frac{\partial f}{\partial n}\right)\mathrm{d}S\,\mathrm{d}\tau, \qquad (3.262)$$

where

$$E(x, y, z, t) = \begin{cases} 1, & (x, y, z) \text{ outside } S, \quad t > 0 \\ 0 & \text{otherwise} \end{cases} \qquad (3.263)$$

Substitution of Eq. (3.262) into the transonic governing equation, Eq. (3.244), yields the following:

$$\nabla^2\hat{\Phi} - \frac{1}{a_\infty^2}\frac{\partial^2\hat{\Phi}}{\partial t^2} = \hat{\chi}, \qquad (3.264)$$

where

$$\hat{\chi} = E_\chi + \nabla E \cdot \nabla\Phi + \nabla\cdot(\Phi\nabla E) - \frac{1}{a_\infty^2}\left[\frac{\partial E}{\partial \tau}\frac{\partial \Phi}{\partial \tau} + \frac{\partial}{\partial \tau}\left(\Phi\frac{\partial E}{\partial \tau}\right)\right] \qquad (3.265)$$

The solution to Eq. (3.264) is finally expressed as follows by Green's identity:

$$\hat{\Phi} = \int_{-\infty}^{t}\iiint_{V} G(\mathbf{X} - \mathbf{X}*, \tau - \tau*)\hat{\chi}(\mathbf{X}, \tau)\mathrm{d}V\,\mathrm{d}\tau, \qquad (3.266)$$

where $\mathbf{X} = (x, y, z)^T$, and G is the elementary solution (Green's function) associated with disturbance applied at $(\mathbf{X}*, \tau*)$ of the wave equation:

$$\nabla^2 G - \frac{1}{a_\infty^2}\frac{\partial^2 G}{\partial t^2} = \delta(\mathbf{X} - \mathbf{X}*)\delta(\tau - \tau*) \qquad (3.267)$$

In three-dimensional flow, the Green's function is expressed as the following retarded potential solution:

$$G(\mathbf{X} - \mathbf{X}*, \tau - \tau*) = -\frac{1}{4\pi R}\delta\left(\tau - \tau* + \frac{R}{R_\infty}\right) \qquad (3.268)$$

where $R = |\mathbf{X} - \mathbf{X}*|$. We have already utilized this result for the linear wave equation, i.e., subsonic and supersonic flows.

3.8.2   *Transonic Doublet-Lattice Method*

Consider Φ to be the perturbation velocity potential over that of the freestream, rendered nondimensional by dividing by $U_\infty b$, where b is a characteristic length.

Then the velocity components nondimensionalized by the freestream velocity are the following:

$$\frac{u}{U_\infty} = \Phi_x; \qquad \frac{v}{U_\infty} = \Phi_y; \qquad \frac{w}{U_\infty} = \Phi_z, \tag{3.269}$$

where the subscripts denote partial derivatives with respect to spatial coordinates, which are rendered nondimensional by dividing by b. With the small-perturbation assumptions,

$$\Phi_x << 1; \qquad \Phi_y << 1; \qquad \Phi_z << 1, \tag{3.270}$$

and the time nondimensionalized by b/U_∞, the FPE is approximated by the following nondimensional TSD equation (see above):

$$\left[1 - M_\infty^2 - (\gamma + 1)M_\infty^2 \Phi_x\right] \Phi_{xx} + \Phi_{yy} + \Phi_{zz} = 2M_\infty^2 \Phi_{xt} + M_\infty^2 \Phi_{tt}. \tag{3.271}$$

The freestream Mach number is subsonic, $M_\infty < 1$, but the local flow can be mixed (subsonic, sonic, and supersonic). The TSD equation is essentially nonlinear and thus capable of capturing weak shock waves associated with the mixed regions, but is theoretically simpler to solve than the full-potential equation, because the solid boundary conditions can be applied on the mean surface of the thin wing. If we change the definition of the non-dimensional coordinates in the y and z directions by dividing them by $b\beta$, where

$$\beta = \sqrt{1 - M_\infty^2},$$

we obtain the following form of the TSD equation used by Lu and Voss [105]:

$$(1 - K\Phi_x)\Phi_{xx} + \Phi_{yy} + \Phi_{zz} = 2\frac{M_\infty^2}{\beta^2}\Phi_{xt} + \frac{M_\infty^2}{\beta^2}\Phi_{tt}, \tag{3.272}$$

where

$$K = (\gamma + 1)M_\infty^2/\beta^2.$$

The main assumption required by the TDLM approach is the concept of time linearization. This approximation involves expressing the total perturbation Φ as a superimposition of the first harmonic component Φ^1 over that of the mean steady flow Φ^0, and is expressed as follows:

$$\Phi(x, y, z, t) = \Phi^0(x, y, z) + \mathrm{Re}[\Phi^1(x, y, z)e^{ikt}], \tag{3.273}$$

where $k = \omega b/U_\infty$ is the reduced frequency. It is therefore assumed that only the first harmonic component of a Fourier series expansion is retained, which is tantamount to linearization in time. This assumption has been verified in experimental investigations over oscillating airfoils with small amplitude motion [176]. Since the harmonic perturbation amplitude Φ^1 is considered small, its products can be neglected, resulting in the following low-frequency TSD equation:

$$\left(1 - K\Phi_x^0\right)\Phi_{xx}^1 + \Phi_{yy}^1 + \Phi_{zz}^1 - (2i\epsilon + K\Phi_{xx}^0)\Phi_x^1 + k\epsilon\Phi^1 = 0, \tag{3.274}$$

where

$$\epsilon = kM_\infty^2/\beta^2$$

The spatial nonlinearity is thus preserved, while the time dependence is rendered linear. However, there is an inherent coupling between the steady and unsteady flow components. Such a coupling could cause a nonuniformity of the coefficients over the flowfield, especially in the presence of supersonic bubbles and shock waves. In fact, the coefficent $(1 - K\Phi_x^0)$ changes sign when the local speed changes from subsonic to supersonic (and vice versa). Since it is expected that mixed subsonic/supersonic flow is present in the transonic flowfield, the low-frequency TSD equation has mixed elliptic/hyperbolic nature, which can cause convergence issues in a finite-difference/finite-volume computational scheme.

The boundary condition on the wing S_b is the following:

$$\frac{\bar{w}}{U_\infty} = \Phi_z^1 = \frac{\partial h}{\partial x} + ikh, \tag{3.275}$$

while that on the wake S_w is given by

$$\Delta\Phi_x^1 + ik\Delta\Phi^1 = 0. \tag{3.276}$$

In order to obtain the solution by Green's integral formula, the following change of perturbation variables is adopted:

$$\phi = \Phi^1 e^{-i\epsilon x}. \tag{3.277}$$

$$u = \phi_x + iv\phi, \tag{3.278}$$

where $v = k/\beta^2$. This substitution transforms the low-frequency TSD to the following Helmholtz form :

$$u_{xx} + u_{yy} + u_{zz} + \lambda^2 u = \left(\frac{\partial}{\partial x} + iv\right)\psi, \tag{3.279}$$

where

$$\lambda = kM_\infty/\beta^2$$

$$\psi = \left(\frac{\partial}{\partial x} + i\epsilon\right)\sigma, \tag{3.280}$$

with

$$\sigma = K\Phi_x^0(\phi_x + i\epsilon\phi) \tag{3.281}$$

For a source at (ξ, η, ζ), the perturbation at (x, y, z) satisfying the homogeneous Helmholtz equation

$$G_{xx} + G_{yy} + G_{zz} + \lambda^2 G = -\delta(x - \xi, y - \eta, z - \zeta), \tag{3.282}$$

and the far-field (Sommerfeld) condition, is given by

$$G(r) = \frac{e^{-i\lambda r}}{4\pi r} \tag{3.283}$$

where $r = \sqrt{(x - \xi)^2 + (y - \eta)^2 + (z - \zeta)^2}$. With the Green's identity applied to the flowfield (Fig. 3.45),

$$\iiint_V \left(u\nabla^2 G - G\nabla^2 u\right) dV = \iint_S \left(u\frac{\partial G}{\partial n} - G\frac{\partial u}{\partial n}\right) dS \tag{3.284}$$

we have

$$u(x, y, z) = -\iiint_V G\left(\frac{\partial}{\partial\xi} + iv\right)\psi dV - \iint_{S_b+S_w} \left(\Delta u\frac{\partial G}{\partial n} - G\frac{\partial\Delta u}{\partial n}\right) dS \tag{3.285}$$

where $S_b + S_w$ is the wing plus wake region and Δ represents the difference between upper and lower surfaces, $\Delta u = u_u - u_\ell$. For a thin wing and wake, this reduces to the following:

$$u(x, y, z) = -\iiint_V G\left(\frac{\partial}{\partial\xi} + iv\right)\psi dV - \iint_{S_b} \Delta u\frac{\partial G}{\partial n} dS \tag{3.286}$$

The solution is now split into the surface integral (u_1) computed by a doublet lattice grid, and the volume integral (u_2) evaluated over the field V:

$$u_1 = -\iint_{S_b} \Delta u\frac{\partial G}{\partial n} dS$$

$$u_2 = -\iiint_V G\left(\frac{\partial}{\partial\xi} + iv\right)\psi dV \tag{3.287}$$

This yields a system of the following coupled integrodifferential equations [105] :

$$\frac{\bar{w}}{U_\infty}(x, y, z) = \frac{1}{8\pi}\iint_{S_b} \frac{\bar{\Delta p}}{q_\infty} K(x_0, y_0, k, M_\infty) dS - \beta e^{i\epsilon x}\iiint_V \sigma\left(G_{\xi\xi} - i\epsilon G_\zeta\right) dV \tag{3.288}$$

and

$$\phi(x, y, z) = \frac{e^{-i\epsilon x}}{8\pi b} \iint_{S_b} \frac{\bar{\Delta p}}{q_\infty} K(x_0, y_0, k, M_\infty) \mathrm{d}S + \iiint_V \sigma \left(G_\xi - i\epsilon G\right) \mathrm{d}V$$

$$(3.289)$$

Here $K(x_0, y_0, k, M_\infty)$ is the familiar subsonic kernel function.

Lu and Voss suggest a solution procedure by discretizing the wing plan form into trapezoidal panels (like the doublet-lattice method), and the flowfield into volume (field) elements. Each panel element is assumed to have a constant pressure difference $\bar{\Delta p}$ on it, whereas each field element has constant value of σ. These assumptions yield the following linear set of equations, to be solved for $\bar{\Delta p}$ and ϕ:

$$\frac{\bar{\mathbf{w}}}{U_\infty} = [A]\frac{\bar{\Delta \mathbf{p}}}{q_\infty} + [B]\boldsymbol{\sigma}$$

$$\boldsymbol{\phi} = [C]\frac{\bar{\Delta \mathbf{p}}}{q_\infty} + [D]\boldsymbol{\sigma}, \qquad (3.290)$$

where the elements of the respective coefficient matrices are given by

$$A_{ij} = \frac{1}{8\pi} \iint_{S_j} K[(x_i - \xi), (y_i - \eta), k, M_\infty] \mathrm{d}\xi \mathrm{d}\eta$$

$$B_{ij} = -\beta e^{i\epsilon x_i} \iiint_{V_j} \left(G_{\xi\xi} - i\epsilon G_\zeta\right) \mathrm{d}\xi \mathrm{d}\eta \mathrm{d}\zeta$$

$$C_{ij} = \frac{e^{-i\epsilon x_i}}{8\pi b} \iint_{S_j} K[(x_i - \xi), (y_i - \eta), k, M_\infty] \mathrm{d}\xi \mathrm{d}\eta$$

$$D_{ij} = \iiint_{V_j} \left(G_\xi - i\epsilon G\right) \mathrm{d}\xi \mathrm{d}\eta \mathrm{d}\zeta.$$

It is to be noted here that the Green's function is evaluated in the field panel integrals by using $r = \sqrt{(x_i - \xi)^2 + (y_i - \eta)^2 + (z_i - \zeta)^2}$. Noting the relationship between σ and ϕ (Eq. (3.281)), Lu and Voss suggest a finite-difference approximation of the derivatives of ϕ, including an artificial viscosity [120] term for numerical convergence. Solution of the linear equations is through the upwash boundary condition specified on the wing, and $\phi = 0$ on the field elements. However, instead of direct inversion of the aerodynamic influence coefficients, it is recommended [105] that an iterative scheme be adopted in which subsonic and supersonic field elements are separately inverted, based upon the flow velocity $\sqrt{U^2 + \phi_x^2 + \phi_y^2 + \phi_z^2}$ in each element. Numerical results reported in the paper of Lu and Voss include fighter and transport aircraft wings, over which unsteady shock waves are successfully captured in comparison with experimental data. Thus a relatively simple TSD method can yield reasonable results for use in aeroservoelastic analysis.

Chapter 4
Finite-State Aeroelastic Modeling

4.1 Finite-State Unsteady Aerodynamics Model

Dynamic aeroelastic applications—such as the flutter analysis—have traditionally incorporated unsteady aerodynamics models based upon simple harmonic motion of lifting surfaces. The previous chapter was devoted to techniques wherein linearized, integral equation formulations were employed to develop frequency-domain aerodynamic forces and moments. The approximation of harmonic motion to aeroelastic modeling is equivalent to applying Fourier transform to the equations of motion in the time domain. This traditional approach is covered in many aeroelasticity textbooks, cf. Bisplinghoff et al. [21], Fung [55], and Scanlan and Rosenbaum [147].

Consider a linear aeroelastic system with the following governing equation of motion:

$$\mathsf{M}\ddot{\mathbf{q}} + \mathsf{K}\mathbf{q} = \mathbf{Q}, \tag{4.1}$$

where the structural damping is neglected for the convenience of discussion. Here, $\mathbf{q}$ is the vector of generalized coordinates representing the structural vibration, M is the generalized mass matrix, K is the generalized stiffness matrix, and $\mathbf{Q}$ is the vector of generalized aerodynamic forces arising due to structural motion. As discussed in the beginning of Chap. 3, the unsteady aerodynamic forces depend upon structural motion coordinates through a relationship given by

$$\mathbf{D}\left(\mathbf{Q}\right) = \mathbf{F}\left(\mathbf{q}\right), \tag{4.2}$$

where $\mathbf{D}(.)$ is a linear differential operator, and $\mathbf{F}(.)$ is a functional operator. The solution for the motion coordinates, $\mathbf{q}(t)$, requires a simultaneous solution to structural dynamics equations, Eq. (4.1), and unsteady aerodynamic field equations, Eq. (4.2), which could be linear, partial differential equations. The coupled integration of such fluid-structure equations requires iterative solution in the time domain, which is not very amenable to aeroelastic computations.

© Springer Science+Business Media, LLC 2015

A. Tewari, *Aeroservoelasticity*, Control Engineering,

DOI 10.1007/978-1-4939-2368-7_4

4.1.1   Traditional Flutter Analysis

In a flutter (or other dynamic aeroelastic) analysis, one is primarily interested in deriving the boundary between stable (exponentially decaying) and unstable (exponentially growing) structural motion, given by

$$\mathbf{q} = \bar{\mathbf{q}}e^{i\omega t}, \tag{4.3}$$

$\bar{\mathbf{q}}$ is the amplitude (or eigenvector) of structural motion and ω is the vibration frequency. This very boundary, for the linear case, brings us to the simple harmonic (Fourier or Laplace transform) analysis, for which many aerodynamic tools are readily available. The harmonic airloads on a structure can be represented in a transfer matrix form as follows:

$$\mathbf{Q} = \mathbf{G}(i\omega)\bar{\mathbf{q}}e^{i\omega t}, \tag{4.4}$$

where $\mathbf{G}(s)$ is the unsteady aerodynamic transfer matrix in the harmonic limit ($s = i\omega$), and is a function of flow properties, namely the airspeed, U, and atmospheric density, ρ. Substituting Eq. (4.4) into Eq. (4.1), we have

$$\left(\mathbf{K} - \omega^2\mathbf{M}\right)\bar{\mathbf{q}}e^{i\omega t} = \mathbf{G}(i\omega)\bar{\mathbf{q}}e^{i\omega t}, \tag{4.5}$$

which cannot be satisfied for any real value of the frequency ω. If a fictitious, viscous damping term, $\mathbf{C}\dot{\mathbf{q}} = g\mathbf{K}\dot{\mathbf{q}}$ with g being a real constant, is added to the left-hand side of Eq. (4.1), then Eq. (4.5) is modified as follows:

$$\left[(1 + ig)\mathbf{K} - \omega^2\mathbf{M}\right]\bar{\mathbf{q}}e^{i\omega t} = \mathbf{G}(i\omega)\bar{\mathbf{q}}e^{i\omega t}. \tag{4.6}$$

The flutter analyst can now seek a real pair, (g, ω), for which Eq. (4.6) is satisfied. This essentially requires checking the damping parameter, g, for a range of flow parameters (U, ρ), which influence the aerodynamic transfer matrix. For $g < 0$, energy must be continuously fed into the system to keep it oscillating at constant amplitude, hence the system is stable. If a condition exists where $g > 0$, then the flutter boundary is said to have been crossed, because energy must be extracted from the flow to maintain harmonic motion. The traditional $U - g - \omega$ method of flutter analysis thus involves plotting g and ω for various values of airspeed, U (or density, ρ), and noting the airspeed and corresponding frequency for which g becomes zero.

While the simple harmonic flutter condition can be investigated in this manner, there is no possibility of applying the traditional frequency domain analysis to subcritical (below flutter speed), and supercritical (above flutter speed) cases. The non-harmonic (convergent or divergent) motion involves a change of motion amplitude with time, therefore its stability analysis requires either Laplace transform, or time-domain methods. Such a method would not only be able to analyze the transient aeroelastic response for any given flight condition, but also predict the transition from stable to unstable motion encountered close to the flutter point. This latter is invaluable in designing a control system for a stable aeroservoelastic (ASE) response at the open-loop flutter speed. Hence, derivation of a state-space model is necessary for ASE applications. However, deriving a time-domain equation of motion, Eq. (4.1), requires representation of the generalized aerodynamics forces vector, $\mathbf{Q}(t)$, for any arbitrary motion, which is a formidable task.

4.1.2   Unsteady Aerodynamics in Time Domain

With the exception of transonic and high-angle-of-attack-flow regimes, it is possible to approximate the unsteady aerodynamics forces to depend linearly on the exciting structural motion. This important step enables the linear superposition of simple aerodynamic solutions for approximating an arbitrary motion. Consider an airfoil experiencing a sudden (impulsive) change in the angle of attack at $t = 0$, approximated by the unit impulse (Dirac delta) function,

$$\alpha_i(t) = \delta(t) = \begin{cases} \infty & (t = 0) \\ 0 & (t \neq 0) \end{cases}. \tag{4.7}$$

The corresponding change in the two-dimensional lift is given by the unit impulse response function,

$$\ell_i(t) = h(t). \tag{4.8}$$

Assuming the aerodynamics to be linear and time-invariant (LTI), the lift due to an arbitrary change in the angle of attack, $\alpha(t)$, is given by the following convolution integral:

$$\ell(t) = \int_{-\infty}^{t} \alpha(\tau)h(t - \tau)\mathrm{d}\tau = \int_{-\infty}^{t} \alpha(t - \tau)h(\tau)\mathrm{d}\tau, \tag{4.9}$$

the existence of which requires that the impulse response of the system, $h(t)$, should vanish for all previous times, $(t < 0)$, when no input was applied (i.e., the system is causal). Furthermore, it is also necessary that $h(t)$ should be finite, which is a property of a stable system. Therefore, the lift of a causal and stable system in response to an arbitrary change in the angle of attack, which begins to act at $t = 0$ is given by

$$\ell(t) = \int_{0}^{t} \alpha(\tau)h(t - \tau)\mathrm{d}\tau. \tag{4.10}$$

For example, consider the following unit step (indicial) change in the angle of attack:

$$\alpha_s(t) = u_s(t) = \begin{cases} 1 & (t > 0) \\ 0 & (t \leq 0) \end{cases}. \tag{4.11}$$

The corresponding indicial lift is then given by

$$\ell_s(t) = \int_0^t \alpha_s(t - \tau)h(\tau)d\tau = \int_0^t \ell_i(\tau)d\tau. \tag{4.12}$$

Hence, the indicial response of a causal, stable LTI system is the time-integral of its impulsive response. Another useful input is the unit harmonic change in the angle of attack, which begins to act at $t = 0$ and is given by

$$\alpha(t) = e^{i\omega t}u_s(t), \tag{4.13}$$

resulting in the following harmonic lift:

$$\ell(t) = \int_0^t e^{i\omega\tau}h(t - \tau)h(\tau)d\tau = e^{i\omega t}\int_0^t e^{-i\omega\tau}h(\tau)d\tau. \tag{4.14}$$

The last equation can be expressed as follows for a causal, stable LTI system:

$$\ell(t) = e^{i\omega t}\left[\int_0^\infty e^{-i\omega\tau}h(\tau)d\tau - \int_t^\infty e^{-i\omega\tau}h(\tau)d\tau\right], \tag{4.15}$$

which for large times ($t \to \infty$) is approximated by

$$\ell(t) = G(i\omega)e^{i\omega t}, \tag{4.16}$$

where

$$G(i\omega) = \int_0^\infty e^{-i\omega t}h(t)dt \tag{4.17}$$

is the Fourier transform of the impulse response function, and is called the frequency response of the lifting system. The existence of the Fourier transform requires that the second integral in Eq. (4.15) should vanish, implying that the impulse response $h(t)$ should tend to zero as $t \to \infty$. This is the property of an *asymptotically stable* system. Hence, harmonic (frequency response) analysis is possible only for causal, LTI, asymptotically stable systems. By taking the Fourier transform of the lift, we have the following result:

$$\begin{aligned}
\ell(i\omega) &= \int_0^\infty e^{-i\omega t}\ell(t)dt \\[2mm]
&= \int_0^\infty e^{-i\omega t}\int_0^t \alpha(t - \tau)h(\tau)d\tau\,dt \\[2mm]
&= \int_0^\infty e^{-i\omega t}h(t)dt \cdot \int_0^\infty e^{-i\omega t}\alpha(t)dt \\[2mm]
&= G(i\omega)\alpha(i\omega).
\end{aligned} \tag{4.18}$$

Thus, the lift is linearly related to the angle-of-attack via the frequency response function in the frequency domain. Here, it is to be noted that both $\ell(i\omega)$ and $G(i\omega)$

are complex functions, implying that the lift can be different in both magnitude and phase from the applied simple harmonic input, $\alpha(i\omega)$.

Proceeding to more general inputs given by

$$\alpha(t) = \alpha_0 e^{(\sigma+i\omega t)} u_s(t), \tag{4.19}$$

where σ is a real number, the concept of Fourier response is extended to Laplace transform, $\mathcal{L}\{h(t)\}$, for a causal, asymptotically stable system merely by equating $s = i\omega$ in the following defining relationship:

$$\ell(s) = G(s)\alpha(s), \tag{4.20}$$

where

$$G(s) = G(\sigma + i\omega) = \mathcal{L}\{h(t)\} = \int_0^\infty e^{-st} h(t) dt, \tag{4.21}$$

is called the aerodynamic transfer function between $\alpha(s)\mathcal{L}\{\alpha(t)\}$ and $\ell(s) = \mathcal{L}\{\ell(t)\}$, the Laplace transforms of the angle-of-attack and lift, respectively, for zero initial condition $(\alpha(0) = \dot{\alpha}(0) = \ddot{\alpha}(0) = \ldots = 0$, etc.). For the existence of transfer function, $G(s)$, the requirement of stability is removed, since the integral of Eq. (4.21) can exist even for unstable systems (i.e., $h(t)$ need not be finite)[1]. Clearly, the extension from the Laplace domain to frequency domain (and vice versa) is not possible if the system is either acausal or *unstable*. This fact is commonly missed. The relationship among indicial response of a linear, time-invariant, stable system, and its transfer function is given by the following inverse Laplace transform:

$$\ell_s(t) = \mathcal{L}^{-1}\left\{\frac{G(s)}{s}\right\}, \tag{4.22}$$

comparing which with Eqs. (4.12) and (4.20) reveals that the time integration is equivalent to the division by the Laplace variable in Laplace domain. Stability analysis by transfer function $G(s)$ requires that the latter should be expressible as a rational function of the Laplace variable s, which is possible only for a linear, time-invariant system

$$G(s) = \frac{N(s)}{D(s)} = \frac{b_m s^m + b_{m-1} s^{m-1} + \cdots + b_1 s + b_0}{s^n + a_{n-1} s^{n-1} + \cdots + a_1 s + a_0}. \tag{4.23}$$

For a proper transfer function $G(s)$, the degree, m, of the numerator polynomial, must not be greater than that of the denominator polynomial, n. The roots of the numerator polynomial, $N(s) = 0$, are called the zeroes of the transfer function, while those of the denominator, $D(s) = 0$, are its poles. Stability requires that no pole of $G(s)$ must be in the right-half s-plane, whereas asymptotic stability implies that all the poles are strictly in the left-half s-plane. Please consult a textbook basic controls [168] for linear systems representation by transfer function, and the associated stability analysis.

---

[1] For the necessary conditions for the existence, and properties of Laplace transform of a function, refer to a textbook on engineering mathematics [88].

4.2   Transient Aerodynamics in Two-Dimensions

Historically, the problem of representing two-dimensional, unsteady aerodynamics in the time domain has received much attention, beginning with the early analytical developments by Birnbaum [20], Wagner [189], Küssner [90], Glauert [63], von Karman and Sears [187], and Theodorsen [173]. For a thin airfoil of semi-chord b undergoing simple harmonic oscillation in the angle of attack in an incompressible, two-dimensional freestream of speed U, Theodorsen [173] expressed the harmonic lift as a sum of circulatory ($\ell_c(ik)$) and non-circulatory ($\ell_{nc}(ik)$) parts:

$$\ell(ik) = \ell_c(ik) + \ell_{nc}(ik), \tag{4.24}$$

where

$$k = \frac{\omega b}{U}$$

is the non-dimensional *reduced frequency*, which is a ratio of the airfoil's forward advance in one period of oscillation to the semi-chord. For a thin airfoil, the non-circulatory part of lift is much smaller in magnitude than the circulatory part, and can be regarded as only a slight modification in the mass, damping, and stiffness of the structure[2]. In contrast, the circulatory part is much more significant, being responsible for aerodynamic lag associated with the wake. Theodorsen used a distribution of sources and sinks to relate the circulatory lift to the upwash, $w(ik)$, induced by the wake,

$$\ell_c(ik) = C(ik)w(ik), \tag{4.25}$$

where the *Theodorsen function*, $C(ik)$, is given by (see Chap. 3 for derivation)

$$C(ik) = \frac{\int_1^\infty \frac{x}{\sqrt{x^2-1}} e^{-ikx} \mathrm{d}x}{\int_1^\infty \frac{x+1}{\sqrt{x^2-1}} e^{-ikx} \mathrm{d}x}, \tag{4.26}$$

is the frequency response function representing the circulatory phase lag due to the oscillating wake operating linearly on $w(ik)$. The infinite integrals for Theodorsen function are evaluated by *Hankel functions of the second kind* [6], $H_0^{(2)}(k)$, $H_1^{(2)}(k)$, that are of orders zero and one, respectively, as follows:

$$C(ik) = \frac{H_1^{(2)}(k)}{H_1^{(2)}(k) + i H_0^{(2)}(k)}. \tag{4.27}$$

---

[2] The noncirculatory lift is mainly responsible for a change in the airfoil's effective mass due to a layer of the fluid being vertically displaced. This effect is termed the *apparent inertia*. The noncirculatory effects for a gas on stiffness and damping are usually negligible.

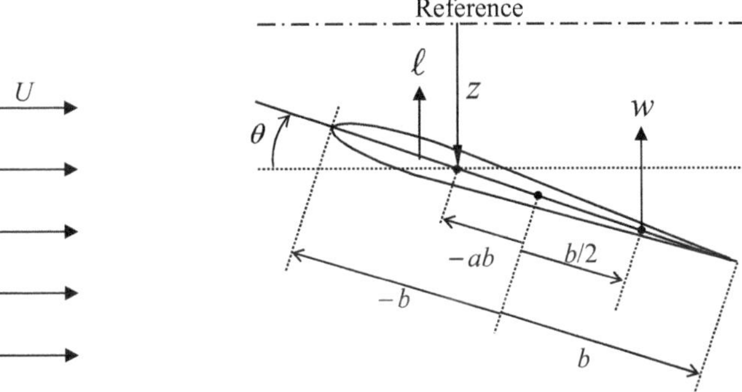

Fig. 4.1 Thin airfoil with pitch ($\theta(t)$) and plunge ($z(t)$) displacements

If the airfoil is undergoing simultaneous oscillations in pitch, $\theta(t) = \theta_0 e^{ikt}$, and plunge, $z(t) = z_0 e^{ikt}$, (see Fig. 4.1), the resulting upwash is given in the time domain by

$$w(t) = \dot{z} + U\theta + b\left(\frac{1}{2} - a\right)\dot{\theta}, \tag{4.28}$$

and in the frequency domain by

$$w(ik) = ikz_0 + \left[U + ikb\left(\frac{1}{2} - a\right)\right]\theta_0. \tag{4.29}$$

This is regarded as the input to the linear lifting system, with the corresponding change in the angle of attack expressed as follows:

$$\alpha(ik) = \frac{w(ik)}{U} = ik\frac{z_0}{U} + \left[1 + ikb\left(\frac{1}{2} - a\right)\right]\frac{\theta_0}{U}. \tag{4.30}$$

4.2.1 Rational Function Approximation

Since the Theodorsen function, $C(ik)$, represents the frequency response of a circulatory wake-induced upwash, it can be generalized as a linear aerodynamic operator for general motions. Such an extension (called *analytic continuation*) from the simple harmonic limit to a general transient motion is justified, because a step change of the upwash (angle of attack) results in an asymptotically stable (decaying) pressure magnitude (lift) response, which is related to the inverse Fourier transform of the Theodorsen function as follows [56]:

$$\Phi(\tau) = 1 + \frac{2}{\pi}\int_0^\infty \frac{C(ik) - 1}{ik}e^{ik\tau}\mathrm{d}k. \tag{4.31}$$

Here $\tau = Ut/b > 0$ is the nondimensional time representing the time required to travel one semichord, and $\Phi(\tau)$ is the *Wagner function* [189] representing the indicial response of circulatory lift to a unit step change in the angle of attack. When comparing Eq. (4.31) with Eq. (4.22), the relationship between Wagner function and Theodorsen function is seen to be identical to that between the indicial response and the frequency response, respectively, of the circulatory lift:

$$\frac{C(ik) - 1}{ik} = \int_0^\infty [\Phi(\tau) - 1]e^{-ik\tau}\, d\tau. \tag{4.32}$$

Jones [80] derived the following approximate expression for $\Phi(\tau)$ from a curve fit with Wagner's numerical data:

$$\Phi(\tau) \simeq 1 - 0.165e^{-0.0455\tau} - 0.335e^{-0.3\tau}, \tag{4.33}$$

which is, however, not accurate enough for practical applications. Jones' method of using a series of decaying exponential functions in time domain to approximate the effect of circulation on transient aerodynamic response, by fitting the frequency response to Theodorsen function at selected frequency points can be generalized. Herein, the process of analytic continuation is brought into effect by adopting an asymptotically stable transfer function representation of $C(ik)$, wherein ik is replaced by the nondimensional Laplace variable $s = \sigma + ik$ and results in the following rational function approximation (RFA):

$$C_a(s) = \frac{N(s)}{D(s)}. \tag{4.34}$$

Such a representation of $C(s)$ is by a series of first order, asymptotically stable poles was given by Dowell [40] as follows:

$$C_a(s) = a_0 + \sum_{n=1}^{N} \frac{a_n s}{s + b_n}, \tag{4.35}$$

where $b_n > 0, n = 1, \ldots, N$. The real coefficients a_0, a_n, b_n can be determined by fitting $C_a(ik) = C(ik)$ at a discrete set of reduced frequencies k. This corresponds to the approximate indicial response,

$$\Phi(\tau) \simeq a_0 + \sum_{n=1}^{N} a_n e^{-b_n \tau}. \tag{4.36}$$

The curve fitting is practically carried out by using a least squares process, where the squared fit error over m selected reduced frequencies is given by

$$\epsilon^2 = \sum_{j=1}^{m} \left[\bar{C}_a(ik_j) - \bar{C}(ik_j)\right] \left[C_a(ik_j) - C(ik_j)\right], \tag{4.37}$$

where the bar represents the complex conjugate. Let $\mathbf{F}$ be the matrix of the following frequency data:

$$\mathbf{F} = \begin{pmatrix} \dfrac{ik_1}{ik_1+b_1} & \dfrac{ik_1}{ik_1+b_2} & \cdots & \dfrac{ik_1}{ik_1+b_n} \\[2ex] \dfrac{ik_2}{ik_2+b_1} & \dfrac{ik_2}{ik_2+b_2} & \cdots & \dfrac{ik_2}{ik_2+b_n} \\[2ex] \vdots & \vdots & \vdots & \vdots \\[2ex] \dfrac{ik_m}{ik_m+b_1} & \dfrac{ik_m}{ik_m+b_2} & \cdots & \dfrac{ik_m}{ik_m+b_n} \end{pmatrix}, \tag{4.38}$$

and $\mathbf{a} = (a_0, a_1, a_2, \ldots, a_n)^T$ be the vector of the unknown numerator coefficients to be determined from the fitting process. The squared fit error can be expressed as follows:

$$\epsilon^2 = \left(\bar{\mathbf{C}}^T - \mathbf{a}^T \bar{\mathbf{F}}^T \right) (\mathbf{C} - \mathbf{F}\mathbf{a}), \tag{4.39}$$

where $\mathbf{C}$ is the vector of Theodorsen function evaluated at the selected frequency points,

$$\mathbf{C} = [C(ik_1), \ \ C(ik_2), \ldots, \ \ C(ik_m)]^T .$$

When the fit error is minimized with respect to the numerator coefficients, we have

$$\frac{\partial \epsilon^2}{\partial \mathbf{a}} = -\bar{\mathbf{F}}^T (\mathbf{C} - \mathbf{F}\mathbf{a}) - \mathbf{F}^T \left(\bar{\mathbf{C}}^T - \mathbf{a}^T \bar{\mathbf{F}}^T \right)^T = 0, \tag{4.40}$$

or

$$\mathbf{a} = \left(\bar{\mathbf{F}}^T \mathbf{F} + \mathbf{F}^T \bar{\mathbf{F}} \right)^{-1} \left(\bar{\mathbf{F}}^T \mathbf{C} + \mathbf{F}^T \bar{\mathbf{C}} \right) \tag{4.41}$$

Dowell [40], following the method of Jones, used $a_0 = 1$, and curve-fitted $C_a(ik) - 1$ in order to determine the remaining linear coefficients, $a_n, n = 2, \ldots, N$. In Dowell's method, the denominator coefficients (poles), $b_n, n = 1, \ldots, N$, are held constant in the minimization process. However, these parameters can be selected by an optimization process for minimizing the total least-squared fit error summed over a range of reduced frequencies, ϵ^2. Peterson and Crawley [128] carried out such an optimization for Dowell's RFA by a gradient-based method, while Eversman and Tewari [50] employed a nongradient (Simplex) optimization for the poles of $C_a(s)$. Table 4.1 compares the results of the two methods for up to 3 lag parameters ($N = 3$) for the Theodorsen function. It was shown [50] that the optimum fit accuracy with the Thoedorsen function can be remarkably improved by using a multiple-pole approximation:

$$C_a(s) = a_0 + \sum_{n=1}^{r} \sum_{p=1}^{m_n} \frac{a_{n_p} s}{(s + b_n)^p}, \tag{4.42}$$

Table 4.1 Rational function approximations for the theodorsen function

$$k_j = (0,\ 0.025,\ 0.05,\ 0.1,\ 0.2,\ 0.3,\ 0.4,\ 0.5,\ 0.6,\ 0.8,\ 1)$$

	Optimized Dowell's Method [128]	Eversman and Tewari [50]
$(N = 1)$		
	$a_0 = 1$	$a_0 = 0.9672$
	$a_1 = -0.4542$	$a_1 = -0.4299$
	$b_1 = 0.1660$	$b_1 = 0.1851$
	$\epsilon = 0.0215$	$\epsilon = 0.0072$
$(N = 2)$		
	$a_0 = 1$	$a_0 = 0.9962$
	$a_1 = -0.4027$	$a_1 = -0.1667$
	$b_1 = 0.1297$	$b_1 = 0.0553$
	$a_2 = -0.1343$	$a_2 = -0.3119$
	$b_2 = 1.2660$	$b_2 = 0.2861$
	$\epsilon = 0.0109$	$\epsilon = 0.0005608$
$(N = 3)$		
	$a_0 = 1$	$a_0 = 0.9994$
	$a_1 = -0.1524$	$a_1 = -0.1055$
	$b_1 = 0.0490$	$b_1 = 0.0371$
	$a_2 = -0.2212$	$a_2 = -0.2879$
	$b_2 = 0.2385$	$b_2 = 0.1859$
	$a_3 = -0.1088$	$a_3 = -0.1003$
	$b_3 = 0.3576$	$b_3 = 0.5886$
	$\epsilon = 0.00102$	$\epsilon = 0.0002034$

where

$$\sum_{n=1}^{r} m_n = N,$$

which has the advantage of reducing the total number of poles N required for a given curve-fit accuracy.

A variation of the least squares method is the Padé approximation by Vepa [185], which uses a common denominator polynomial, $D(s) = (s+b_1)(s+b_2)\cdots(s+b_N)$, in all the terms of the series, resulting in

$$C_a(s) = a_0 + \frac{1}{D(s)} \sum_{n=1}^{N} a_n s\, D(s), \tag{4.43}$$

which enables a determination of the poles, b_n (along with the numerator coefficients, a_n), by the least squares fitting process, thereby removing the ambiguity in their selection. However, this method gives undue weightage to the fit-error at higher values of k, which degrades accuracy at smaller frequencies. Furthermore, the poles could migrate into the right-half Laplace domain during the curve-fitting process.

Another interesting RFA for the Theodorsen function is the following continued fraction approximation by Desmarais [38]:

$$C_a(s) = 1 + \cfrac{-1/2}{1 + \cfrac{1}{4s + \cfrac{1}{1 + \cfrac{3}{4s + \cfrac{3}{1 + \cfrac{5}{4s + \cfrac{5}{1 + \cdots}}}}}}} \tag{4.44}$$

This can be truncated to give a desired order of transfer function by a recursive method. A disadvantage of this RFA is the denser clustering of the poles as the order is increased. This implies that the order of approximation is unnecessarily increased beyond what is required by the conventional approximations of Eq. (4.35) and Eq. (4.42) for a given fit accuracy.

It is possible to extend the two-dimensional (2D) incompressible RFA modeling approach to compressible, subsonic, and supersonic flows. In such a case, the Theodorsen function $C(ik)$ is replaced by a more general frequency response function, $G(ik)$, and the harmonic data is obtained from a suitable method. The subsonic integral equation formulation of Possio [134] can be employed for determining the unknown harmonic lift distribution by enforcing upwash boundary condition at specific collocation points on the airfoil. Similar 2D subsonic integral equation–solution techniques are the iterative kernel evaluation methods of Dietze [39] and Turner [179], and the kernel function expansion methods of Schade [148, 149] and Fettis [52]. A direct solution procedure that does not employ the integral equation at all is the boundary-value solution by Mathieu functions developed by Haskind [69], Reissner [139], Timman [177, 178], and Küssner [94].

For the 2D supersonic case, the available frequency domain methods that can provide the harmonic data for the least squares curve fitting, are based upon a modification of the acceleration potential integral equation employed by Possio for subsonic flow. The modification essentially consists of limiting the integration to the area inside the Mach cone emanating from the given load point, (ξ, ζ), and changing β to $\beta = \sqrt{M_\infty^2 - 1}$ in the kernel function. Alternative formulation is the velocity potential source distribution proposed by Possio [133] and developed into a numerical procedure by Garrick and Rubinow [57]. It is interesting to note that the supersonic integral formulation based upon velocity potential is valid also for arbitrary (non-harmonic) motions. Stewartson [164] showed that the harmonic solution for the velocity potential can be directly derived by taking Laplace transform of the governing wave equation. Apparently, the presence of Mach cones allows the small pressure disturbances to attenuate in the far field, thereby enabling a process of analytic continuation from Laplace to frequency domain. It is to be further noted that the linearized supersonic aerodynamics is valid only for thin airfoils undergoing infinitesimal oscillations. For a thick airfoil undergoing finite amplitude oscillations, the Mach waves are replaced by oblique shock waves whose analysis is essentially nonlinear. This results in a reduction of supersonic damping, which can potentially cause flutter, as investigated by Van Dyke [183]. However, at large reduced frequencies and reasonably large supersonic-hypersonic Mach numbers, a linear approximation called *piston theory* becomes valid even for thick airfoils. Lighthill [101] and Ashley and Zartarian [8] developed the piston theory into a systematic numerical procedure for flutter calculation, based upon Hayes [71] equivalence principle, which states that a steady hypersonic flow on a slender body is equivalent to an unsteady flow in one fewer space dimensions, and can include real gas effects.

4.2.2 Indicial Admittance by Duhamel's Integral

The alternative approach to RFA for deriving unsteady aerodynamics representation in the time domain for a general transient motion is by employing linear superposition of the prescribed geometric upwash via Duhamel's integral. In this way, the indicial admittance is directly evaluated by an integral. For the incompressible case, this approach gives the unsteady lift by Wagner function as follows:

$$\ell(t) = 2b\pi\rho U \left[\Phi(t)w(\xi,0) + \int_0^t \Phi(t-\tau)\frac{\partial}{\partial\tau}w(\xi,\tau)d\tau \right]. \qquad (4.45)$$

Peloubet et al. [126] developed a finite series of Tschebychev polynomials to approximate the real part of the unsteady aerodynamic transfer function in this manner. This approximation is integrated in a closed form to yield the indicial admittance in terms of a finite series of constants and inverse tangent functions in time. Hassig [70] applied the integral method to determine the indicial response of circulatory aerodynamics in incompressible flow. The Laplace transform of indicial admittance derived by this method gives the unsteady aerodynamics transfer function at discrete points in the Laplace domain, which is fitted by a cubic spline to frequency response data at selected frequencies. Hassig demonstrated a good correlation with the work of Luke and Dengler [106] in deriving a *generalized* Theodorsen type function for complex reduced frequency, which was valid for growing (unstable) oscillations. However, the spline fit requires harmonic data at a much wider range of frequencies than is possible to generate either theoretically or experimentally. Furthermore, the approach is invalid for decaying (convergent) oscillations. Apart from the tables of the generalized Theodorsen function provided by Luke and Dengler [106], an early systematic attempt to extend the Theodorsen function for decaying oscillations was by Fraeys de Veubeke [54].

Mazelsky and Drischler [112] derived indicial responses for the compressible subsonic case, and used reciprocal relationship between the pitch and angle of attack changes on unsteady lift and pitching moment. They also proposed that the circulatory lag effect is a property of the fluid in the linear subsonic case, and is independent of airfoil boundary conditions. The indicial response technique was adopted by Leishman [100] for the unsteady lift and moment of an airfoil with an arbitrary flap motion, and was shown to have a reasonable comparison with experimental results for the compressible subsonic case.

4.2.3   Transient Aerodynamics in Three-Dimensions

The problem of representing unsteady aerodynamics for transient structural response of three-dimensional lifting surfaces has been the topic of research for the last five decades. Incompressible flutter calculations of finite-span wings were routinely carried out in World War II era by a strip-wise application of the Theodorsen function (cf. Smilg and Wasserman [159]), followed by subsonic, supersonic, and hypersonic flutter calculations in 1950s and 60s by the modified strip theory of Yates [196]. However, the earliest transient application was the simple extension from two-dimensions proposed by Rodden and Stahl [142] in their subsonic strip theory, wherein the wing span is broken up into a number of chordwise rigid strips (Fig. 4.2). Each strip has essentially the degrees of freedom of pitch and plunge (plus a control surface rotation, if present) like an airfoil, and can be regarded as being connected to neighboring strips by linear and rotary springs, whose stiffnesses are the bending and torsional stiffnesses of the wing at the given spanwise station (Chap. 2). On each strip, the subsonic transient aerodynamic response is calculated by applying a two-term rational polynomial approximation to the Wagner function, which is based upon W.P. Jones' [81] improvement over the approximation of R.T. Jones, Eq. (4.33). The upwash (or angle of attack) boundary condition is enforced at a selected chordwise location, x_i, on the ith strip, based upon its vertical deflection, z_i, and slope, θ_i:

$$w_i = \dot{z}_i + U\theta_i, \tag{4.46}$$

while the lift, L_i, and pitching moment, M_i, are calculated at the aerodynamic center. Since these are now functions of the spanwise variable, y_i, locating the midspan of each strip (Fig. 4.2), the degrees of freedom of the structure are equal to $2n$ (plus the number of control surface rotations, if any), where n is the number of strips. Control surface degrees of freedom, β_i, if present, can be added to each strip, with the corresponding hinge moment, H_i, added to the vector of generalized aerodynamic forces. The strip method is made more accurate by using a compressible, two-dimensional subsonic model (such as those discussed in the last subsection) for generating the harmonic aerodynamic data, rather than the Wagner function. Obvious limitations of the strip method are its inapplicability to low and medium aspect-ratio wings, and its inability to account for loads due to chordwise bending deformations.

In order to overome the deficiencies of the strip theory, the method of RFA in Laplace domain for three-dimensional lifting surfaces was developed in the 1970s. This method required curve fitting with frequency domain data, which can be generated using doublet-lattice, kernel collocation, and double-point methods covered in Chap. 3. For the subsonic case, the double-lattice method of Albano and Rodden [7], and its nonplanar extension by Giesing et al. [61], became standard and

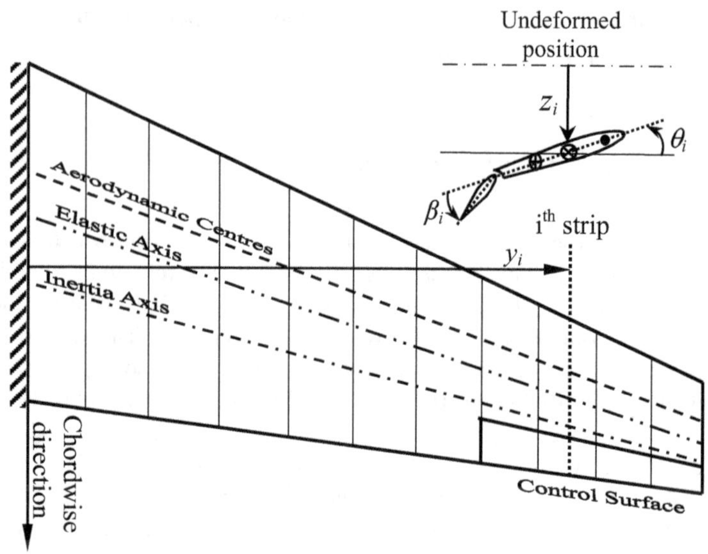

Fig. 4.2 Strip idealization of an aircraft wing

was incorporated into industrial software used by the major aircraft companies (e.g.,
NASTRAN, STABCAR, ASTROS, H7WC, N5KM). However, the problem of deriving a suitable RFA for transient modeling remained open, and became the subject
of active research in the 1975–1995 era.

Inspired by the linear transform relationship between the Theodorsen and Wagner
functions in two-dimensional incompressible flow, Richardson [140] was the first to
suggest deriving a finite-state aerodynamic model for a three-dimensional aircraft
wing based upon a exponential series type approximation (similar to that of R.T.
Jones [80]), which could be combined with a series approximation of the continuous
structure by only a finite number of modes. This immediately led to a finite order,
linear unsteady aerodynamic transfer matrix in the Laplace domain, given by the
following relationship

$$\mathbf{Q}(s) = \mathsf{G}(s)\mathbf{q}(s), \qquad (4.47)$$

where $\mathbf{Q}(s)$ is the unsteady aerodynamic generalized force vector, $\mathsf{G}(s)$ is the unsteady aerodynamic transfer matrix, and $\mathbf{q}(s)$ is the vector of generalized motion
coordinates based upon a finite number, n, of structural degrees of freedom. For in
vacuo structural response ($\mathbf{Q} = \mathbf{0}$), the structural modes $(\lambda_i, \bar{\mathbf{q}}_i), i = 1,\ldots,n$ are
identified from the solution of the following eigenvalue problem:

$$\left(\mathsf{K} + \lambda_i^2 \mathsf{M}\right)\bar{\mathbf{q}}_i = \mathbf{0} \qquad (i = 1,\ldots,n), \qquad (4.48)$$

and result in the following structural modal matrix:

$$\varPhi = \left(\bar{\mathbf{q}}_1, \ \bar{\mathbf{q}}_2,\ldots,\bar{\mathbf{q}}_n\right). \qquad (4.49)$$

Motivated by having to accurately predict the flutter envelope of highly maneuverable fighter aircraft being prototyped at that time, as well as by the need to design active flutter suppression systems for transport/bomber type aircraft, many workers simultaneously developed the RFA for finite-span wings and wing–tail combinations. Sevart [152] was the first to apply an RFA-based model with a second-order polynomial with simple poles (lag parameters) using a curve fit with doublet-lattice data. This was followed by Roger et al. [143, 144], Vepa [186], Edwards [44], and Abel et al. [2–4]. While the RFAs developed by Sevart, Roger et al., and Abel et al. were of simple pole type, where each element of $G(s)$ has the same denominator polynomial, the matrix Padé RFA of Vepa and Edwards had a different denominator polynomial for each column of $G(s)$. The latter approach was expected to reduce the number of state variables in the aeroelastic model required for representing the lag effect of the wake (called lag states), which are approximated by the denominator polynomial of the transfer matrix. Since the Padé approximation does not employ simple poles, but rather a common denominator polynomial (see Eq. (4.43)), and applies least squares curve fit to simultaneously determine the numerator and denominator coefficients, it cannot guarantee that the poles do not migrate into the right-half s-plane for some reduced frequencies during the curve-fitting process. This was quite a major limitation, and was only addressed later by imposing optimization constraints (at the cost of fit accuracy). Other disadvantages of the Padé approximant were its giving undue weightage to higher frequencies in the fitting process, and needing as many frequency points as the order of the aerodynamic influence coefficients (AIC) matrix (thereby requiring an interpolation of the frequency domain database). Consequently, a minor reduction in the number of lag states is offset by an increased computational complexity for obtaining a given fit accuracy, when compared with the method of Sevart, Roger, and Abel et al.

A variation to matrix Padé approximation was the twin transfer functions model of Burkhart [27], wherein the aerodynamic stiffness and damping terms were approximated by rational functions of second degree numerator polynomial and first order denominator, and each element of the transfer matrix employed different poles (which dramatically increased the number of lag states). The poles were constrained to be stable, and a gradient-indexgradient optimization based optimizer (rather than least-squares fitting) determined both numerator and denominator coefficients. Hence, Burkhart was the first to employ optimization in the determination of RFA. However, his model produced a much larger state-space model as well as computation cost than those of the other schemes, for a given accuracy.

In the second wave of RFA developments, there was an attempt to improve the curve-fit accuracy with the frequency domain data by employing nonlinear optimization of the lag parameters (poles) of the transfer matrix. Dunn [43] introduced lag parameter optimization in the RFA of Vepa [186] and Edwards [44], calling it the modified matrix Padé approximation. He used a gradient-based optimizer for the purpose and reported considerable improvement in the fit-accuracy over the unoptimized method. Karpel [85] derived a minimum-state RFA, which yielded the smallest number of aerodynamic states in the aeroelastic model for a given fit accuracy. This was achieved by reversing the traditional approach of first finding the RFA and then the state-space model, and used a constrained, gradient optimization for the lag parameters based upon variable cost metric. The main advantage of Karpel's method was the independence of the number of aerodynamic states from the order of the structural plant, which had the potential for a large reduction in the overall order of the ASE plant. Tiffany and Adams [175] introduced a nongradient optimization method for lag minimization, and applied it to improve the methods of Karpel, Dunn, and Roger. The new optimizer was more efficient and robust with respect to starting conditions, when compared to the previously used gradient-based schemes. Tiffany and Adams [175] compared the various RFA formulations extended by them and reported that it was possible to achieve a similar fit accuracy with the previous (unextended) methods, but with up to 63 % reduction in the number of aerodynamic states. This was taken as an inspiration by Eversman and Tewari [49], who devised their own consistent RFA based upon non-gradient optimization of the aerodynamic states, and reported a remarkable reduction in the number of lag parameters to be optimized for a given curve-fit accuracy. The formulation of Eversman and Tewari [49] used multiple poles, which meant only a single lag parameter needed to be optimized for a comparable accuracy with Tiffany and Adam's optimized least squares RFA of several simple poles. This resulted in a great reduction in the order (and cost) of nonlinear optimization.

Our focus here is on the technique devised by Sevart [152], Roger [144], and Abel et al. [2], and pole optimized by Tiffany and Adams [175] and Eversman and Tewari [49], which will be referred to as the least squares method. Here, the RFA is of the following type:

$$G(s) = A_0 + A_1 s + A_2 s^2 + \sum_{j=1}^{N} A_{j+2} \frac{s}{s + b_j}, \tag{4.50}$$

where $b_j > 0$, $j = 1, \ldots, N$. The numerator coefficient matrices,

$$A_0, A_1, A_2, \ldots, A_{N+2},$$

are determined by fitting $G(ik)$ to the doublet-lattice (or doublet-point) data, $D(ik)$, at a discrete set of reduced frequencies k. The curve fitting is practically carried out by using a least-squares process, where the squared fit error[3] over M-selected reduced frequencies is given by

$$\epsilon^2 = \sum_{i=1}^{n} \sum_{j=1}^{n} \sum_{m=1}^{M} \left[\bar{g}_{ij}(ik_m) - \bar{d}_{ij}(ik_m) \right] \left[g_{ij}(ik_m) - d_{ij}(ik_m) \right], \tag{4.51}$$

where g_{ij} is the (i, j) element of G, and d_{ij} is the (i, j) element of D. The order of the aerodynamic transfer function, n, must be same as the number of structural degrees of freedom.

---

[3] The fit error, ϵ, is divided by the number of elements n^2 as well as by the number M over which the curve fit is carried out, in order to yield an average fit error per element and per frequency point.

For deriving the frequency domain, generalized aerodynamic matrix, D, the following procedure is adopted. After generating the discretized wing geometry data as box corner points, the coordinates and sweep angles required for calculating the aerodynamic influence coefficients (AICs) by either the subsonic doublet-lattice, the supersonic Mach box, or the subsonic/supersonic doublet-point method are computed and stored. Then the AIC computation begins by enforcing the upwash boundary condition at the selected chordwise location, x_i (3/4-chord in doublet-lattice/subsonic doublet-point and mid-chord in Mach box and supersonic doublet-point), on the ith box, based upon its vertical deflection, z_i, and slope, $(dz/dx)_i$, which in turn, are determined by the normalized, in vacuo structural mode shapes (including control surfaces, if any), $\bar{\mathbf{q}}_i, i = 1, \ldots, n$:

$$w_i = \dot{z}_i + U(dz/dx)_i. \tag{4.52}$$

The resulting aerodynamic influence coefficients matrix, AIC, is inverted and pre- and postmultiplied by the structural modal matrix to yield the generalized (and normalized) aerodynamic matrix:

$$\mathsf{D} = \phi^T \{\mathsf{AIC}\}^{-1}\phi. \tag{4.53}$$

If the nondimensional form of AIC matrix, $\overline{\mathsf{AIC}}$, is available (such as the one calculated by doublet-lattice method of Chap. 3), it requires conversion to the dimensional form as follows, before being used in the RFA model:

$$\Delta\mathbf{p} = \{\mathsf{AIC}\}^{-1}\mathbf{w} = \frac{1}{2}\rho U^2\{\overline{\mathsf{AIC}}\}^{-1}\frac{\mathbf{w}}{U} = \frac{1}{2}\rho U\{\overline{\mathsf{AIC}}\}^{-1}\mathbf{w} \tag{4.54}$$

This implies a multiplication by the factor $1/2\rho U$.

For the curve-fitting process, let F be the matrix of the following frequency data:

$$\mathsf{F} = \begin{pmatrix} 1 & ik & (ik)^2 & \frac{ik_1}{ik_1+b_1} & \frac{ik_1}{ik_1+b_2} & \cdots & \frac{ik_1}{ik_1+b_N} \\ 1 & ik & (ik)^2 & \frac{ik_2}{ik_2+b_1} & \frac{ik_2}{ik_2+b_2} & \cdots & \frac{ik_2}{ik_2+b_N} \\ \vdots & \vdots & \vdots & \vdots & & & \\ 1 & ik & (ik)^2 & \frac{ik_M}{ik_M+b_1} & \frac{ik_M}{ik_M+b_2} & \cdots & \frac{ik_M}{ik_M+b_N} \end{pmatrix}, \tag{4.55}$$

and

$$\mathbf{A} = \left(\mathsf{A}_0^T, \mathsf{A}_1^T, \mathsf{A}_2^T, \ldots, \mathsf{A}_{N+2}^T \right)^T$$

be the $[(N+3) \times n \times n]$ array of the unknown numerator coefficients to be determined from the fitting process. The squared fit error matrix is then given by

$$\mathsf{E}^2 = \left(\bar{\mathbf{D}}^T - \mathbf{A}^T\bar{\mathsf{F}}^T \right)(\mathbf{D} - \mathsf{F}\mathbf{A}), \tag{4.56}$$

where $\mathbf{D}$ is the ($M \times n \times n$) array of the generalized aerodynamics data, D, evaluated at the selected frequency points,

$$\mathbf{D} = \left[\mathsf{D}^T(ik_1),\ \mathsf{D}^T(ik_2),\dots,\ \mathsf{D}^T(ik_M)\right]^T.$$

When the fit error is minimized with respect to the numerator coefficients, we have

$$\frac{\partial \mathsf{E}^2}{\partial \mathbf{A}} = -\bar{\mathsf{F}}^T\left(\mathbf{D} - \mathsf{F}\mathbf{A}\right) - \mathsf{F}^T\left(\bar{\mathbf{D}}^T - \mathbf{A}^T\bar{\mathsf{F}}^T\right)^T = 0, \tag{4.57}$$

or

$$\mathbf{A} = \left(\bar{\mathsf{F}}^T\mathsf{F} + \mathsf{F}^T\bar{\mathsf{F}}\right)^{-1}\left(\bar{\mathsf{F}}^T\mathbf{D} + \mathsf{F}^T\bar{\mathbf{D}}\right) \tag{4.58}$$

The total order of the aeroelastic system produced by the least squares RFA is $n(2 + N)$, where N is the number of lag parameters (poles) in the aerodynamic transfer matrix. If a suitable optimization can be performed to reduce the number of aerodynamic (lag) states, then the overall size of the aeroelastic plant can be significantly reduced. This is the objective of the nonlinear optimization method of Tiffany and Adams [175] and Eversman and Tewari [50]. The latter observed the tendency of some of the RFA poles to coalesce together, which indicated the need of a multiple-pole RFA that is consistent with the optimization process. Otherwise, using simple poles that have nearly identical values results in the inconsistent (ill-conditioned, or nearly singular) state-space representation. Table 4.2 shows some results of the optimizations reported by the latter [50] for a high aspect-ratio wing with six structural modes retained in the modal matrix, at a Mach number of 0.9 and fitted to doublet-lattice data at 11 reduced frequencies. It is observed in the table that a given fit accuracy can be achieved by using multiple poles of the same order as simple poles in the conventional RFA[4]. This has the advantage of significant reduction in the size of the nonlinear optimization problem to be solved, although the order of the resulting state-space model remains unchanged from that of simple poles.

The methodology of RFA derivation is schematically shown in Fig. 4.3. Note the necessity of selecting the lag parameters, $b_i, i = 1,\dots, N$, by a suitable optimization process. The options are either the usage of a constrained minimization by gradient based optimizer (such as CONMIN [184] and the function *fmincon* of MATLAB's

---

[4] The RFA of Eversman and Tewari [50] is slightly different from Eq. (4.50) in that the lag terms do not have the Laplace variable in their numerators:

$$\mathsf{G}(s) = \mathsf{A}_0 + \mathsf{A}_1 s + \mathsf{A}_2 s^2 + \sum_{j=1}^{N} \mathsf{A}_{j+2}\frac{1}{s + b_j},$$

An effect of this change is that the coefficient matrix A_0 can no longer be regarded as the "aerodynamic stiffness", which was the case in the original RFA of Sevart [152] and Roger [144].

Table 4.2 Multiple-pole rational function approximations (RFA) [50] for a wing at mach number 0.9

$$k_j = (0,\ 0.05,\ 0.1,\ 0.125,\ 0.175,\ 0.2,\ 0.3,\ 0.5,\ 0.7,\ 0.9,\ 1.2)$$

	Simple pole RFA	Multiple pole RFA
$(N = 2)$		
	$b_1 = 0.21822$	$b_1 = 0.2185$ (Double pole)
	$b_2 = 0.21854$	
	$\epsilon = 0.04197$	$\epsilon = 0.041967$
$(N = 3)$		
	$b_1 = 0.142086$	$b_1 = 0.1425$ (Triple pole)
	$b_2 = 0.142831$	
	$b_3 = 0.142031$	
	$\epsilon = 0.03259$	$\epsilon = 0.03257$
$(N = 4)$		
	$b_1 = 0.172794$	$b_1 = 0.175295$ (Triple pole)
	$b_2 = 0.17281$	$b_2 = 0.58698$ (Simple pole)
	$b_3 = 0.17733$	
	$b_4 = 0.603264$	
	$\epsilon = 0.006112$	$\epsilon = 0.006107$

Optimization Toolbox), or a direct search method without the need for computing gradients, such as the simple but versatile simplex algorithm of Nelder and Mead [121]. A gradient-based scheme requires extensive user experience, because it can get stuck into convergence problems, especially if the number of lag parameters is larger than 2 or 3, and can also have starting problems if the initial guess is outside the feasible space. In contrast, the simplex method is easier to use and is also blessed with good convergence properties, but can quickly get out of the feasible space because it is unconstrained. A way of adding hard constraints to the simplex method is through wall-penalty functions, such as when a lag parameter crosses into the negative region, the objective function is made to assume a large value. The author had little difficulty with the simplex algorithm as a graduate student in the early days of personal computers, and thus would expect anybody with a basic programming experience to easily write a simplex code nowadays.

For an illustration of the RFA method, let us consider a high-aspect-ratio wing modeled after NASA-Langley's *Drone for Aeroelastic Testing*, (DAST-ARW1) [34] (see Chap. 3) with a doublet-lattice gridding of 20 spanwise and 10 chordwise boxes. The DAST-ARW1 wing is modified for ease of aerodynamic modeling by removing the root leading-edge extension of the original wing, which results in a trapezoidal plan form (see Chap. 3), and assuming that this does not result in a significant change in the mass and structural parameters. The in vacuo structural vibration modes for this drone were experimentally investigated by Cox and Gilyard [34] using a ground vibration analysis, and the natural frequencies and damping ratios for the first six

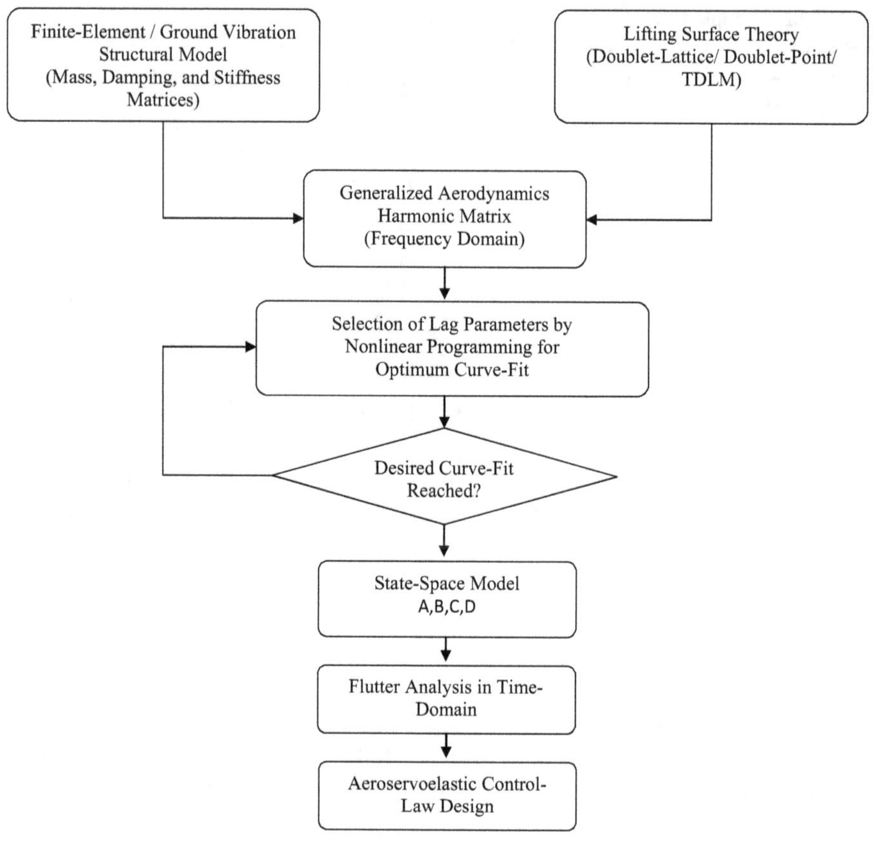

Fig. 4.3 Schematic diagram showing the process of rational function approximation (RFA)

wing modes reported by them are listed in Table 4.3. The mode shapes for the symmetric and anti-symmetric[5] modes are plotted in Figs. 4.4–4.9.

In order to generate RFA for the given set of symmetric and anti-symmetric wing modes, consider the simplified doublet-lattice grid of one chordwise and six span-wise boxes for the optimization of the lag parameters. Such a coarse grid, while being insufficient for an accurate computation of aerodynamic influence coefficients and pressure distribution, is useful in reducing the size of the nonlinear programming problem of lag parameters selection. Once the lag parameters are determined, the doublet-lattice grid can be refined for a more accurate fitting. Using the simplex nongradient optimizer of Nelder and Mead [121] as coded in MATLAB's Optimization Toolbox function *fminsearch*, optimization is carried out for the least

---

[5] Antisymmetric modes have mutually opposite signs of deflections at either side of the wing, and result in a modification of the aerodynamic influence coefficients, such as those computed by the doublet-lattice method (Chap. 3).

Table 4.3 Structural vibration modes of DAST-ARW1 Wing [34]

Natural frequency (Hz)	Damping ratio	Mode type
9.3	0.00588	Symmetric bending
13.56	0.00882	Antisymmetric bending
30.30	0.00937	Symmetric bending/
		Torsion
32.72	0.01943	Antisymmetric bending/
		Torsion
38.96	0.01447	Symmetric torsion
48.91	0.02010	Antisymmetric torsion

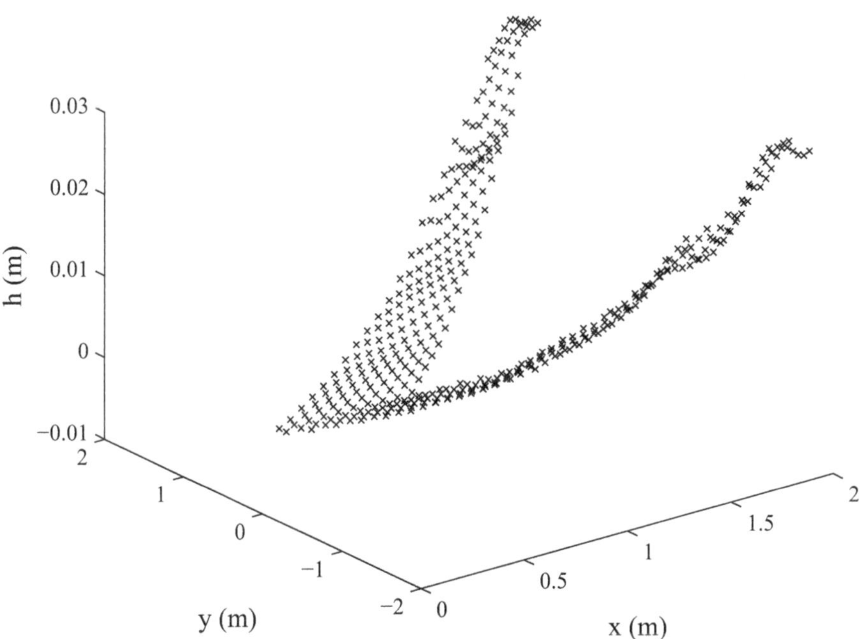

First symmetric bending mode (9.3 Hz)

Fig. 4.4 Vertical deflection mode shape of the first symmetric bending mode of natural frequency 9.3 Hz

squares RFA poles (Eq. (4.50)) $b_j, j = 1, \ldots, N$ where the number of poles N is varied from 2 to 6. The objective function for minimization is the curve-fit error, ϵ, without any weighting for the frequencies. The optimized (1x6) RFAs are generated for the exact harmonic data from the doublet-lattice code at 30 frequency points (selected to cover the natural frequencies of the structural modes), and a flight Mach number of $M = 0.807$ (corresponding to a flight speed of 250 m/s

First anti–symmetric bending mode (13.56 Hz)

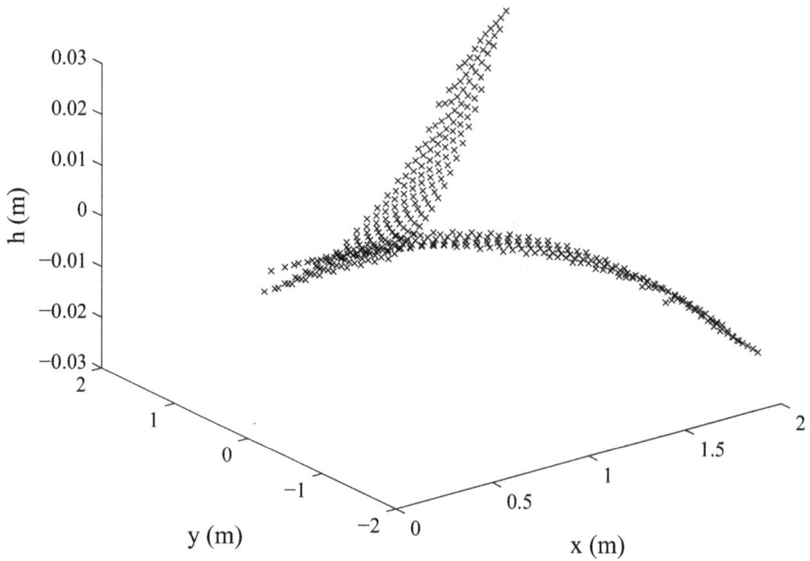

Fig. 4.5 Vertical deflection mode shape of the first antisymmetric bending mode of natural frequency 13.56 Hz

Second symmetric bending/torsion mode (30.3 Hz)

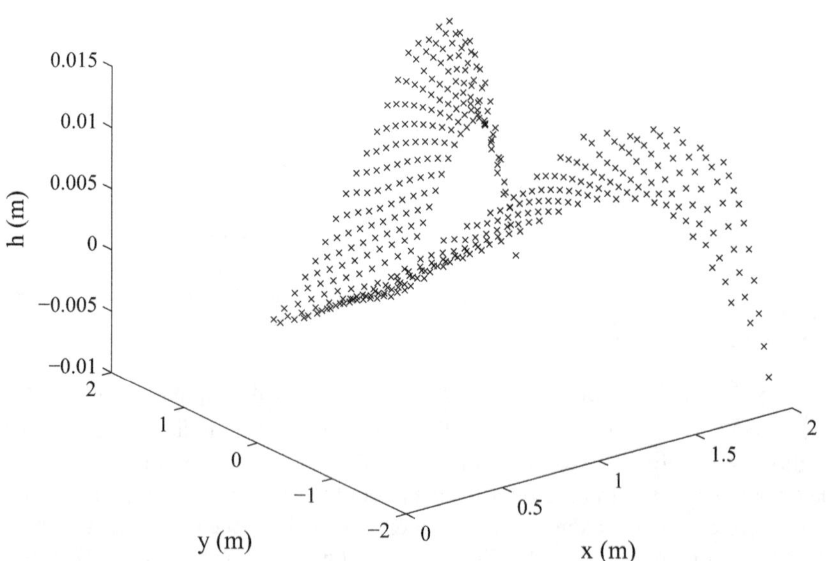

Fig. 4.6 Vertical deflection mode shape of the second symmetric bending/torsion mode of natural frequency 30.3 Hz

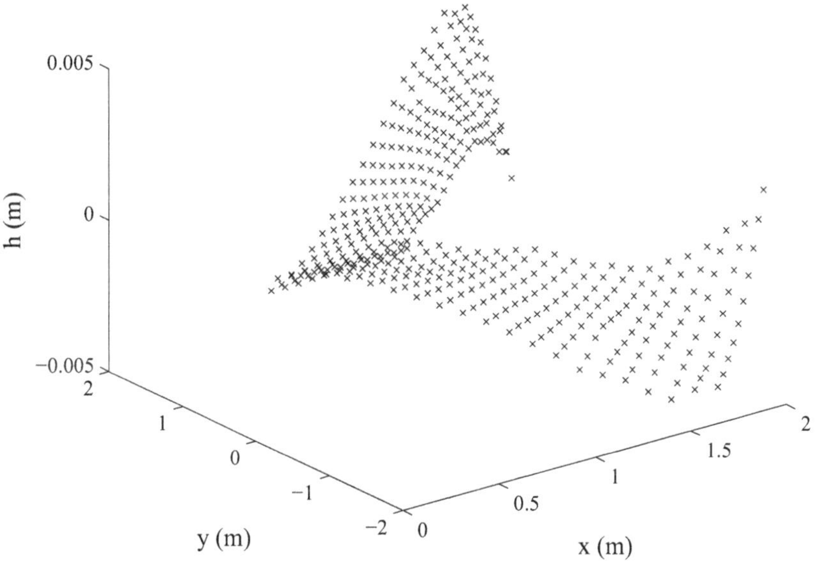

Fig. 4.7 Vertical deflection mode shape of the second antisymmetric bending/torsion mode of natural frequency 32.72 Hz

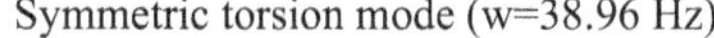

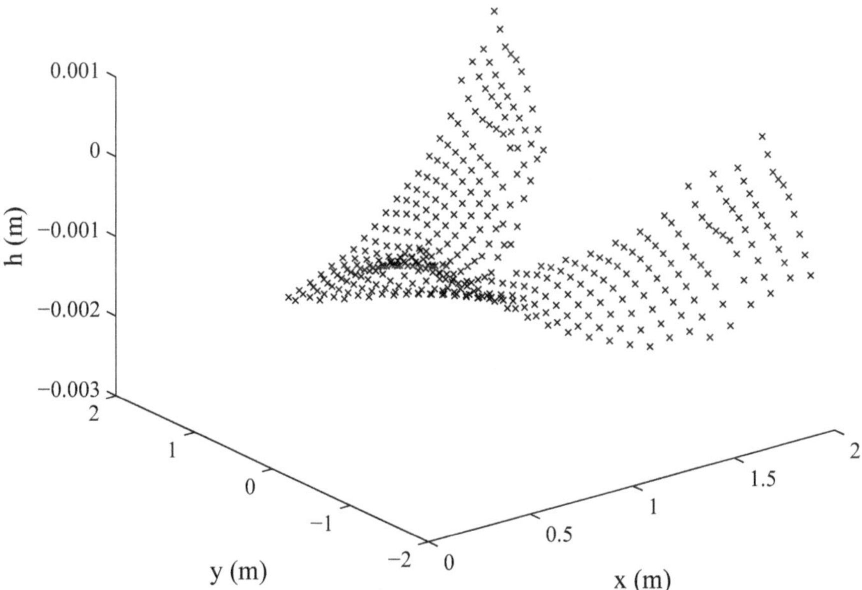

Fig. 4.8 Vertical deflection mode shape of the symmetric torsion mode of natural frequency 38.96 Hz

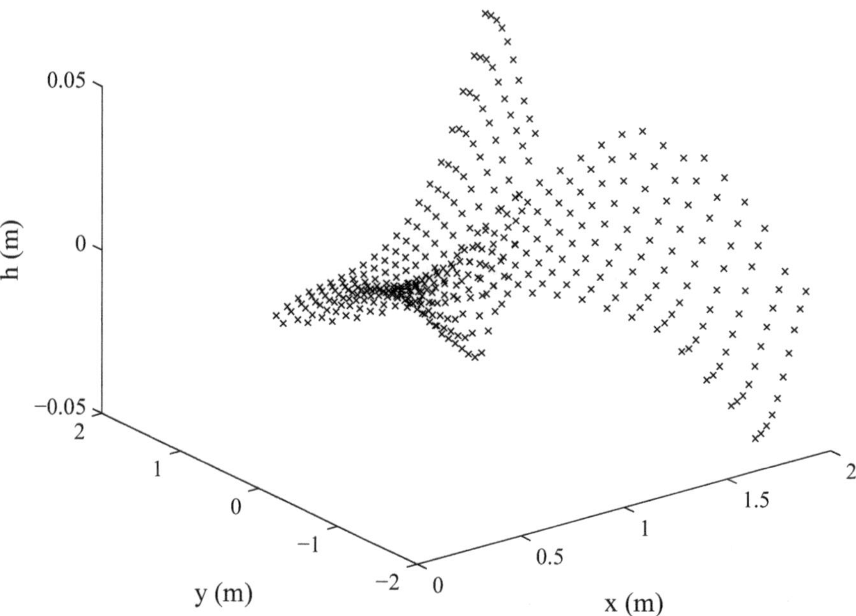

Fig. 4.9 Vertical deflection mode shape of the antisymmetric torsion mode of natural frequency 48.91 Hz

at a standard altitude of 7.6 km):

$$
\omega = \begin{bmatrix}
0 & 0.05 & 0.1 & 0.2 & 0.5 & 1 & 1.5 & 2 & 2.5 & 3 \\
4 & 5 & 7 & 10 & 12 & 15 & 17 & 20 & 22 & 25 \\
27 & 30 & 32.5 & 35 & 37 & 40 & 42 & 45 & 47 & 50
\end{bmatrix} \quad \text{(Hz)}
$$

Curve fits for sample elements of the RFA matrix are shown in Figs. 4.10–4.14, while the optimized values of the lag parameters are listed in Table 4.4. It is to be noted that while 2- or 3-pole RFA can give a rough fit for some elements, they are inadequate for other elements, for which the use of either 4- or 6-pole RFAs shows a significant improvement in the fit accuracy. However, there is a tendency for the term in Eq. (4.58) involving the frequency weighting matrix F to become nearly singular for some values of the lag parameters when closely space frequency points are taken. An obvious solution to this problem is to take fewer frequency points for the curve fit, and then applying a spline interpolation for intermediate frequencies. The higher order RFAs give a better fit at higher values of reduced frequency, which indicates that a further optimization is possible by employing frequency weighting functions, which can reduce the number of poles required for a given accuracy. The harmonic aerodynamic transfer-matrix data used in this example for carrying out the curve fits are listed in Appendix C.

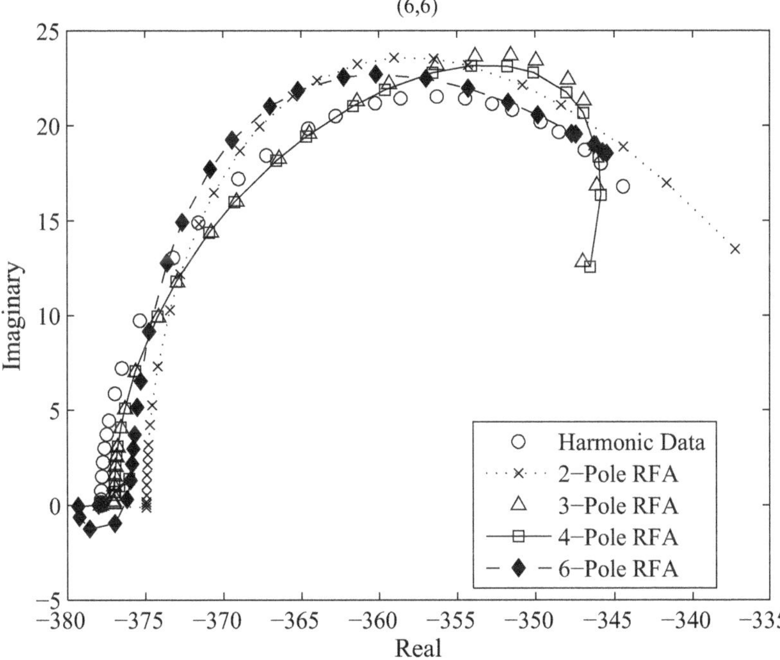

Fig. 4.10 Curve fitting for element (6,6) of generalized aerodynamic transfer matrix of modified DAST-ARW1 wing with 6 symmetric and anti-symmetric modes and (1×6) doublet-lattice grid at $M = 0.807$

4.2.4   Alternative Methods for 3D Transient Aerodynamics

Alternative techniques have been presented for bypassing the rational function approximations (RFA) method, and to derive the unsteady aerodynamics transfer matrix directly from a generalized integral equation. The first such method was by Cunningham and Desmarais [36] who derived the generalized kernel function for the subsonic lifting surfaces, for validity in both converging and diverging oscillations. They compared their results with an RFA model for the convergent case, and observed a looping and spiraling of the generalized aerodynamic forces in the complex plane for high-reduced frequencies. Blair and Williams [24] presented a time-domain panel method in which the subsonic integral equation is solved by a variation of the indicial (Duhamel's integral) method based on calculating and storing the time-history of doublet strengths. Cho and Williams [32] devised a Laplace domain evaluation of the kernel function, which could be applied to both subsonic and supersonic regimes, and obtained good agreement with doublet-lattice and doublet-point methods for oscillatory response. Unfortunately, neither Blair and Williams [24] nor Cho and Williams [32] compare their data with an RFA model for the transient case, and hence it is difficult to comment on their applicability to an aeroservoelastic plant model. Furthermore, they

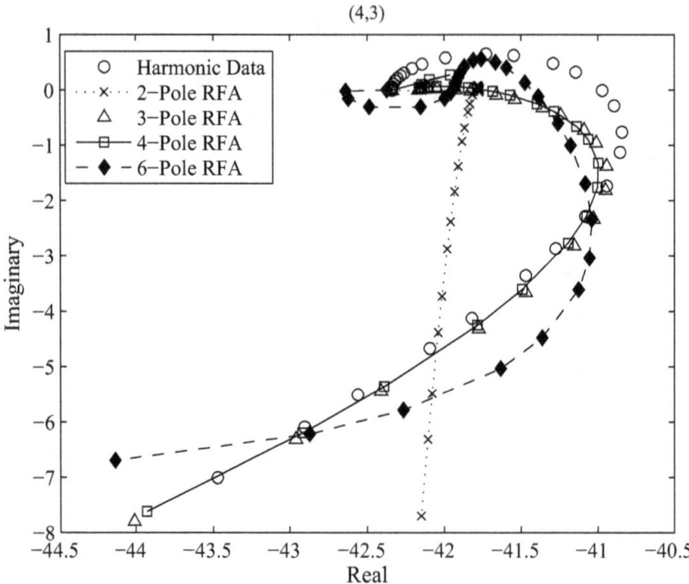

Fig. 4.11 Curve fitting for element (4,3) of generalized aerodynamic transfer matrix of modified DAST-ARW1 wing with 6 symmetric and antisymmetric modes and (1×6) doublet-lattice grid at $M = 0.807$

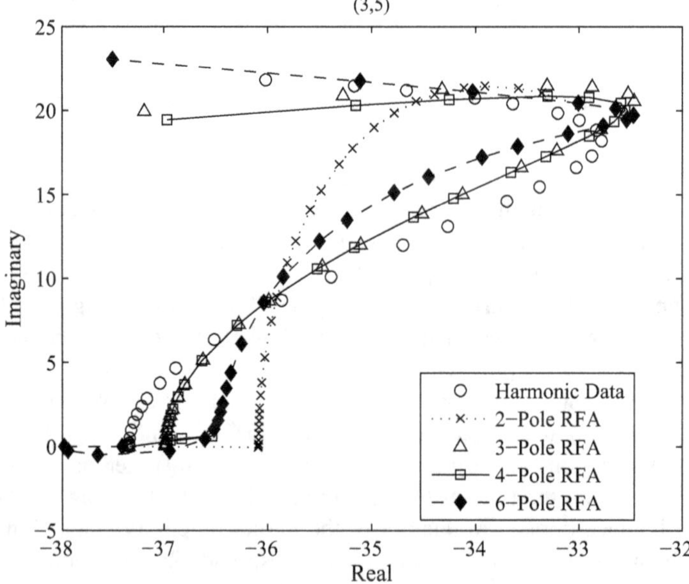

Fig. 4.12 Curve fitting for element (3,5) of generalized aerodynamic transfer matrix of modified DAST-ARW1 wing with 6 symmetric and antisymmetric modes and (1×6) doublet-lattice grid at $M = 0.807$

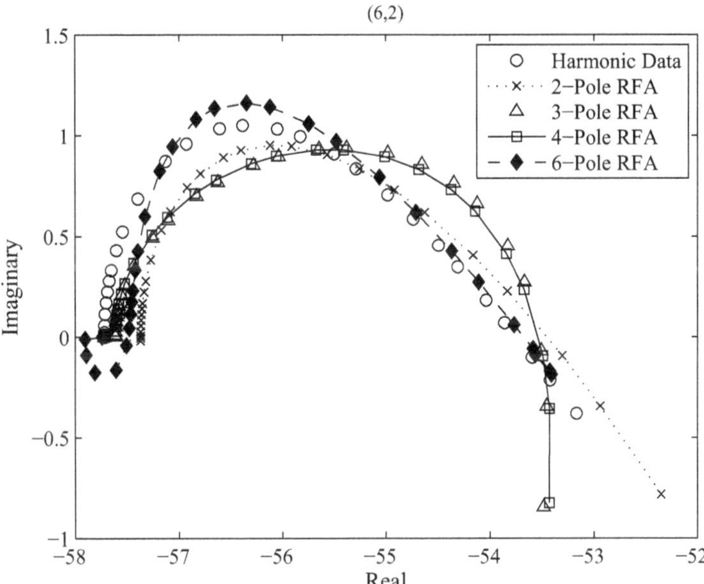

Fig. 4.13 Curve fitting for element (6,2) of generalized aerodynamic transfer matrix of modified DAST-ARW1 wing with 6 symmetric and antisymmetric modes and (1×6) doublet-lattice grid at $M = 0.807$

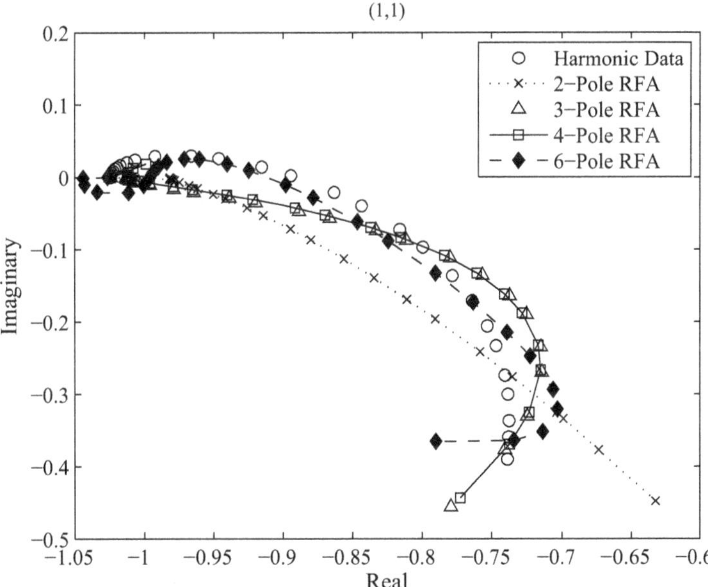

Fig. 4.14 Curve fitting for element (1,1) of generalized aerodynamic transfer matrix of modified DAST-ARW1 wing with 6 symmetric and antisymmetric modes and (1×6) doublet-lattice grid at $M = 0.807$

Table 4.4 Optimized values of lag parameters for modified DAST-ARW1 wing at $U = 250$ m/s

N	b_j	ϵ
2	0.020077, 414503.75	0.00159
3	6.9187×10^{-5}, 3403.8425	0.00093
	5202.256	
4	0.001969, 0.230118	0.0008935
	7869.2315, 7782.0475	
6	0.238246, 0.05157	0.00074
	2.03284, 813.363	
	8570.242, 7980.97	

seem to require much higher data storage than an RFA-based model, which indicates that they suffer from a similar disadvantage as the CFD-based methods for routine ASE applications.

4.2.5   Direct Integration of Governing Equations

With the advent of high-speed digital computers in the 1980s, it became possible to directly solve the governing partial differential equations of unsteady aerodynamics by an appropriate discretization procedure, such as finite-difference, finite-volume, and finite-element method. The resulting field came to be known as computational fluid dynamics (CFD), and has been employed in many practical applications, partly covered in Chap. 3. The early developments at NASA-Langley for a transonic small-disturbance (CAP-TSD) code [14], and its applications to realistic configurations are noteworthy. The main utility of CFD is in calculating the nonlinear aerodynamic and aeroelastic response associated with unsteady flow separation and oscillation of strong shock waves, typically in the transonic regime (Chap. 8). In such an application, the aerodynamic loads vector, $\mathbf{Q}(t)$, is related to the structural response, $\mathbf{q}(t)$, by a nonlinear relationship,

$$\mathbf{Q} = \mathsf{F}(\mathbf{q}, \dot{\mathbf{q}}), \tag{4.59}$$

where $\mathsf{F}(.)$ can be regarded as a nonlinear operator representing the CFD solution, with $\mathbf{q}(t), \dot{\mathbf{q}}(t)$ acting as the time-dependent boundary conditions. The CFD solution is then coupled to the essentially linear structural dynamics model, Eq. (4.1), resulting in the following aeroelastic plant model:

$$\mathbf{M}\ddot{\mathbf{q}} + \mathbf{K}\mathbf{q} = \mathsf{F}(\mathbf{q}, \dot{\mathbf{q}}). \tag{4.60}$$

The coupled CFD-structural dynamics system much be solved iteratively for a time-accurate solution, which is a computationally intensive process, and cannot be conducted routinely as part of an ASE stability analysis. Another drawback of the CFD procedure that it often does not lead to any physical insight into the aeroelastic problem that could possibly be employed in a control system design. However, the CFD method can be used to simulate the ASE response, thereby verifying an already designed control system.

4.3   State-Space Representation

The basic aeroelastic plant is governed by linear ordinary differential equations, which can be written in the following matrix form:

$$\mathsf{M}\ddot{\mathbf{q}} + \mathsf{C}\dot{\mathbf{q}} + \mathsf{K}\mathbf{q} = \mathbf{Q} + \mathbf{Q}_c, \tag{4.61}$$

where $\mathbf{q}(t)$ is the generalized coordinate vector corresponding to the n degrees of structural freedom, $\mathsf{M}, \mathsf{C},$ and K are the corresponding generalized mass, damping, and stiffness matrices of the structure respectively, and $\mathbf{Q}_c(t)$ is an n-dimensional vector of generalized aerodynamic control forces acting as inputs to the system. The generalized aerodynamic force vector, $\mathbf{Q}(t)$, is assumed to be linearly related to $\mathbf{q}(t)$, $\dot{\mathbf{q}}(t)$, and $\ddot{\mathbf{q}}(t)$, as well as to gust inputs, regarded as the *generalized gust state* vector, $\mathbf{x}_g(t)$, and also to certain additional state variables collected into the *aerodynamic lag state* vector, $\mathbf{x}_a(t)$, which is required for modeling the aerodynamic lag caused by a circulatory wake. Such a relationship is enabled by an RFA for the unsteady aerodynamics transfer matrix in the Laplace domain, as discussed in the last section, resulting in the following finite-state approximation for the aerodynamic force vector:

$$\mathbf{Q} = \mathsf{M}_\mathsf{a}\ddot{\mathbf{q}} + \mathsf{C}_\mathsf{a}\dot{\mathbf{q}} + \mathsf{K}_\mathsf{a}\mathbf{q} + \mathsf{N}_\mathsf{g}\mathbf{x}_g + \mathsf{N}_\mathsf{a}\mathbf{x}_a, \tag{4.62}$$

where $\mathsf{M}_\mathsf{a}, \mathsf{C}_\mathsf{a},$ and K_a are the generalized aerodynamic inertia, aerodynamic damping, and aerodynamic stiffness matrices, respectively, N_g is the gust coefficient matrix, and N_a is the *aerodynamic lag* coefficient matrix associated with the time lag due to a circulatory wake. Substitution of Eq. (4.62) into Eq. (4.61) yields

$$(\mathsf{M} - \mathsf{M}_\mathsf{a})\ddot{\mathbf{q}} + (\mathsf{C} - \mathsf{C}_\mathsf{a})\dot{\mathbf{q}} + (\mathsf{K} - \mathsf{K}_\mathsf{a})\mathbf{q} = \mathsf{N}_\mathsf{g}\mathbf{x}_g + \mathsf{N}_\mathsf{a}\mathbf{x}_a + \mathbf{Q}_c. \tag{4.63}$$

The gust states are governed by the following state equations:

$$\dot{\mathbf{x}}_g = \mathsf{A}_\mathsf{g}\mathbf{x}_g + \mathbf{\Gamma}_\mathsf{g} \left\{ \begin{array}{c} \mathbf{q} \\ \dot{\mathbf{q}} \end{array} \right\}, \tag{4.64}$$

where $\mathsf{A}_\mathsf{g} \in \mathbf{R}^{n_g \times n_g}, \mathbf{\Gamma}_\mathsf{g} \in \mathbf{R}^{n_g \times 2n}$ are the coefficient matrices corresponding to aerodynamic effects of the gust.

The aerodynamic lag states are assumed to be governed by the following linear state equations:

$$\dot{\mathbf{x}}_a = \mathsf{A}_\mathsf{a}\mathbf{x}_a + \mathbf{\Gamma}_\mathsf{a} \left\{ \begin{array}{c} \mathbf{q} \\ \dot{\mathbf{q}} \end{array} \right\}, \tag{4.65}$$

where $A_a \in \mathbf{R}^{n_a \times n_a}$, $\Gamma_a \in \mathbf{R}^{n_a \times 2n}$ are the aerodynamic coefficient matrices corresponding to noncirculatory (apparent aerodynamic inertia, stiffness, and damping) and circulatory lag effects. By collecting the structural and aerodynamic state vectors into an *augmented state vector*, $\mathbf{x}(t) : \mathbf{R} \rightarrow \mathbf{R}^{2n+n_a}$,

$$\mathbf{x} = \left\{ \begin{array}{c} \mathbf{q} \\ \dot{\mathbf{q}} \\ \mathbf{x}_g \\ \mathbf{x}_a \end{array} \right\}, \tag{4.66}$$

we have the following augmented state equations of the linear aeroelastic plant:

$$\dot{\mathbf{x}} = \mathbf{A}\mathbf{x} + \mathbf{B}\mathbf{Q}_c, \tag{4.67}$$

where

$$\mathbf{A} = \begin{pmatrix} 0 & \mathsf{I} & 0 & 0 \\ -\bar{\mathsf{M}}^{-1}\bar{\mathsf{K}} & -\bar{\mathsf{M}}^{-1}\bar{\mathsf{C}} & \bar{\mathsf{M}}^{-1}\mathsf{N}_g & \bar{\mathsf{M}}^{-1}\mathsf{N}_a \\ \Gamma_g & & \mathsf{A}_g & 0 \\ \Gamma_a & & 0 & \mathsf{A}_a \end{pmatrix}, \tag{4.68}$$

and

$$\mathbf{B} = \begin{pmatrix} 0 \\ \bar{\mathsf{M}}^{-1}\mathsf{I} \\ 0 \\ 0 \end{pmatrix}, \tag{4.69}$$

where $\bar{\mathsf{M}} = \mathsf{M} - \mathsf{M}_a$, $\bar{\mathsf{C}} = \mathsf{C} - \mathsf{C}_a$, $\bar{\mathsf{K}} = \mathsf{K} - \mathsf{K}_a$ are the generalized mass, damping, and stiffness matrices, respectively, of the aeroelastic system, and 0 and I represent the null and identity matrices, respectively, of appropriate dimensions.

An output equation is necessary for the aeroelastic plant before control can be applied to it, and can be expressed as follows:

$$\mathbf{y} = \mathbf{C}\mathbf{x} + \mathbf{D}\mathbf{Q}_c, \tag{4.70}$$

where the output variables, $\mathbf{y}$, can consist of a set of normal accelerations measured by accelerometers at selected locations, and/or optically (laser) sensed vertical deflections. The sensor locations must be selected such that the resulting plant is observable [171] with the given combination of the coefficient matrices, A, C.

The most common output for an aeroelastic wing is the normal acceleration, $\mathbf{y} = \ddot{\mathbf{q}}_i$, measured at selected points by accelerometers. Let the coordinates of the

Table 4.5 State-space variables for aeroservoelastic system

Symbol	Name	Dimension
$\mathbf{q}$	Generalized coordinates	$n \times 1$
$\mathbf{x}_a$	Aerodynamic states	$n_a \times 1$
$\mathbf{x}_g$	Gust states	$n_g \times 1$
$\mathbf{x}_c$	Actuators' states	$n_c \times 1$
$\mathbf{u}$	Actuators' torques	$m \times 1$
$\mathbf{Q}_c$	Generalized aerodynamic forces produced by control surfaces	$n \times 1$
$\mathbf{y}$	Sensors' outputs	$n_s \times 1$

sensor locations correspond to the indices $\mathbf{i}_s$ of the doublet-lattice grid. Then the normal acceleration output vector picked up by the sensors is the following:

$$\mathbf{y} = -(\bar{\mathbf{M}}^{-1}\mathbf{K})_{\mathbf{i}_s,j}\mathbf{q} - (\bar{\mathbf{M}}^{-1}\mathbf{C})_{\mathbf{i}_s,j}\dot{\mathbf{q}}$$
$$+ (\bar{\mathbf{M}}^{-1}\mathbf{N}_a)_{\mathbf{i}_s,j} + (\bar{\mathbf{M}}^{-1}\mathbf{N}_g)_{\mathbf{i}_s,j} + \bar{\mathbf{M}}^{-1}_{\mathbf{i}_s,j}\mathbf{Q}_c, \quad j = 1,\ldots,n \qquad (4.71)$$

where $\mathbf{A}_{\mathbf{i}_s,j}$, $j = 1,\ldots,n$ represents the submatrix constructed out of $\mathbf{i}_s$ rows of the original matrix $\mathbf{A}$. Thus, the output coefficient matrices are given by

$$\mathbf{C} = \left[-(\bar{\mathbf{M}}^{-1}\mathbf{K})_{\mathbf{i}_s,j}, \ -(\bar{\mathbf{M}}^{-1}\mathbf{C})_{\mathbf{i}_s,j}, \ (\bar{\mathbf{M}}^{-1}\mathbf{N}_a)_{\mathbf{i}_s,j}, \ (\bar{\mathbf{M}}^{-1}\mathbf{N}_g)_{\mathbf{i}_s,j} \right]$$
$$\mathbf{D} = \bar{\mathbf{M}}^{-1}_{\mathbf{i}_s,j}, \qquad j = 1,\ldots,n \qquad (4.72)$$

The control surfaces are governed by separate subsystems called *actuators* whose linearized dynamical model can be represented by the following state-space model:

$$\dot{\mathbf{x}}_c = \mathbf{A}_c\mathbf{x}_c + \mathbf{B}_c\mathbf{u}, \qquad (4.73)$$

$$\mathbf{Q}_c = \mathbf{C}_c\mathbf{x}_c + \mathbf{D}_c\mathbf{u}, \qquad (4.74)$$

where $\mathbf{x}_c$ is the actuator state vector of order n_c, $\mathbf{u}$ is the vector of m control torques applied by the actuators as inputs to drive the control surfaces, and $\mathbf{A}_c \in \mathbf{R}^{n_c \times n_c}$, $\mathbf{B}_c \in \mathbf{R}^{n_c \times m}$, $\mathbf{C}_c \in \mathbf{R}^{n \times n_c}$, $\mathbf{D}_c \in \mathbf{R}^{n \times m}$ are the actuator coefficient matrices. Table 4.5 lists the important variables of the aeroservoelastic state-space plant model.

4.3.1 Typical Section Model

When the typical section model proposed by Theodorsen and Garrick [174] is applied to an aircraft wing of reasonably high aspect ratio, the resulting aerodynamics is of two-dimensional in nature. Such a model is essentially based upon two degrees of freedom for the airfoil, plunge, $z(t)$, and pitch, $\theta(t)$, and an additional degree of freedom for control surface rotation, $\beta(t)$. The control surface hinge moment is regarded as the single control input, $u(t)$, resulting in the following structural equations of motion:

$$M\ddot{q} + C\dot{q} + Kq = Q + (0,\ 1)^T u, \qquad (4.75)$$

where $q = (z/b, \theta, \beta)^T$ is the generalized coordinates vector, $Q = (-L, M_\theta, M_\beta)^T$ is the generalized air loads vector with b being a characteristic length (usually the semi-chord) used for nondimensionalizing the distances, and $M, C,$ and K are the generalized mass, damping, and stiffness matrices, respectively, of the structure. If the total mass of the wing per unit span is m, and total mass of the control surface per unit span is m_c, then the nondimensional distance of the center of mass of the wing behind the pitch axis is x_θ, the nondimensional distance of the control surface hinge line behind the pitch axis is x_c, and the nondimensional distance of the control surface center of mass behind its own hinge line is x_β (Fig. 4.1). The structural stiffnesses in plunge, pitch, and control surface rotation are k, k_θ, k_β, respectively, which denote the bending, torsion, and control stiffnesses, respectively. The generalized mass and stiffness matrices of the structure are given by:

$$M = \begin{pmatrix} m & mx_\theta & m_c x_\beta \\ mx_\theta & I_\theta & m_c x_c \left(x_\beta - x_c \right) \\ m_c x_\beta & m_c x_c \left(x_\beta - x_c \right) & I_\beta \end{pmatrix} \qquad (4.76)$$

$$K = \begin{pmatrix} k & 0 & 0 \\ 0 & k_\theta & 0 \\ 0 & 0 & k_\beta \end{pmatrix}. \qquad (4.77)$$

The generalized structural damping matrix, C, is difficult to model and is thus commonly neglected.

For aerodynamic modeling, the typical airfoil section of semichord b is idealized as a flat plate, placed in uniform flow of speed U and density ρ, and oscillating with reduced frequency, $k = \omega b/U$. In Theodorsen's [173] incompressible flow model, which is used as a basis for most typical section applications, the plunge is taken about the mid-chord point, and pitching is about an axis located a distance ab aft of the mid-chord point. The trailing edge control surface has its hinge line located a distance bc aft of the mid-chord point. The two-dimensional lift L, pitching moment M_θ, and control surface hinge moment M_β in such a case are given as follows [173]:

$$L = \rho b^2 \left(\pi \ddot{z} + \pi U \dot{\theta} - \pi a b \ddot{\theta} - U T_4 \dot{\theta} - T_1 b \ddot{\theta} \right) + 2\pi \rho U b \Psi \qquad (4.78)$$

$$M_\theta = -\rho b^2 \left[-\pi a b \ddot{z} + \pi U b \left(\frac{1}{2} - a \right) \dot{\theta} + \pi b^2 \left(\frac{1}{8} + a^2 \right) \ddot{\theta} + (T_4 + T_{10}) U^2 \beta \right.$$

$$+ Ub \left\{ T_1 - T_8 - (c - a)T_4 + \frac{1}{2}T_{11} \right\} \dot{\beta} - b^2 \left\{ T_7 + (c - a)T_1 \right\} \ddot{\beta} \right] \quad (4.79)$$

$$+ 2\pi\rho Ub^2 \left(a + \frac{1}{2} \right) \Psi$$

$$M_\beta = -\rho b^2 \left[-T_1 b\ddot{z} + Ub \left\{ -2T_9 - T_1 + T_4 \left(a - \frac{1}{2} \right) \right\} \dot{\theta} \right.$$

$$+ 2T_{13}\ddot{\theta} + \frac{1}{\pi}(T_5 - T_4 T_{10})U^2\beta - \frac{1}{2\pi}UbT_4 T_{11}\dot{\beta}$$

$$\left. - \frac{1}{\pi}b^2 T_3 \ddot{\beta} \right] + 2\pi\rho Ub^2 \left(a + \frac{1}{2} \right)$$

$$- \rho Ub^2 T_{12}\Psi \quad (4.80)$$

Here, $\Psi(k)$ is the lag effect of circulation due to the wake, and can be regarded as the frequency response of the upwash w induced by the wake at the 3/4-chord location as follows:

$$\Psi(k) = C(ik)w(k)$$

$$= C(ik) \left\{ \dot{z} + U\theta + b \left(\frac{1}{2} - a \right) \dot{\theta} + \frac{1}{\pi}U T_{10}\beta + \frac{b}{2\pi}T_{11}\dot{\beta} \right\}, \quad (4.81)$$

$C(ik)$ is the Theodorsen function, and the geometric coefficients, T_i, $i = 1 \ldots 14$ employed in Eqs. (4.78)–(4.81) are listed in Theodorsen's report [173]. The generalized loads vector, $\mathbf{Q} = (-L, M_\theta, M_\beta)^T$, contains both circulatory and noncirculatory terms. It is to be noted here that the noncirculatory terms in Eqs. (4.78)–(4.80) (those not involving Ψ) are clubbed together with the mass, stiffness, and damping coefficients of the structure, resulting in the generalized mass, damping, and stiffness matrices of the modified structural system, $\bar{M}$, $\bar{C}$, and $\bar{K}$, respectively, which are expressed in a nondimensional form as follows:

$$\bar{M} = \begin{pmatrix} \kappa + 1 & x_\theta - a\kappa \\ x_\theta - a\kappa & r_\theta^2 + \kappa \left\{ \frac{1}{8} + a^2 \right\} \\ x_\beta - T_1 \frac{\kappa}{\pi} & r_\beta^2 + (c - a)x_\beta - \frac{\kappa}{\pi}\left\{ T_7 + T_1(c - a) \right\} \end{pmatrix}$$

$$\begin{pmatrix} x_\beta - T_1 \frac{\kappa}{\pi} \\ r_\beta^2 + (c - a)x_\beta - \frac{\kappa}{\pi}\left\{ T_7 + T_1(c - a) \right\} \\ r_\beta^2 - \frac{\kappa}{\pi^2}T_3 \end{pmatrix}$$

$$\bar{C} = \begin{pmatrix} 0 & \kappa & -\frac{\kappa}{\pi}T_4 \\ 0 & \kappa\left(\frac{1}{2} - a \right) & \frac{\kappa}{\pi}\left\{ 4T_9 - \left(a + \frac{1}{2} \right) T_4 \right\} \\ 0 & -\frac{\kappa}{\pi}\left\{ T_1 + 2T_9 - \left(a - \frac{1}{2} \right) T_4 \right\} & -\frac{\kappa}{2\pi^2}T_4 T_{11} \end{pmatrix}$$

$$\bar{K} = \begin{pmatrix} \omega_z^2 & 0 & 0 \\ 0 & \omega_\theta^2 r_\theta^2 & \frac{\kappa}{\pi}\{T_4 + T_{10}\} \\ 0 & 0 & \omega_\beta^2 r_\beta^2 + \frac{\kappa}{\pi^2}\{T_5 - T_4 T_{10}\} \end{pmatrix},$$

where

$$\kappa = \frac{\pi \rho b^2}{m}$$

is the nondimensional mass parameter representing the ratio of mass of a cylinder of air per unit span with diameter equal to the chord to the wing mass per unit span. The nondimensional radii of gyration in pitch and control rotation are given by

$$r_\theta = \sqrt{\frac{I_\theta}{mb^2}}, \qquad r_\beta = \sqrt{\frac{I_\beta}{mb^2}},$$

respectively, while the non-dimensional structural frequencies of plunge, pitch, and control surface modes are respectively the following:

$$\omega_z = \frac{b}{U}\sqrt{\frac{k}{m}}, \qquad \omega_\theta = \frac{b}{U}\sqrt{\frac{k_\theta}{I_\theta}}, \qquad \omega_\beta = \frac{b}{U}\sqrt{\frac{k_\beta}{I_\beta}}.$$

For the transient aeroelastic plant, the Theodorsen function is approximated in the Laplace domain by a rational function approximation, as explained in the previous section, resulting in the unsteady aerodynamics transfer matrix, $G(s)$,

$$Q(s) = G(s)q(s). \tag{4.82}$$

The equations of motion of the aeroelastic plant, Eq. (4.75), can now be written in the non-dimensional Laplace domain for zero initial conditions:

$$\left(s^2 \bar{M} + s\bar{C} + \bar{K}\right) q(s) = G(s)q(s) + (0, 0, 1)^T u(s). \tag{4.83}$$

The order of the plant is $6 + \ell$, where ℓ is the total number of aerodynamic states (the number of poles of the RFA, $C_a(s)$). A possible choice of the augmented state vector, $x(t)$, is the following:

$$x = \left\{ \begin{matrix} q \\ \dot{q} \\ x_a \end{matrix} \right\}, \tag{4.84}$$

where the aerodynamic states, $x_a(t)$, are derived from the transfer-function relationship derived from analytic continuation of Eq. (4.81), and the overdot represents differentiation with respect to the non-dimensional time, $\tau = tU/b$:

$$\Psi(s) = C_a(s)w(s) = C_a(s)\left\{s, \quad 1 + \left(\tfrac{1}{2} - a\right)s, \quad \tfrac{1}{\pi}\left(T_{10} + \tfrac{1}{2}T_{11}s\right)\right\} q$$

and an RFA $C_a(s)$ such as Eq. (4.35) or Eq. (4.42). The aerodynamic state equations can thus be expressed as follows:

$$\dot{\mathbf{x}}_a = \mathsf{A}_a \mathbf{x}_a + \Gamma_a \begin{Bmatrix} \mathbf{q} \\ \dot{\mathbf{q}} \end{Bmatrix}, \tag{4.85}$$

where $\mathsf{A}_a \in \mathbf{R}^{\ell \times \ell}$, $\Gamma_a \in \mathbf{R}^{\ell \times 6}$ are the aerodynamic coefficient matrices. The structural equations of motion in non-dimensional time are then written as follows:

$$\bar{\mathsf{M}}\ddot{\bar{q}} + \bar{\mathsf{C}}\dot{\bar{q}} + \bar{\mathsf{K}}\bar{q} = \mathsf{C}_\ell \dot{\mathbf{q}} + \mathsf{K}_\ell \mathbf{q} + \mathsf{N}_a \mathbf{x}_a + (0, 0, 1)^T u, \tag{4.86}$$

where the aerodynamic coefficient matrix $\mathsf{N}_a \in \mathbf{R}^{3 \times \ell}$, and the circulatory damping and stiffness matrices, $\mathsf{C}_\ell \in \mathbf{R}^{3 \times 3}$ and $\mathsf{K}_\ell \in \mathbf{R}^{3 \times 3}$, respectively, arise from the aerodynamic transfer matrix $\mathsf{G}(s)$. The matrices $\mathsf{C}_\ell, \mathsf{K}_\ell$ must be subtracted from the corresponding structural matrices, $\bar{\mathsf{C}}, \bar{\mathsf{K}}$, respectively.

For example, if the simple-pole RFA of Eq. (4.35) is adopted for $C_a(s)$, we have,

$$\mathsf{N}_a = -2\kappa \begin{Bmatrix} -1 \\ \frac{1}{2} + a \\ -\frac{1}{2\pi} T_{12} \end{Bmatrix} \begin{Bmatrix} a_1 b_1 & a_2 b_2 & \cdots & a_\ell b_\ell \end{Bmatrix} \tag{4.87}$$

$$\mathsf{K}_\ell = 2\kappa (a_0 + a_1 + \cdots + a_\ell) \begin{Bmatrix} -1 \\ \frac{1}{2} + a \\ -\frac{1}{2\pi} T_{12} \end{Bmatrix} \begin{Bmatrix} 0 & 1 & \frac{1}{\pi} T_{10} \end{Bmatrix} \tag{4.88}$$

$$\mathsf{C}_\ell = 2\kappa (a_0 + a_1 + \cdots + a_\ell) \begin{Bmatrix} -1 \\ \frac{1}{2} + a \\ -\frac{1}{2\pi} T_{12} \end{Bmatrix} \begin{Bmatrix} 1 & (\frac{1}{2} - a) & \frac{1}{2\pi} T_{11} \end{Bmatrix} \tag{4.89}$$

The final state equations of the aeroelastic system are as follows:

$$\dot{\mathbf{x}} = \mathsf{A}\mathbf{x} + \mathsf{B}u, \tag{4.90}$$

where

$$\mathsf{A} = \begin{pmatrix} 0 & \mathsf{I} & 0 \\ -\bar{\mathsf{M}}^{-1}\hat{\mathsf{K}} & -\bar{\mathsf{M}}^{-1}\hat{\mathsf{C}} & -\bar{\mathsf{M}}^{-1}\mathsf{N}_a \\ & \Gamma_a & & \mathsf{A}_a \end{pmatrix}, \tag{4.91}$$

and

$$\mathsf{B} = (0_{1\times 5}, 1, 0_{1\times \ell})^T , \tag{4.92}$$

where $\hat{\mathsf{C}} = \bar{\mathsf{C}} - \mathsf{C}_\ell$, $\hat{\mathsf{K}} = \bar{\mathsf{K}} - \mathsf{K}_\ell$. The controllability of the plant with the control surface torque input $u(t)$ can be verified by the rank of the following controllability test matrix [168]:

$$\mathsf{P} = \left(\mathsf{B},\ \mathsf{AB},\ \mathsf{A}^2\mathsf{B},\ \mathsf{A}^3\mathsf{B},\ \mathsf{A}^4\mathsf{B},\ \mathsf{A}^5\mathsf{B},\ \mathsf{A}^6\mathsf{B}\right). \tag{4.93}$$

Since B is a column vector, P is a square. The determinant $|\,\mathsf{P}\,|$ is always nonzero, which signifies an unconditionally controllable plant.

The gust states can be added to the foregoing plant by suitably augmenting the state vector, as discussed above. For compressible flows, the aerodynamic force vector is derived in a similar manner, but without the recourse to the closed-form expressions (Eqs. (4.78)–(4.81)). The RFA of $\mathsf{G}(s)$ in that case involves curve-fitting with the frequency domain data that is available only in a tabular form for compressible subsonic and supersonic flows, in the same manner as carried out for the three-dimensional case discussed below.

4.3.2 Three-Dimensional Wing Model

By substituting the RFA $\mathsf{G}(s)$ for a three-dimensional wing, into the Laplace transform of structural dynamics equations with zero initial conditions,

$$\left(\mathsf{M}s^2 + \mathsf{C}s + \mathsf{K}\right)\mathbf{q}(s) = \mathsf{G}(s)\mathbf{q}(s) + \mathbf{Q}_c(s), \tag{4.94}$$

we have the following state equations:

$$\begin{Bmatrix} \mathbf{q} \\ \dot{\mathbf{q}} \end{Bmatrix} = \begin{pmatrix} 0 & \mathsf{I} \\ -\bar{\mathsf{M}}^{-1}\bar{\mathsf{K}} & -\bar{\mathsf{M}}^{-1}\bar{\mathsf{C}} \end{pmatrix} \begin{Bmatrix} \mathbf{q} \\ \dot{\mathbf{q}} \end{Bmatrix} + \begin{pmatrix} 0 \\ \bar{\mathsf{M}}^{-1}\mathsf{N}_{\mathsf{a}} \end{pmatrix} \mathbf{x}_a$$

$$+ + \begin{pmatrix} 0 \\ \bar{\mathsf{M}}^{-1}\mathsf{N}_{\mathsf{g}} \end{pmatrix} \mathbf{x}_g \begin{pmatrix} 0 \\ \bar{\mathsf{M}}^{-1}\mathsf{I} \end{pmatrix} \mathbf{Q}_c, \tag{4.95}$$

and

$$\dot{\mathbf{x}}_a = \mathsf{A}_{\mathsf{a}}\mathbf{x}_a + \Gamma_{\mathsf{a}} \begin{Bmatrix} \mathbf{q} \\ \dot{\mathbf{q}} \end{Bmatrix}, \tag{4.96}$$

$$\dot{\mathbf{x}}_g = \mathsf{A}_\mathsf{g}\mathbf{x}_g + \Gamma_\mathsf{g}\begin{Bmatrix}\mathbf{q}\\\dot{\mathbf{q}}\end{Bmatrix}, \tag{4.97}$$

where $\bar{\mathsf{M}}, \bar{\mathsf{K}}, \bar{\mathsf{C}}$ denote the generalized mass, stiffness, and damping matrices, respectively, of the aeroelastic system (each of size $n \times n$) derived by clubbing the relevant terms of the RFA transfer matrix with the corresponding structural matrices, $\mathsf{M}, \mathsf{K}, \mathsf{C}$, respectively. The number n_a of the additional aerodynamic state variables, $\mathbf{x}_a$, and the dimensions of their state-space coefficient matrices, $\mathsf{A}_\mathsf{a}, \mathsf{N}_\mathsf{a}, \Gamma_\mathsf{a}$, depend on the type of the RFA employed. A comparison of the augmented state dimensions for the selected RFAs is shown in Table 4.6. The gust influence on the unsteady aerodynamic forces is modeled by a total number n_g of gust states, $\mathbf{x}_g$, with corresponding gust coefficient matrices, $\mathsf{A}_\mathsf{g}, \mathsf{N}_\mathsf{g}, \Gamma_\mathsf{g}$.

The state equations of the overall aeroelastic plant can be expressed as follows:

$$\dot{\mathbf{x}} = \mathsf{A}\mathbf{x} + \mathsf{B}\mathbf{Q}_c, \tag{4.98}$$

where

$$\mathbf{x} = \left(\mathbf{q}^T, \dot{\mathbf{q}}^T, \mathbf{x}_a^T\right)^T,$$

$$\mathsf{A} = \begin{pmatrix} 0 & \mathsf{I} & 0 & 0 \\ -\bar{\mathsf{M}}^{-1}\bar{\mathsf{K}} & -\bar{\mathsf{M}}^{-1}\bar{\mathsf{C}} & \bar{\mathsf{M}}^{-1}\mathsf{N}_\mathsf{a} & \bar{\mathsf{M}}^{-1}\mathsf{N}_\mathsf{g} \\ \Gamma_\mathsf{a} & \mathsf{A}_\mathsf{a} & & 0 \\ \Gamma_\mathsf{g} & 0 & & \mathsf{A}_\mathsf{g} \end{pmatrix}, \tag{4.99}$$

and

$$\mathsf{B} = \begin{pmatrix} 0 \\ \bar{\mathsf{M}}^{-1}\mathsf{I} \\ 0 \end{pmatrix}. \tag{4.100}$$

When the least squares RFA of Eq. (4.50) is introduced (without considering any gust states for the moment), we have the following expression for the generalized aerodynamic force (GAF) vector:

$$\mathbf{Q}(t) = \mathsf{A}_0\mathbf{q} + \mathsf{A}_1\dot{\mathbf{q}} + \mathsf{A}_2\ddot{\mathbf{q}} + \mathsf{N}_\mathsf{a}\mathbf{x}_a, \tag{4.101}$$

where

$$\mathsf{N}_\mathsf{a} = (\mathsf{A}_3,\ \mathsf{A}_4,\ \mathsf{A}_5,\ \ldots,\ \mathsf{A}_{N+2}) \tag{4.102}$$

Table 4.6 Augmented state dimensions for various RFA formulations

Method	Lag parameters	n_a
Least-squares [143]	$\sum_{k=1}^{N} \{a_{k+2}\}_{ij}\, \frac{1}{s+b_k}$	$n_a = nN$
Matrix-Padé [185]	$\sum_{k=1}^{N_j} \{a_{k+2}\}_{ij}\, \frac{1}{s+\{b_j\}_k}$	$n_a = \sum_{j=1}^{n+n_g} N_j$
Minimum-state [85]	$\bar{D}(s\mathsf{I} - \mathsf{R})^{-1}\bar{E}$	$n_a = N$

$$[\mathsf{R} = \text{diag.}(b_1, b_2, \ldots, b_N) \text{ and } \bar{D}, \bar{E}$$
$$\text{to be derived by nonlinear optimization}]$$

is the coefficient matrix of dimension $n \times nN$ and $\mathbf{x}_a$ is the $nN \times 1$-dimensional aerodynamic state vector satisfying the following state equation:

$$\dot{\mathbf{x}}_a = \mathsf{A}_a \mathbf{x}_a + \Gamma_a \dot{\mathbf{q}}, \tag{4.103}$$

with

$$\mathsf{A}_a = \begin{pmatrix} -b_1 \mathsf{I}_n & 0 & 0 & \cdots & 0 \\ 0 & -b_2 \mathsf{I}_n & 0 & \cdots & 0 \\ \vdots & \vdots & \vdots & \vdots & \vdots \\ 0 & 0 & 0 & \cdots & -b_N \mathsf{I}_n \end{pmatrix}, \tag{4.104}$$

$$\Gamma_a = \begin{pmatrix} \mathsf{I}_n \\ \mathsf{I}_n \\ \vdots \\ \mathsf{I}_n \end{pmatrix}, \tag{4.105}$$

where I_n denotes the identity matrix of dimension n (order of structural system), $b_j > 0, j = 1, \ldots, N$ are the lag parameters (or aerodynamic poles), A_a has dimension $nN \times nN$, and Γ_a is of dimension $nN \times n$. When substituted into Eq. (4.67), this results in the following state-space model:

$$\mathsf{A} = \begin{pmatrix} \mathsf{0}_n & \mathsf{I}_n & \mathsf{0}_{n \times nN} \\ -\bar{\mathsf{M}}^{-1}\bar{\mathsf{K}} & -\bar{\mathsf{M}}^{-1}\bar{\mathsf{C}} & \bar{\mathsf{M}}^{-1}\mathsf{N}_a \\ \mathsf{0}_{nN \times n} & \Gamma_a & \mathsf{A}_a \end{pmatrix}, \tag{4.106}$$

where $\bar{\mathsf{M}} = \mathsf{M} - \mathsf{A}_2$, $\bar{\mathsf{C}} = \mathsf{C} - \mathsf{A}_1$, $\bar{\mathsf{K}} = \mathsf{K} - \mathsf{A}_0$, and single subscript n indicates a square matrix. The control coefficient matrix is the following:

$$\mathsf{B} = \begin{pmatrix} \mathsf{0}_{n \times m} \\ \bar{\mathsf{M}}^{-1}\mathsf{I}_n \\ \mathsf{0}_{nN \times m} \end{pmatrix}. \tag{4.107}$$

Modeling of Control Surfaces

Modeling of control surfaces requires additional state variables, which can generate the generalized aerodynamic control force vector, $\mathbf{Q}_c$. The control inputs can either take the form of control hinge moments, $\mathbf{u}$, applied by the actuators, which need to be modeled separately, or control surface deflections, δ, which are treated as separate generalized coordinates. The latter approach of treating control surface deflections as control inputs is tantamount to a quasi-steady approximation, in which the unsteady aerodynamic inertia and lag effect of the wake are neglected. Many flight dynamics textbooks adopt this approach, but we will not use it here due to its obvious limitations. Instead, full account is given of the aerodynamic noncirculatory and circulatory effects of control surfaces by using a rational function approximation. Thus a control surface can be regarded as a separate lifting surface with its own degrees of freedom and aerodynamics.

The most accurate model of a control surface would include the bending deformation of the hinge line, a rigid rotation of the surface about the hingeline, as well as spanwise bending and twisting of the surface. However, due to its generally small size in comparison to the wing, the bending and twisting deformations can be often neglected, resulting in the approximation of a rigid rotation, δ, of the control surface about the hinge-line. This allows a significant reduction in the degrees of freedom of the overall structural system, as well as an ease of control law development, without any great penalty on modeling accuracy.

Consider a control surface (assumed to be rigid for simplicity) with the following actuator dynamics:

$$I_\delta \ddot{\delta} + c_\delta \dot{\delta} + k_\delta \delta = u + H, \tag{4.108}$$

where $\delta(t)$ is the control surface deflection about the hinge line, I_δ, c_δ, and k_δ the corresponding moment of inertia, damping, and stiffness of the control surface respectively, $H(t)$ is the unsteady aerodynamic hinge moment acting on the control surface, and $u(t)$ is the hinge moment control input applied by the actuator. The aerodynamic hinge moment $H(t)$, and the generalized aerodynamic force vector created on the wing by the control surface deflection, $\mathbf{Q}_c(t)$, are assumed to be linearly related to $\delta(t)(t)$, $\dot{\delta}(t)(t)$, and $\ddot{\delta}(t)(t)$, as well as to certain additional state variables collected into the control surface aerodynamic lag state vector, $\mathbf{x}_c(t)$. Such a relationship can be represented by the following least squares RFA:

$$\mathbf{Q}_c(t) = \mathbf{B}_0 \delta + \mathbf{B}_1 \dot{\delta} + \mathbf{B}_2 \ddot{\delta} + \mathsf{N}_c \boldsymbol{\xi}_c, \tag{4.109}$$

where

$$\mathsf{N}_c = (\mathbf{B}_3, \quad \mathbf{B}_4, \quad \mathbf{B}_5, \quad \ldots, \quad \mathbf{B}_{N+2}) \tag{4.110}$$

is the coefficient matrix of dimension $n \times N$ and $\boldsymbol{\xi}_c$ is the $N \times 1$-dimensional control aerodynamic state vector satisfying the following state equation:

$$\dot{\boldsymbol{\xi}}_c = \Lambda_c \boldsymbol{\xi}_c + \boldsymbol{\Gamma}_c \dot{\delta}, \tag{4.111}$$

with

$$
\Lambda_c =
\begin{pmatrix}
-b_1 & 0 & 0 & \cdots & 0 \\
0 & -b_2 & 0 & \cdots & 0 \\
\vdots & \vdots & \vdots & \vdots & \vdots \\
0 & 0 & 0 & \cdots & -b_N
\end{pmatrix},
\tag{4.112}
$$

$$
\Gamma_c =
\begin{pmatrix}
1 \\
1 \\
\vdots \\
1
\end{pmatrix},
\tag{4.113}
$$

Similarly, the aerodynamic hinge moment can be expressed by

$$
H(t) = a_0 \delta + a_1 \dot{\delta} + a_2 \ddot{\delta} + \mathbf{N}_h^T \boldsymbol{\xi}_c,
\tag{4.114}
$$

where

$$
\mathbf{N}_h^T = (a_3, \quad a_4, \quad a_5, \quad \ldots, \quad a_{N+2})
\tag{4.115}
$$

The coefficients sets $(\mathbf{B}_0, \mathbf{B}_1, \mathbf{B}_2, \mathbf{B}_3, \ldots, \mathbf{B}_N)$ and $(a_0, a_1, a_2, a_3, \ldots, a_N)$ are the numerator coefficients of the RFAs for the unsteady aerodynamic-generalized forces and the hinge moment, respectively, with N lag parameters determined by a least squares curve fit with the harmonic aerodynamic data.

By defining the control surface state vector as follows:

$$
\mathbf{x}_c = \left(\delta, \dot{\delta}, \boldsymbol{\xi}_c^T \right)^T,
$$

we can write the control surface state equation by

$$
\dot{\mathbf{x}}_c = \mathbf{A}_c \mathbf{x}_c + \mathbf{B}_c u,
\tag{4.116}
$$

and the output equation as the following:

$$
\mathbf{Q}_c = \mathbf{C}_c \mathbf{x}_c + \mathbf{D}_c u,
\tag{4.117}
$$

where

$$
\mathbf{A}_c =
\begin{pmatrix}
0 & 1 & \mathbf{0}_{1 \times N} \\
\dfrac{a_0 - k_\delta}{I_\delta - a_2} & \dfrac{a_1 - c_\delta}{I_\delta - a_2} & \dfrac{1}{I_\delta - a_2} \mathbf{N}_h^T \\
0 & \Gamma_c & \Lambda_c
\end{pmatrix},
\tag{4.118}
$$

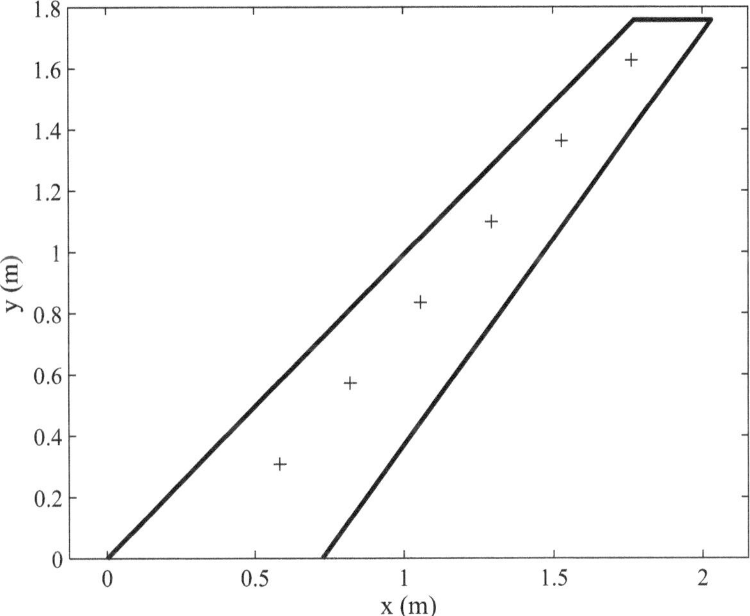

Fig. 4.15 Six selected spanwise locations on modified DAST-ARW1 wing for vertical deflection mode shapes

$$\mathbf{B_c} = \begin{pmatrix} 0 \\ \frac{1}{I_\delta - a_2} \\ \mathbf{0}_{N \times 1} \end{pmatrix}, \tag{4.119}$$

$$\mathbf{C_c} = \left(\mathbf{B}_0 + \frac{(a_0 - k_\delta)}{(I_\delta - a_2)}\mathbf{B}_2, \ \ \mathbf{B}_1 + \frac{(a_1 - c_\delta)}{(I_\delta - a_2)}\mathbf{B}_2, \ \ \frac{1}{(I_\delta - a_2)}\mathbf{N_c} \right), \tag{4.120}$$

and

$$\mathbf{D_c} = \mathbf{B}_2 / (I_\delta - a_2). \tag{4.121}$$

4.3.3   *Illustrative Example*

For illustration of the state-space modeling method, consider the modified DAST-ARW1 wing for which doublet-lattice computations were reported in Chap. 3, and rational function approximations were derived in the previous section. The vertical deflection mode shapes of the six structural modes (Table 4.3) at six selected spanwise locations (Fig. 4.15) listed column-wise are the following:

$$\{\mathbf{z}_i\} = \begin{pmatrix} -0.0029 & -0.1437 & -0.0030 & -0.0793 & 0.0691 & -0.0418 \\ 0.0442 & -0.1775 & 0.2304 & 0.1076 & -0.3412 & -0.3048 \\ 0.1595 & -0.1179 & 0.4772 & 0.2880 & -0.5993 & -0.5098 \\ 0.2879 & 0.0763 & 0.6381 & 0.5552 & -0.5854 & -0.6427 \\ 0.4864 & 0.3901 & 0.5500 & 0.6535 & -0.3532 & -0.4567 \\ 0.8081 & 0.8809 & 0.0980 & 0.4048 & 0.2284 & 0.1542 \end{pmatrix},$$

while the corresponding mode shapes for the deflection slopes are the following:

$$\{dz/dx_i\} = \begin{pmatrix} 0.0199 & 0.0064 & -0.0211 & -0.0265 & 0.0004 & -0.0821 \\ 0.1017 & 0.0676 & -0.0162 & -0.0934 & -0.2300 & -0.2363 \\ 0.2032 & 0.1949 & -0.0832 & -0.1066 & -0.3250 & -0.3518 \\ 0.4406 & 0.4023 & -0.2004 & -0.0414 & -0.0047 & -0.3758 \\ 0.7306 & 0.6987 & -0.4231 & -0.0319 & 0.5398 & -0.4374 \\ 0.4692 & 0.5544 & -0.8793 & -0.9882 & -0.7417 & -0.6936 \end{pmatrix},$$

These are combined to yield the modal matrix for aerodynamic influence coefficients by Eq. (4.53):

$$\phi = i\omega\{\mathbf{z}_i\} + U\{dz/dx_i\}.$$

The modal masses, stiffnesses, and damping coefficients are derived in order to fit the flight flutter test data for this wing reported by Bennett and Abel [19], and are the following:

$$M = \text{diag.}(2.0967,\ 1.9164,\ 1.8036,\ 1.7360,\ 1.6007,\ 1.4429)\ \text{kg}$$

$$K = 10^5 \begin{pmatrix} 1.7861 & 1.9831 & 0.2042 & 1.1739 & 0.5544 & 0.4213 \\ 1.4578 & 2.3976 & -2.9320 & -2.4571 & 3.5898 & 3.5093 \\ -1.3326 & -4.6836 & 5.8085 & 2.6710 & -8.3447 & -6.9302 \\ -1.4480 & -1.4594 & 1.9777 & 2.2929 & -1.8597 & -2.4083 \\ -0.0841 & -0.1959 & 0.4750 & 0.3269 & -0.4011 & -0.5714 \\ -1.6214 & -3.8540 & 4.8391 & 3.1149 & -6.5868 & -5.7113 \end{pmatrix}\ \text{N/m}$$

$$C = \begin{pmatrix} 3.9877 & 18.0085 & -35.1898 & -19.0859 & 40.9523 & 42.3896 \\ 75.3328 & 50.9966 & 72.4875 & 66.8918 & -54.2124 & -63.2161 \\ -56.3060 & -60.5175 & -13.7980 & -35.6402 & -14.8539 & 2.2176 \\ -60.9295 & -38.6040 & -51.3028 & -42.4412 & 41.6701 & 42.9690 \\ 4.3764 & -0.8387 & 17.5343 & 14.1483 & -15.3260 & -18.6483 \\ -89.2669 & -66.3578 & -78.0079 & -80.2630 & 48.9853 & 66.8707 \end{pmatrix}\ \text{N s/m}$$

The state–space model for a particular flight condition ($U = 250\ m/s$, $M = 0.807$, altitude 7.6 km) requires aerodynamic representation. The unsteady harmonic aerodynamic data for this flight condition are listed in Appendix C. For brevity, we will consider the RFA derived above for this condition with only two lag paramaters, whose values are selected as follows:

$$b_1 = 10^{-6}(b/U), \qquad b_2 = 1.85(b/U)$$

Such a choice allows for easily varying the lag parameters with a changing flight speed U—which is necessary for conducting a flutter analysis—without having to conduct a nonlinear optimization at every speed. The aerodynamic coefficient matrices for the given flight condition are as follows:

$$\mathsf{K}_a = \mathsf{A}_0 = \begin{pmatrix} -0.0439 & -1.8487 & -1.3809 & 3.9490 & 13.1090 & 10.1600 \\ -1.0560 & -9.4088 & -12.9280 & -4.3663 & 10.8807 & -17.4327 \\ -2.3908 & -16.3010 & -24.9145 & -19.3065 & -7.4486 & -50.4676 \\ -4.0111 & -15.5809 & -28.4172 & -44.9707 & -66.6647 & -69.2923 \\ -6.5650 & -11.3063 & -29.2660 & -82.7733 & -162.4312 & -98.8830 \\ -10.5992 & -49.2321 & -85.0047 & -100.0319 & -123.2653 & -310.7789 \end{pmatrix}\ \text{N/m}$$

$$\mathsf{C}_a = \mathsf{A}_1 = \begin{pmatrix} 0.0520 & 0.2624 & 0.5311 & 0.8836 & 1.3630 & 1.8723 \\ 0.0576 & 0.4082 & 0.8074 & 1.2481 & 1.8042 & 2.4824 \\ 0.0643 & 0.5062 & 0.9901 & 1.4725 & 2.0781 & 3.0577 \\ 0.0793 & 0.5296 & 1.0416 & 1.5725 & 2.2962 & 3.6159 \\ 0.1049 & 0.5011 & 1.0036 & 1.6079 & 2.5485 & 4.2165 \\ 0.1318 & 0.4444 & 0.9058 & 1.5382 & 2.6537 & 4.7587 \end{pmatrix}\ \text{N s/m}$$

$$\mathsf{M}_a = \mathsf{A}_2 = 10^{-3} \begin{pmatrix} -0.030532 & -0.15834 & -0.31831 & -0.52236 & -0.80225 & -1.142 \\ -0.037911 & -0.22624 & -0.44583 & -0.69237 & -1.0269 & -1.5394 \\ -0.046541 & -0.2666 & -0.52024 & -0.78915 & -1.1747 & -1.914 \\ -0.059263 & -0.27506 & -0.53894 & -0.83841 & -1.3151 & -2.2701 \\ -0.077774 & -0.27169 & -0.53823 & -0.88757 & -1.5016 & -2.6505 \\ -0.095264 & -0.28659 & -0.56147 & -0.93688 & -1.6284 & -2.9442 \end{pmatrix}\ \text{kg}$$

$$\mathsf{A}_3 = \begin{pmatrix} -0.93425 & -4.3053 & -8.5849 & -13.636 & -21.377 & -37.71 \\ -1.0965 & -5.0894 & -10.107 & -16.025 & -25.031 & -43.459 \\ -1.2479 & -5.8489 & -11.583 & -18.324 & -28.489 & -48.958 \\ -1.4115 & -6.7693 & -13.283 & -20.635 & -31.657 & -55.203 \\ -1.5679 & -7.6391 & -14.872 & -22.774 & -34.585 & -60.877 \\ -1.6774 & -8.1089 & -15.734 & -24.037 & -36.443 & -63.821 \end{pmatrix}$$

$$A_4 = \begin{pmatrix} -97.801 & -508.18 & -1032.1 & -1717.8 & -2646.2 & -3603.9 \\ -105.13 & -773.73 & -1549.3 & -2437.5 & -3563 & -4743 \\ -115.29 & -949.42 & -1888.5 & -2880.3 & -4142.5 & -5880.7 \\ -145.1 & -980.48 & -1966.5 & -3042.1 & -4539.5 & -7127.2 \\ -198.92 & -905.88 & -1853.1 & -3030.2 & -4923.9 & -8493.1 \\ -245.95 & -854.23 & -1735.1 & -2903.1 & -5038.3 & -9282.9 \end{pmatrix}$$

Flutter Analysis

When the aeroelastic data given above are substituted in Eq. (4.106) and an eigenvalue analysis is carried out by solving the characteristic equation,

$$|\, s\mathsf{I} - \mathsf{A}\, | = 0$$

the open-loop poles of the aeroelastic system at $M = 0.807$ and altitude 7.6 km are the following:

$$
\begin{array}{rrcr}
 & -0.00097699 & + & 0i \\
5\times & -0.0010173 & + & 0i \\
 & -0.31449 & \pm & 57.797i \\
 & -1.6542 & \pm & 100.35i \\
 & -0.4830 & \pm & 181.35i \\
 & -5.5118 & \pm & 219.83i \\
 & -3.0523 & \pm & 237.59i \\
 & -6.0303 & \pm & 305.03i \\
 & -1875.1 & + & 0i \\
 & -1881.6 & + & 0i \\
2\times & -1882 & + & 0i \\
 & -1882 & \pm & 0.019859i.
\end{array}
$$

The aeroelastic modes and the aerodynamic lag parameters can be clearly identified in this list. All the stable, real poles result from the aerodynamic lag states, while the complex conjugate pairs (except the last one) are the aeroelastic modes. When flight speed is increased, flutter is experienced by one of the aeroelastic modes becoming unstable. This is shown in Figs. 4.16–4.18, which show the variation of the natural frequencies and damping ratios of the aeroelastic modes in the speed range 250–295 m/s at 7.6 km standard altitude. The second symmetric bending/torsion mode of in vacuo natural frequency 30.3 Hz is seen, in Fig. 4.16, to become unstable at flutter speed of 284.7 m/s, which corresponds to a Mach number of 0.9192 at the given altitude. The flutter frequency for this mode is 28.691 Hz (181.272 rad/s). This is in agreement with the flight flutter test [19], which indicates a flutter Mach number of 0.92 for the same mode. At a higher velocity, close to the sonic speed, the

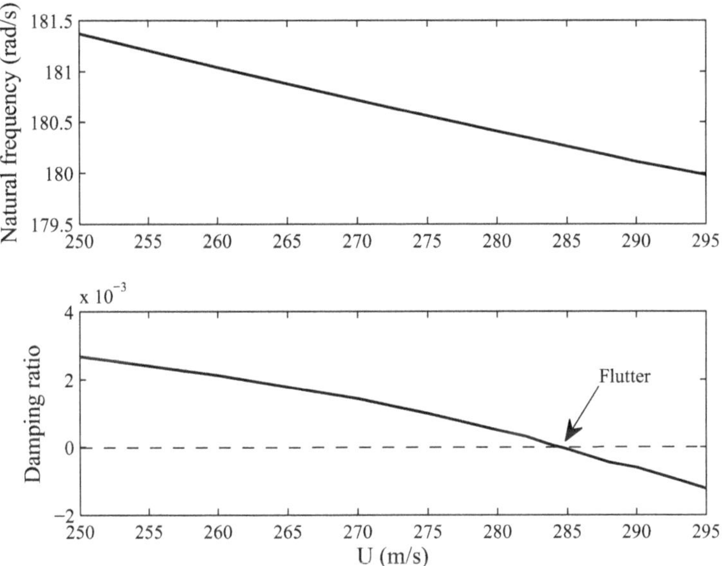

Fig. 4.16 Open-loop flutter analysis of modified DAST-ARW1 wing at 7.6 km standard altitude (second symmetric bending/torsion mode of in vacuo natural frequency 30.3 Hz)

symmetric torsion mode (in vacuo natural frequency 38.96 Hz) becomes unstable, as indicated by Fig. 4.17. The natural frequencies and damping ratios of the stable aeroelastic modes are plotted in Fig. 4.18. In order to ensure that the curve fit does not deteriorate badly at higher Mach number, the average fit error is plotted in Fig. 4.19. The error does not increase very much as the speed is varied, which shows that the results of flutter analysis can be considered reliable.

Control Surface Effects

For the determination of the controls coefficient matrix, $\mathbf{B}$, we require the generalized aerodynamics forces (GAF), as well as the hinge moment due to control surface. Here, we derive the GAF contribution of a single trailing edge flap of 40 % chord flap on the outboard 76–98 % semi-span location for the modified DAST-ARW1 wing[6]: (Fig. 4.20). The moment of inertia of the flap about its hinge line is $I_\delta = 0.1$ kg m^2 while its rotational stiffness about the hinge line is $k_\delta = 100$ N m/rad, which implies a flap mode of 31.623 rad/s.

The effect of an oscillating flap on the pressure distribution on DAST-ARW1 wing is shown in Figs. 4.21 and 4.22, which plots the real and imaginary parts of the pressure coefficient for the flap oscillating with $k = 0.8$ at $M = 0.8$. The flap motion causes large and discontinuous pressure magnitude and phase at outboard locations, and relatively smaller pressure modifications at inboard points. The flap hinge line has a sharp pressure discontinuity in the chordwise direction. Detailed chordwise pressure plots for this case at selected spanwise stations are shown in Appendix-C.

---

[6] The original DAST-ARW1 wing has two trailing edge control surfaces of approximately 20 % chord each, one inboard, and the other outboard. These are replaced by a single outboard flap here.

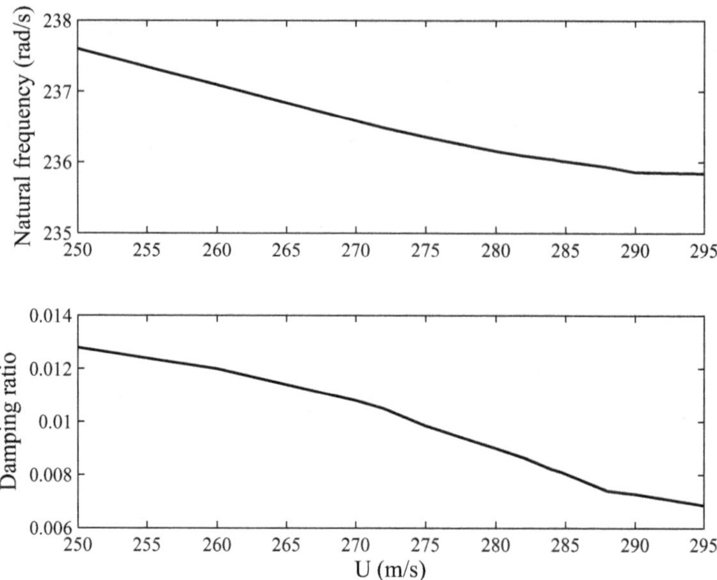

Fig. 4.17 Open-loop flutter analysis of modified DAST-ARW1 wing at 7.6 km standard altitude (symmetric torsion mode of in vacuo natural frequency 38.96 Hz)

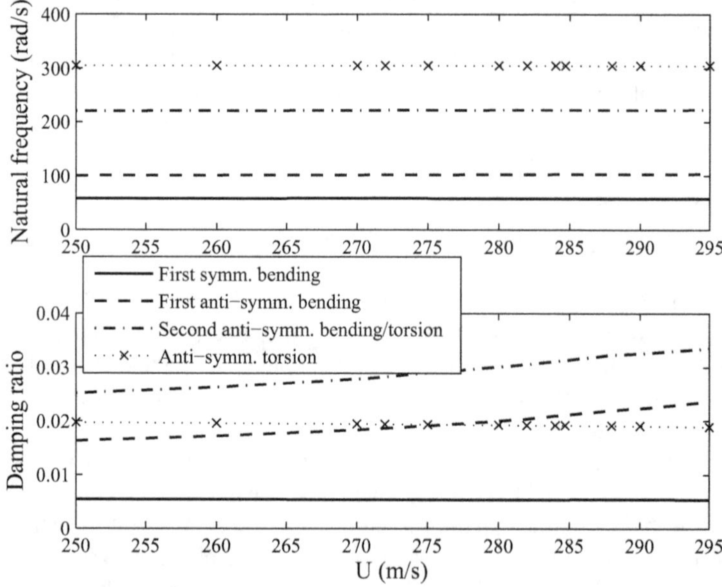

Fig. 4.18 Open-loop flutter analysis of modified DAST-ARW1 wing at 7.6 km standard altitude (stable aeroelastic modes)

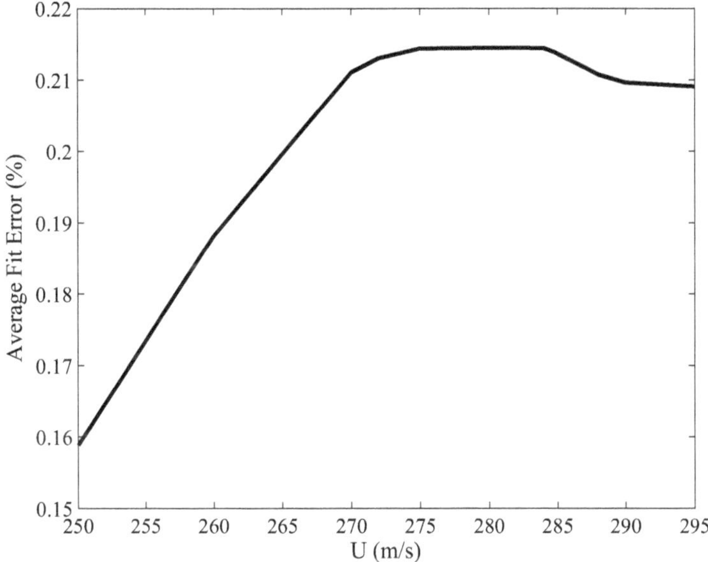

Fig. 4.19 Curve fit error with two lag parameters for flutter analysis of modified DAST-ARW1 wing at 7.6 km standard altitude

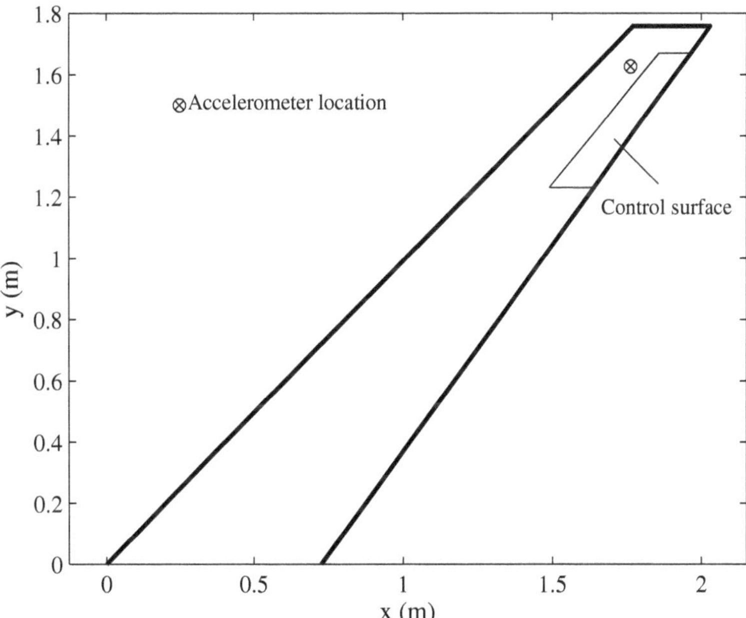

Fig. 4.20 Trailing-edge flap and accelerometer location on modified DAST-ARW1 wing

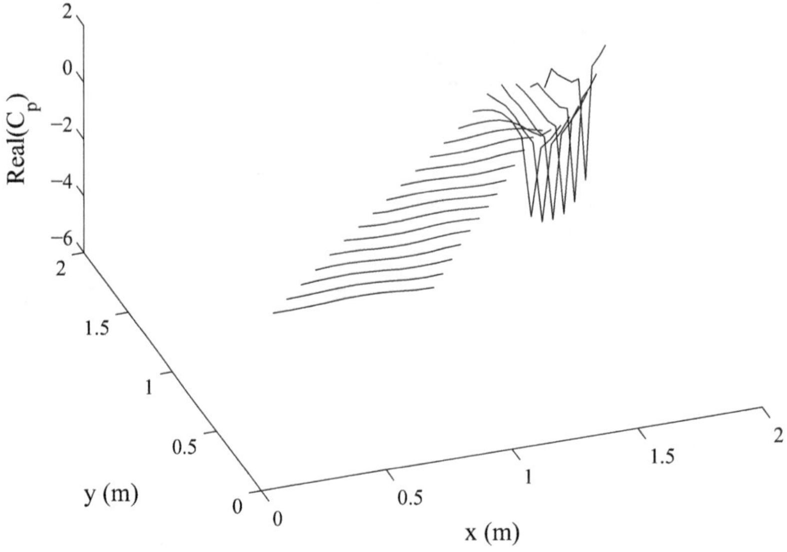

Fig. 4.21 Real part of unsteady pressure distribution on modified DAST-ARW1 wing due to the flap mode at $M = 0.807$ and $k = 0.8$

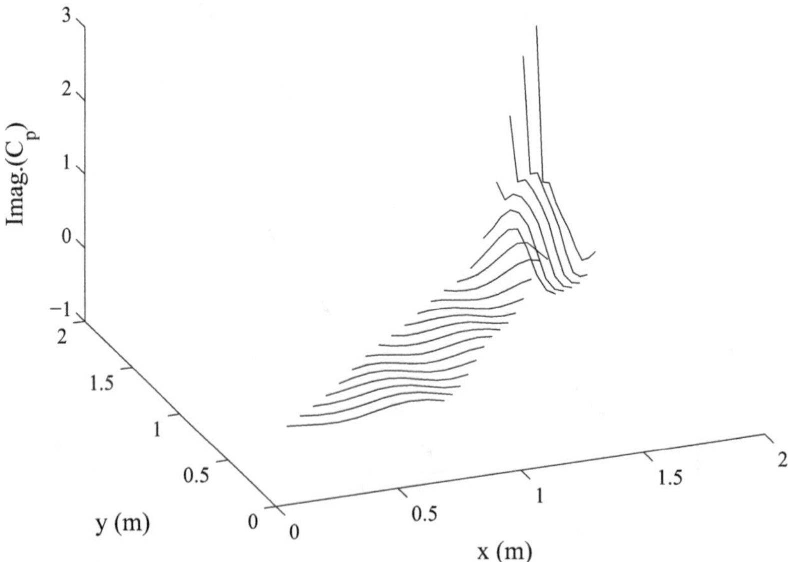

Fig. 4.22 Imaginary part of unsteady pressure distribution on modified DAST-ARW1 wing due to the flap mode at $M = 0.807$ and $k = 0.8$

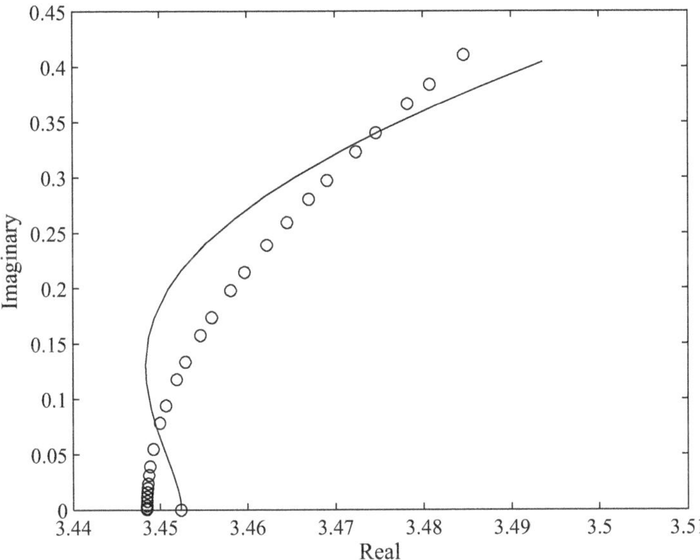

Fig. 4.23 Least squares curve fit of unsteady hinge moment for modified DAST-ARW1 wing at $M = 0.807$ with two lag parameters. (*circle*: harmonic data; *solid line*: RFA)

Due to a pressure discontinuity at the hinge line of an oscillating flap, it is expected that the curve fit for control surface GAF with a given number of lag parameters would be degraded when compared to that of wing without the control surface. This is indeed the fact for the two lag parameters case, wherein the fit error is increased at $M = 0.807$ from 0.18 % for wings with 6 structural modes, to 0.35 % for wings with only the trailing-edge flap mode. The RFA coefficients for the control surface GAF with two lag parameters are derived from the GAF data (Appendix C) to be the following:

$$\mathbf{B}_0^T = (-1.1880, \ -1.5713, \ -2.9118, \ -10.4382, \ -101.6408, \ -120.6015)$$

$$\mathbf{B}_1^T = (1.0286, \ 0.8900, \ 0.4865, \ 0.2060, \ 0.1702, \ 0.2863)$$

$$\mathbf{B}_2^T = 10^{-3} \, (-0.5646, \ -0.5204, \ -0.3229, \ -0.1787, \ -0.1288, \ -0.1616)$$

$$\mathbf{B}_3^T = (0.6953, \ 0.4578, \ 0.3557, \ 0.3885, \ 0.4454, \ 0.3734)$$

$$\mathbf{B}_4^T = 10^3 \, (-1.9612, \ -1.7064, \ -0.9540, \ -0.4534, \ -0.4078, \ -0.5711)$$

The hinge-moment RFA is computed next, with the curve fit shown in Fig. 4.23 using the same two lag parameters as those for the generalized aerodynamic forces. The average fit error per frequency point is only 0.027 %. The numerator coefficients of hinge-moment RFA for $U = 250$ m/s are listed as follows:

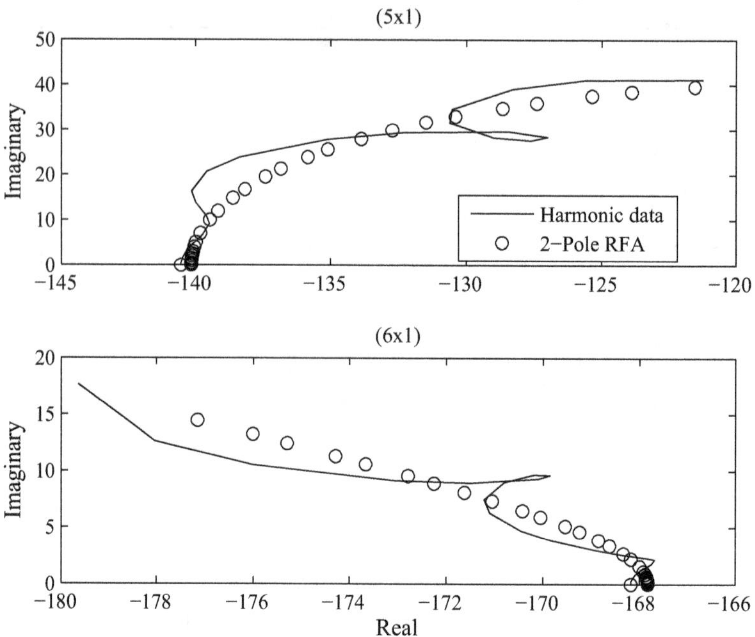

Fig. 4.24 Least squares curve fits for elements (1×5) and (1×6) of generalized aerodynamics forces due to flap mode for modified DAST-ARW1 wing at $M = 0.95$ with two lag parameters.

$$a_0 = 3.4524$$

$$a_1 = -0.0036$$

$$a_2 = 8.6236 \times 10^{-7}$$

$$a_3 = -0.0040$$

$$a_4 = 4.4747$$

When the flight speed is increased beyond the flutter velocity to $U = 295$ m/s, the nature of the RFA undergoes a transformation to a more spiraling character, as shown in Fig. 4.24. This indicates the need for including more intermediate frequency points for a better curve fit. The hinge-moment RFA coefficients at $U = 295$ m/s (listed below) for aerodynamic damping and aerodynamic inertia change in sign, while the lag numerator coefficients are seen to change in sign and increase in magnitude, all of which indicate a stabilizing aerodynamic influence on the flap rotation at supercritical speed (above flutter velocity):

$$a_0 = 4.4327$$

$$a_1 = 0.0047$$

$$a_2 = -4.3595 \times 10^{-6}$$

$$a_3 = 0.0048$$

$$a_4 = -11.1086$$

Finally, we derive the coefficient matrices for the output equation. The sensor location for the modified DAST-ARW1 wing is shown in Fig. 4.20. The outboard selection of accelerometer gives the best resolution of individual contributions from all the relevant modes to the acceleration output. The output coefficients for the given flight condition ($U = 250$ m/s, $M = 0.807$, altitude 7.6 km) are the following:

$$
\mathbf{C}^T = \left\{
\begin{array}{c}
1.1207 \times 10^5 \\
2.6638 \times 10^5 \\
-3.3454 \times 10^5 \\
-2.1536 \times 10^5 \\
4.5524 \times 10^5 \\
3.9455 \times 10^5 \\
61.807 \\
46.18 \\
54.575 \\
56.567 \\
-32.051 \\
-42.966 \\
-1.1581 \\
-5.5986 \\
-10.863 \\
-16.595 \\
-25.16 \\
-44.061 \\
-169.87 \\
-589.51 \\
-1197.4 \\
-2003.7 \\
-3478.1 \\
-6409.9
\end{array}
\right\}
$$

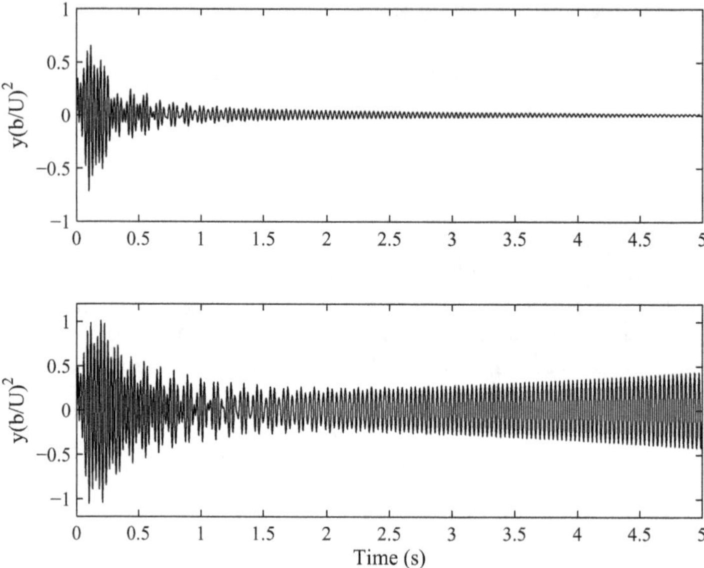

Fig. 4.25 Subcritical (M = 0.8) and supercritical (M = 0.95) accelerometer output for initial unit tip deflection of the modified DAST-ARW1 wing

$$D = (-3.138 \times 10^{-5}, -0.00010323, -0.00021489,$$
$$-0.00037257, -0.00070238, 0.69163)$$

The simulated normal acceleration response to an initial unit tip deflection at a subcritical (below flutter speed) Mach number of 0.8 and a supercritical Mach number of 0.95 at 7.6 km altitude are compared in Fig. 4.25 for the first 5 s. The stable (or decaying) subcritical response and an exponentially growing (unstable) supercritical response are evident.

Chapter 5
Linear Aeroelastic Control

5.1 Introduction

Control problems in aeroservoelasticity (ASE) are broadly categorized into the following two types: (a) avoidance of adverse ASE interactions and (b) design of active aeroelastic control systems. While (a) is necessary for designing all modern aircraft with airspeeds and load factors large enough to bring adverse ASE problems within the flight envelope, and control applications falling under category (b) involve a deliberate use of ASE interactions to improve certain flight characteristics. Category (a) is the area of classical ASE applications that have passive solutions, such as inserting standard filters in an existing flight control system in order to block out (or suppress) problematic aeroelastic modes at specific frequencies. Such solutions were common in early applications of 1950–1970 era before the advent of modern control design techniques. A modern, statically unstable fighter aircraft is an example of category (a), where an attempt to stabilize the rigid flight dynamics can lead to inadvertent destabilization of some aeroelastic modes. Due to its very haphazard nature, the main difficulty associated with the approach of tinkering with the gains of an existing controller can lead to a spillover effect of adversely affecting higher-order (and often unmodeled) dynamics. Furthermore, the analysis of such inadvertent and unmodeled ASE interactions requires extensive experimental work in the form of wind tunnel and flight tests. In contrast, category (b)—consisting of a systematic controller design for achieving a desired set of performance objectives in an optimal manner—takes into account all of the important aeroelastic modes on an ab initio basis, and has less probability of leading to any surprises in actual flight implementation.

Modern control design techniques can be immediately classified as either linear, or nonlinear, based upon their mathematical behavior. While most of the practical ASE control laws are essentially linear, their implementation often requires gain scheduling (or other adaptive techniques) with flight parameters, which make the overall control system a nonlinear one. This chapter presents linear ASE design applications, while Chap. 6 covers nonlinear ASE control analysis and design techniques. The reader is referred to a textbook on modern control design [168] for the

© Springer Science+Business Media, LLC 2015

A. Tewari, *Aeroservoelasticity,* Control Engineering,
DOI 10.1007/978-1-4939-2368-7_5

basic controls concepts covered here. It may also be helpful to see a textbook on automatic flight control systems, such as Tewari [171].

The main problem of active aeroelastic control is the possible lack of robustness of the designed control laws under actual operating conditions. This is the reason why many ASE applications are not implemented on certified civilian aircraft. If it were possible to do so, most commercial airliners could routinely fly beyond the open-loop flutter velocity, thereby increasing their speed and possibly efficiency. Attention is therefore required on designing control laws that can be robust under most operating conditions with some parametric uncertainties. Fortunately, the advancements in control theory in the past 30 years have enabled robust design methodologies for linear systems. The large part of this chapter will focus on such techniques. However, it is hardly possible to do full justice to the robust aeroservoelastic control in a brief chapter, and perhaps a future monograph can be completely devoted to such a topic.

5.2 Linear Feedback Stabilization

Consider the linear aeroservoelastic (ASE) plant derived in Chap. 4 with the state equations,

$$\dot{\mathbf{x}} = \mathbf{A}\mathbf{x} + \mathbf{B}\mathbf{Q}_c, \tag{5.1}$$

$$\dot{\mathbf{x}}_c = \mathbf{A}_c\mathbf{x}_c + \mathbf{B}_c\mathbf{u}, \tag{5.2}$$

with the aeroelastic state vector,

$$\mathbf{x} = \left(\mathbf{q}^T, \dot{\mathbf{q}}^T, \mathbf{x}_g^T, \mathbf{x}_a^T\right)^T,$$

gust state vector, $\mathbf{x}_g$, the actuators state vector,

$$\mathbf{x}_c = \left(\boldsymbol{\delta}^T, \dot{\boldsymbol{\delta}}^T, \boldsymbol{\xi}_c^T\right)^T,$$

control generalized aerodynamics forces vector, $\mathbf{Q}_c$, actuators state vector, $\mathbf{x}_c$, and actuator torque inputs vector, $\mathbf{u}$. The outputs of all control surface actuators, $\mathbf{Q}_c$, are the aeroelastic plant inputs, thereby implying a series connection between the actuators and aeroelastic blocks as shown in Fig. 5.1:

$$\mathbf{Q}_c = \mathbf{C}_c\mathbf{x}_c + \mathbf{D}_c u.$$

The outputs of the aeroelastic plant are accelerometer and laser optic-sensed acceleration and structural displacements signals collected from selected points in the structure and given by the output equation

$$\mathbf{y} = \mathbf{C}\mathbf{x} + \mathbf{D}\mathbf{Q}_c.$$

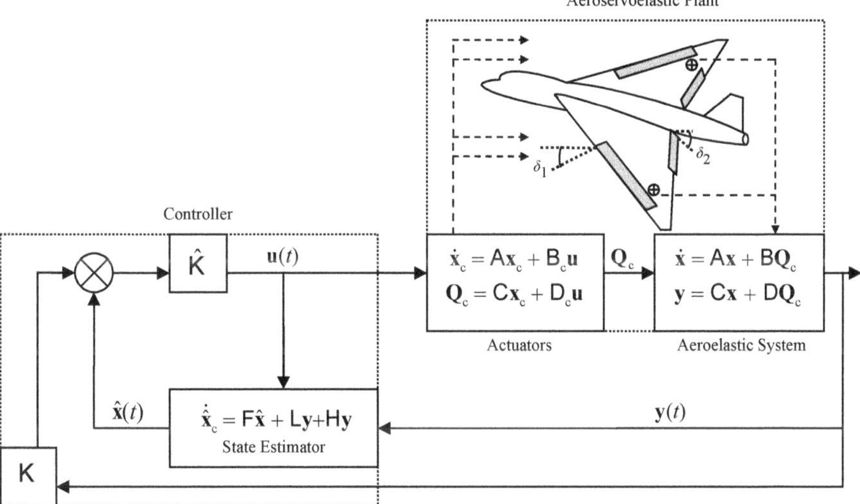

Fig. 5.1 Aeroservoelastic subsystems

Finally, a linear feedback[1] *control law* given by additional controller states, $\hat{\mathbf{x}}$, is necessary to close the loop between the output signals and actuator torque inputs, in order to yield a stable and well-behaved aeroservoelastic system:

$$\mathbf{u} = \hat{\mathsf{K}}\hat{\mathbf{x}} + \mathsf{K}\mathbf{y}, \tag{5.3}$$

where the controller state equations are the following:

$$\dot{\hat{\mathbf{x}}} = \mathsf{F}\hat{\mathbf{x}} + \mathsf{L}\mathbf{y} + \mathsf{H}\mathbf{u}. \tag{5.4}$$

The controller coefficient matrices $\hat{\mathsf{K}}, \mathsf{K}, \mathsf{F}, \mathsf{L}, \mathsf{H}$ must be determined such that the resulting closed-loop system (Fig. 5.1) has desirable properties. A hypothetical case is that of *full-state feedback*, where all the states $\mathbf{x}$ of the aeroelastic plant and the states $\mathbf{x}_c$ of all actuators are available for direct measurement. In such a case, $\mathbf{y}^T = (\mathbf{x}, \mathbf{x}_c)$ and the system can be stabilized with $\hat{\mathsf{K}} = 0$, therefore, no additional controller states are necessary and Eq. (5.4) can be dropped. However, we know that the aerodynamic states required for rational function approximation are not even physical variables that can be measured. Consequently, even if an infinitely large number of sensors can be arranged to measure deflections $\mathbf{q}$ and their rates at virtually every point on the structure, full-state feedback would still be impossible because a large part of the aeroelastic system (namely, the aerodynamic augmented states) cannot be fed

---

[1] Feedback control (or regulation) is part of a more general control scheme called *tracking control*, where in addition to achieving the desired closed-loop system characteristics, it is also required to follow (or track) a time-varying signal.

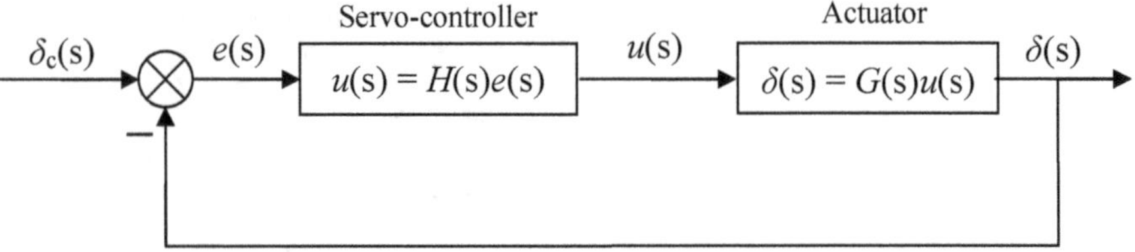

Fig. 5.2 Servo actuator for a control surface

back. Hence, modeling of a controller by Eq. (5.4) is almost always necessary in order to estimate the state variables that cannot be directly measured. The problem of solving Eq. (5.4) such that the solution $\hat{\mathbf{x}}(t)$ yields a good estimate of the actual state variables, based upon a knowledge of the $\mathbf{u}(t)$ and $\mathbf{y}(t)$, is called the *state estimation* (or *observation*) problem. We will return to the problem of state estimation later.

The controller state equation, Eq. (5.4), has another important application. Even if all states were available for feedback, there is no guarantee that the closed-loop response will track a desired step change in the state variables. In order to achieve a good tracking performance, integral action is provided by the addition of the controller states $\hat{\mathbf{x}}(t)$, which can be redefined as the error from a desired (or commanded) state of the ASE plant, $\hat{\mathbf{x}} = \mathbf{x}_c - \mathbf{x}$. This concept is discussed next.

5.2.1 Servo Actuators

The actuators are not entirely free systems in the sense that their response to step commanded control deflections, $\boldsymbol{\delta}_\mathbf{c}$, must be asymptotically stable, with a good transient behavior, and must also be free from any steady-state errors [168]. If these properties are absent, even a well-planned control strategy may fail to stabilize the overall ASE system. We saw in Chap. 4 that a control surface can be considered rigid for modeling purposes, and the rotation of each surface about its hinge line is driven by actuators independently of the other surfaces. Hence, actuators are a collection of second-order subsystems, each of which can be modeled separately of the others.

Let us consider one such actuator with moment of inertia I_δ and rotational stiffness k_δ. The commanded control surface deflection is $\delta_c(t)$, while the actual deflection is $\delta(t)$. The motion of the control surface produces a generalized aerodynamic force vector $\mathbf{Q}_c(t)$ on the wing, which can be regarded as the output vector of the actuator subsystem. The hinge moment $u(t)$ driving the control surface can be regarded as the input to the actuator. If no torque is applied by a control system, we have the free control surface ($u = 0$), which can either stabilize or destabilize the aeroelastic system depending upon its aerodynamic balance. Therefore, a feedback controller called a *servo controller* (or simply servo) is necessary for ensuring system stability. The closed-loop actuator system along with the servo is called the servo actuator, and is depicted by a classical block diagram in Fig. 5.2. Here, $H(s)$ refers to the transfer function of the servo controller, $G(s)$ is that of the actuator, and $e = \delta_c - \delta$ is the error between commanded and actual deflections.

The simplest servo is the proportional-derivative (PD) controller, which applies the control torque as a linear combination of the measured rotation angle, $\delta(t)$, and its rate, $\dot{\delta}(t)$, as follows:

$$u = k_0(\delta_c - \delta) + k_1(\dot{\delta}_c - \dot{\delta}), \tag{5.5}$$

where $\delta_c(t)$ is the commanded deflection, and k_0, k_1 are positive constants called the *gains*. The corresponding servo transfer function is given by

$$H(s) = k_0 + k_1 s.$$

The angle and angular rate can be measured by an optical digital encoder mounted on the hinge line with a high accuracy. Furthermore, the torque output $u(t)$ of the actuator can be calibrated in terms of the measured electric current passing through its driving circuit. Adding the PD control law to the actuator results in the following servo actuator dynamics:

$$(I_\delta - a_2)\ddot{\delta} + (k_1 - a_1)\dot{\delta} + (k_\delta + k_0 - a_0)\delta = k_0\delta_c + k_1\dot{\delta}_c + (a_3, \dots, a_N)\boldsymbol{\xi}_c, \tag{5.6}$$

where $\boldsymbol{\xi}_c$ is the control aerodynamic state vector satisfying the following state equation:

$$\dot{\boldsymbol{\xi}}_c = \Lambda_c \boldsymbol{\xi}_c + \boldsymbol{\Gamma}_c \dot{\delta},$$

with

$$\Lambda_c = \begin{pmatrix} -b_1 & 0 & 0 & \cdots & 0 \\ 0 & -b_2 & 0 & \cdots & 0 \\ \vdots & \vdots & \vdots & \vdots & \vdots \\ 0 & 0 & 0 & \cdots & -b_N \end{pmatrix},$$

$$\boldsymbol{\Gamma}_c = \begin{pmatrix} 1 \\ 1 \\ \vdots \\ 1 \end{pmatrix}.$$

Defining the actuator state vector as follows:

$$\mathbf{x}_c = \left(\delta, \dot{\delta}, \boldsymbol{\xi}_c^T\right)^T,$$

we can write the closed-loop servo actuator state equation by

$$\dot{\mathbf{x}}_c = \mathbf{A}_s \mathbf{x}_c + \mathbf{B}_c(k_0\delta_c + k_1\dot{\delta}_c), \tag{5.7}$$

and the output equation as the following:

$$\mathbf{Q}_c = \mathbf{C_s}\mathbf{x}_c + \mathbf{D_c}(k_0\delta_c + k_1\dot{\delta}_c), \tag{5.8}$$

where

$$\mathbf{A_s} = \begin{pmatrix} 0 & 1 & 0_{1\times N} \\ \dfrac{a_0 - k_\delta - k_0}{I_\delta - a_2} & \dfrac{a_1 - c_\delta - k_1}{I_\delta - a_2} & \dfrac{(a_3,\ldots,a_N)}{I_\delta - a_2} \\ 0 & \mathbf{\Gamma_c} & \mathbf{\Lambda_c} \end{pmatrix}, \tag{5.9}$$

$$\mathbf{B_c} = \begin{pmatrix} 0 \\ \dfrac{1}{I_\delta - a_2} \\ 0_{N\times 1} \end{pmatrix},$$

$$\mathbf{C_s} = \left(\mathbf{B}_0 + \frac{(a_0 - k_\delta - k_0)}{(I_\delta - a_2)}\mathbf{B}_2, \quad \mathbf{B}_1 + \frac{(a_1 - c_\delta - k_1)}{(I_\delta - a_2)}\mathbf{B}_2, \quad \frac{1}{I_\delta - a_2}\mathbf{N_c} \right),$$

$$\mathbf{D_c} = \mathbf{B}_2/(I_\delta - a_2),$$

$$\mathbf{N_c} = (\mathbf{B}_3, \quad \mathbf{B}_4, \quad \mathbf{B}_5, \quad \ldots, \quad \mathbf{B}_{N+2})$$

and $(\mathbf{B}_0, \mathbf{B}_1, \mathbf{B}_2, \mathbf{B}_3, \ldots, \mathbf{B}_N)$, $(a_0, a_1, a_2, a_3, \ldots, a_N)$ are the numerator coefficients of the RFAs with N lag parameters determined by a least squares curve fit with corresponding harmonic aerodynamic data (Chap. 4). The controller gains can be selected in order to yield an asymptotically stable response, $\delta(t)$ for a given excitation $\delta_c(t)$. An example of this fact is shown in Fig. 5.3 for the modified DAST-ARW1 wing at $M = 0.8$ and 7.6 km standard altitude, using actuator constants $I_\delta = 0.1$ kg m^2, $k_\delta = 100$ N m/rad, and servo gains $k_0 = 0.1, k_1 = 10$ for a unit step torque input $u_c(t) = k_0\delta_c + k_1\dot{\delta}_c = 1, t > 0$, and zero initial conditions $\delta(0) = \dot{\delta}(0) = 0$. The servo actuator step response results in a stable normal acceleration output $y(t)$ on the wing, which is plotted in Fig. 5.4 for 2 s.

While PD control gives an acceptable closed-loop servo response, sometimes it may be insufficient for accurately following a commanded step deflection, $\delta_c(t) = 1, t > 0$, because there are no pure integrators in the forward path [168] of the servo actuator (Fig. 5.2). In order to make the steady-state error to a commanded step deflection vanish, integral action must be provided in the servo by adding the controller state variable, $\hat{x}(t)$ as follows:

$$u = k_0(\delta_c - \delta) + k_1(\dot{\delta}_c - \dot{\delta}) + k_2\hat{x}$$

$$\dot{\hat{x}} = \delta_c - \delta. \tag{5.10}$$

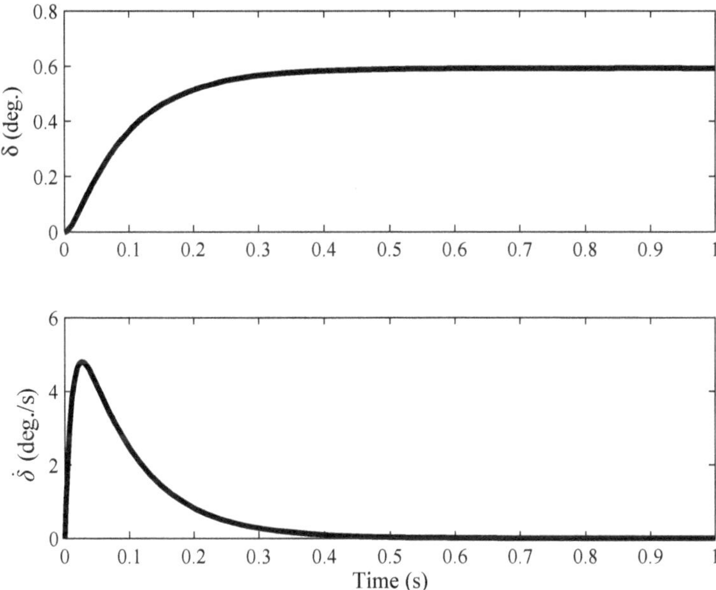

Fig. 5.3 Flap deflection response to a step servo actuator torque input for the modified DAST-ARW1 wing at $M = 0.8$ and 7.6 km standard altitude

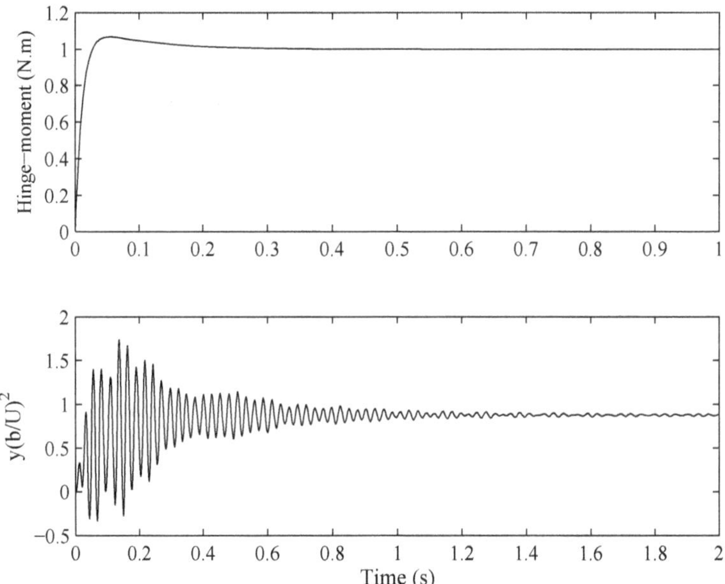

Fig. 5.4 Hinge moment and normal acceleration response to a step servo actuator torque input for the modified DAST-ARW1 wing at $M = 0.8$ and 7.6 km standard altitude

The transfer function of this servo is given by

$$H(s) = k_0 + k_1 s + k_2/s.$$

Such a controller is termed *proportional-integral-derivative* (PID) controller and results in a zero steady-state error to step commands.

The simple feedback technique of PD/PID control used to stabilize a servo actuator can be seen in the context of a larger, systematic design procedure offered by the optimal control theory. We will consider this before proceeding to the control laws for specific ASE applications.

5.3   Optimal Control

The linear optimal control is a special case of optimal control theory and dynamic programming, which has been the driver of control systems technology over the last half-century. The reader can consult the classical texts of Bellman [17], Bryson and Ho [25], and Athans and Falb [11], or Chap. 2 of Tewari [170] for an exposition of the optimal control theory. Here, we will highlight some important concepts and their application to the ASE problem.

Consider a dynamic system with $\mathbf{x}(t)$ as the state vector governed by the following state equation with a known initial condition:

$$\dot{\mathbf{x}} = \mathbf{f}(\mathbf{x}, \mathbf{u}, t), \qquad \mathbf{x}(t_0) = \mathbf{x_0}, \tag{5.11}$$

where $\mathbf{u}(t)$ is the controls vector bounded by constraints in a given interval, $t_0 \leq t \leq t_f$ (called *admissible control input*), and $\mathbf{f}(.)$ is a continuous functional and has a continuous partial derivative with respect to the state $\partial\mathbf{f}/\partial\mathbf{x}$ in the given interval. The optimal control problem is to find $\mathbf{u}(t)$ such that the following *objective function*[2] is minimized subject to the constraint of the dynamic state equation, Eq. (5.11):

$$J = \varphi[\mathbf{x}(t_f), t_f] + \int_{t_0}^{t_f} L[\mathbf{x}(t), \mathbf{u}(t), t]\mathrm{d}t, \tag{5.12}$$

where $L(\mathbf{x}, \mathbf{u}, t)$ is a *Lagrangian function* defining the transient performance objectives to be minimized in the control interval, $t_0 \leq t \leq t_f$, and $\varphi[\mathbf{x}(t_f), t_f]$ is a *terminal cost function* to be minimized at the final time. Minimization of the cost function subject to a full-state feedback control law,

$$\mathbf{u} = \mathbf{g}(\mathbf{x}, t), \tag{5.13}$$

would result in driving the final state vector, $\mathbf{x}(t_f)$, to zero while also minimizing the total excursions of $\mathbf{x}(t)$ from zero (both positive and negative) in the given control

---

[2] The objective function is alternatively called the *performance index* or the *cost function*.

interval $t_0 \leq t \leq t_f$. Derivation of the functional $\mathbf{g}(\mathbf{x}, t)$ for achieving this task is called the optimal regulator problem. There are various alternative formulations of the optimal regulator problem. In our brief discussion, we will focus on the *Hamilton–Jacobi–Bellman* (HJB) formulation given below for its relatively direct nature.

5.3.1 *Hamilton–Jacobi–Bellman Equation*

For an optimal control history, $\tilde{\mathbf{u}}(t)$ and the corresponding optimal trajectory, $\tilde{\mathbf{x}}(t)$, minimizing the performance index given by Eq. (5.12), subject to Eq. (5.11), one can define the following *optimal return function* for the control interval $t_0 \leq t \leq t_f$:

$$\tilde{V}[\tilde{\mathbf{x}}(t), t] = \varphi[\tilde{\mathbf{x}}(t_f), t_f] + \int_t^{t_f} L[\tilde{\mathbf{x}}(\tau), \tilde{\mathbf{u}}(\tau), \tau]\mathrm{d}\tau, \tag{5.14}$$

or

$$\tilde{V}[\tilde{\mathbf{x}}(t), t] = \varphi[\tilde{\mathbf{x}}(t_f), t_f] - \int_{t_f}^t L[\tilde{\mathbf{x}}(\tau), \tilde{\mathbf{u}}(\tau), \tau]\mathrm{d}\tau. \tag{5.15}$$

Differentiating $\tilde{V}$ with time, we have

$$\frac{\mathrm{d}\tilde{V}}{\mathrm{d}t} = -L[\tilde{\mathbf{x}}(t), \tilde{\mathbf{u}}(t), t], \tag{5.16}$$

and also

$$\begin{aligned}
\frac{\mathrm{d}\tilde{V}}{\mathrm{d}t} &= \frac{\partial \tilde{V}}{\partial t} + \frac{\partial \tilde{V}}{\partial \mathbf{x}}\dot{\mathbf{x}} \\
&= \frac{\partial \tilde{V}}{\partial t} + \frac{\partial \tilde{V}}{\partial \mathbf{x}}\mathbf{f}.
\end{aligned} \tag{5.17}$$

By equating Eqs. (5.16) and (5.17), we have the following partial differential equation, called the HJB equation, that must be satisfied by the optimal return function:

$$-\frac{\partial \tilde{V}}{\partial t} = L[\tilde{\mathbf{x}}(t), \tilde{\mathbf{u}}(t), t] + \frac{\partial \tilde{V}}{\partial \mathbf{x}}\mathbf{f}. \tag{5.18}$$

The boundary condition for the optimal return function is the following:

$$\tilde{V}[\tilde{\mathbf{x}}(t_f), t_f] = \varphi[\tilde{\mathbf{x}}(t_f), t_f]. \tag{5.19}$$

A solution for the HJB equation in optimal trajectory space, $\tilde{\mathbf{x}}$, and time, t, for a set of initial states, $\mathbf{x_0}$, results in a field of optimal solutions. A finite set of small initial perturbations about a given optimal path thus produces a set of *neighboring optimal solutions* by the solution of HJB equation. However, a solution of the HJB equation is a formidable task requiring simultaneous integration in space and time for neighboring extremal trajectories with finite-difference (or finite volume) methods. A more practical use of the HJB equation is to provide a sufficient condition for testing the optimality of a given return function.

In an alternative formulation of Euler–Lagrange, the following equivalence of notation exists:

$$-\frac{\partial \tilde{V}}{\partial t} \doteq H[\tilde{\mathbf{x}}(t), \tilde{\boldsymbol{\lambda}}(t), \tilde{\mathbf{u}}(t), t] \tag{5.20}$$

where $H(\boldsymbol{\lambda}, \mathbf{u}, t)$ is the *Hamiltonian function* and

$$\tilde{\boldsymbol{\lambda}}^T = \frac{\partial \tilde{V}}{\partial \mathbf{x}}. \tag{5.21}$$

is the optimal *co-state* (or Lagrange multipliers) vector.

While the HJB equation can be used to provide the necessary and sufficient optimality conditions, its main application lies in deriving nonlinear feedback control laws of the form,

$$\mathbf{u}(t) = \tilde{\mathbf{u}}\,[t, \mathbf{x}(t)]\,, \tag{5.22}$$

from the arguments of the optimal return function, $\tilde{V}[\tilde{\mathbf{x}}(t), t]$.

If the optimal return function, $\tilde{V}[\tilde{\mathbf{x}}(t), t]$, satisfies Lyapunov's theorem (Appendix-D) for global asymptotic stability for all optimal trajectories, $\tilde{\mathbf{x}}(t)$, then the given system is globally asymptotically stable about the origin, $\mathbf{x} = \mathbf{0}$. In this regard, the optimal return function is regarded as a Lyapunov function. The task of a designer is to find a return function, $V(\mathbf{x}, t)$, which is continuously differentiable with respect to the time and the state variables of the system, and satisfies the following sufficient conditions for global asymptotic stability about the equilibrium point at origin:

$$V(\mathbf{0}, t) = 0, \qquad V(\mathbf{x}, t) > 0; \qquad \frac{dV}{dt}(\mathbf{x}, t) < 0; \qquad \text{for all } \mathbf{x} \neq \mathbf{0} \tag{5.23}$$

$$\| \mathbf{x} \| \to \infty \text{ implies } V(\mathbf{x}, t) \to \infty. \tag{5.24}$$

5.3.2 Linear Systems with Quadratic Performance Index

Consider a linear, time-varying system given by

$$\dot{\mathbf{x}} = \mathsf{A}(t)\mathbf{x}(t) + \mathsf{B}(t)\mathbf{u}(t), \qquad \mathbf{x}(t_0) = \mathbf{x}_0, \tag{5.25}$$

for which the following quadratic objective function, representing a regulation of the state and control to zero steady state, is to be minimized:

$$J = \mathbf{x}^T(t_f)\mathsf{Q}_f\mathbf{x}(t_f)$$
$$+ \int_{t_0}^{t_f} \left[\mathbf{x}^T(t)\mathsf{Q}(t)\mathbf{x}(t) + 2\mathbf{x}^T(t)\mathsf{S}(t)\mathbf{u}(t) + \mathbf{u}^T(t)\mathsf{R}(t)\mathbf{u}(t) \right] dt. \tag{5.26}$$

Here, Q_f, $Q(t)$, and $R(t)$ are symmetric (but $S(t)$ could be asymmetric) cost coefficient matrices. Since $J \geq 0$, we also require that Q_f, $Q(t)$, and $R(t)$ be at least positive semidefinite.

$$L[\mathbf{x}(t), \mathbf{u}(t), t] = \mathbf{x}^T(t)Q(t)\mathbf{x}(t) + 2\mathbf{x}^T(t)S(t)\mathbf{u}(t) + \mathbf{u}^T(t)R(t)\mathbf{u}(t), \qquad (5.27)$$

the following quadratic value function with a positive definite, symmetric matrix, $P(t)$, is proposed:

$$V[\mathbf{x}(t), t] = \mathbf{x}^T(t)P(t)\mathbf{x}(t), \qquad (5.28)$$

whose optimal value must satisfy the HJB equation with the terminal boundary condition,

$$\tilde{V}[\tilde{\mathbf{x}}(t_f), t_f] = \mathbf{x}^T(t_f)Q_f\mathbf{x}(t_f). \qquad (5.29)$$

Thus, we have

$$\lambda^T = \frac{\partial V}{\partial \mathbf{x}} = 2\mathbf{x}^T(t)P(t), \qquad (5.30)$$

and the following Hamiltonian:

$$H = \mathbf{x}^T(t)Q(t)\mathbf{x}(t) + 2\mathbf{x}^T(t)S(t)\mathbf{u}(t) + \mathbf{u}^T(t)R(t)\mathbf{u}(t)$$
$$+ 2\mathbf{x}^T(t)P(t)[A(t)\mathbf{x}(t) + B(t)\mathbf{u}(t)]. \qquad (5.31)$$

For deriving the optimal control, we differentiate H with respect to $\mathbf{u}$ and equate the result to zero $H_{\mathbf{u}} = \mathbf{0}$:

$$\tilde{\mathbf{x}}^T(t)S(t) + \tilde{\mathbf{u}}^T(t)R(t) + \tilde{\mathbf{x}}^T(t)\tilde{P}(t)B(t) = \mathbf{0}, \qquad (5.32)$$

or

$$\tilde{\mathbf{u}}(t) = -\mathbf{R}^{-1}(t)[\mathbf{B}^T(t)\tilde{P}(t) + S^T(t)]\tilde{\mathbf{x}}(t). \qquad (5.33)$$

Equation (5.33) is perhaps the most important equation in modern control theory, and is called the *linear optimal feedback control law*. Since the matrix $P(t)$ is chosen to be positive definite, we have $V[\mathbf{x}(t), t] > 0$ for all $\mathbf{x}(t)$, and by substituting the linear, optimal feedback control law into

$$\tilde{H} = -\frac{\partial \tilde{V}}{\partial t} = -\tilde{\mathbf{x}}^T \dot{\tilde{P}} \tilde{\mathbf{x}}, \qquad (5.34)$$

we have the *Legendre–Clebsch* sufficiency condition for optimality, $H_{\mathbf{uu}} > \mathbf{0}$, which implies a negative definite matrix, $\dot{\tilde{P}}(t)$ (i.e., $\dot{V}[\mathbf{x}(t), t] < 0$ for all $\mathbf{x}(t)$). Therefore, the optimal return function given by Eq. (5.28) with a positive definite $\tilde{P}(t)$ is globally asymptotically stable by Lyapunov's theorem. We also note that the Legendre–Clebsch condition also implies that $\mathbf{R}(t)$ is positive definite.

Substitution of Eq. (5.30) into Eq. (5.28) results in the optimal Hamiltonian, which satisfies the HJB equation, Eq. (5.18):

$$\tilde{\mathbf{x}}^T \dot{\tilde{\mathbf{P}}} \tilde{\mathbf{x}} = -\tilde{\mathbf{x}}^T [(\mathbf{A} - \mathbf{B}\mathbf{R}^{-1}\mathbf{S}^T)^T \tilde{\mathbf{P}} + \tilde{\mathbf{P}}(\mathbf{A} - \mathbf{B}\mathbf{R}^{-1}\mathbf{S}^T)$$
$$- \tilde{\mathbf{P}}\mathbf{B}\mathbf{R}^{-1}\mathbf{B}^T\tilde{\mathbf{P}} + \mathbf{Q} - \mathbf{S}\mathbf{R}^{-1}\mathbf{S}^T]\tilde{\mathbf{x}}, \qquad (5.35)$$

and yields the following *matrix Riccati equation* (MRE) to be satisfied by the optimal matrix, $\tilde{\mathbf{P}}$:

$$-\dot{\tilde{\mathbf{P}}} = \mathbf{Q} + (\mathbf{A} - \mathbf{B}\mathbf{R}^{-1}\mathbf{S}^T)^T \tilde{\mathbf{P}} + \tilde{\mathbf{P}}(\mathbf{A} - \mathbf{B}\mathbf{R}^{-1}\mathbf{S}^T)$$
$$- \tilde{\mathbf{P}}\mathbf{B}\mathbf{R}^{-1}\mathbf{B}^T\tilde{\mathbf{P}} - \mathbf{S}\mathbf{R}^{-1}\mathbf{S}^T, \qquad (5.36)$$

which must be solved subject to the boundary condition,

$$\tilde{\mathbf{P}}(t_f) = \mathbf{Q_f}. \qquad (5.37)$$

MRE is fundamental to linear, optimal control and must be integrated backward in time by an appropriate numerical scheme. Being nonlinear in nature, the MRE solution procedure is termed *nonlinear* (or *dynamic*) *programming*. The existence of a unique, positive definite solution, $\tilde{\mathbf{P}}(t)$, is guaranteed if $\mathbf{Q_f}, \mathbf{Q}(t), \mathbf{R}(t)$ are positive definite, although less restrictive conditions on the cost coefficients are possible.

Alternatively, the linear state and co-state equations for the optimal control problem with a quadratic cost function are the following:

$$\dot{\mathbf{x}} = [\mathbf{A}(t) - \mathbf{B}(t)\mathbf{R}^{-1}(t)\mathbf{S}^T(t)]\mathbf{x}(t) - \mathbf{B}(t)\mathbf{R}^{-1}(t)\mathbf{B}^T(t)\lambda(t), \qquad (5.38)$$

$$\dot{\lambda} = -[\mathbf{A}^T(t) - \mathbf{S}(t)\mathbf{R}^{-1}(t)\mathbf{B}^T(t)]\lambda(t) + [\mathbf{S}(t)\mathbf{R}^{-1}(t)\mathbf{S}^T(t) - \mathbf{Q}(t)]\mathbf{x}(t), \qquad (5.39)$$

and must be solved subject to the following boundary conditions,

$$\mathbf{x}(t_0) = \mathbf{x_0} \; ; \qquad \lambda(t_f) = \mathbf{Q_f}\mathbf{x}(t_f) . \qquad (5.40)$$

Linearity of the adjoint system of equations assures the existence of a state transition matrix, $\Phi(t, t_0)$, such that

$$\left\{ \begin{array}{c} \mathbf{x}(t) \\ \lambda(t) \end{array} \right\} = \Phi(t, t_0) \left\{ \begin{array}{c} \mathbf{x}(t_0) \\ \lambda(t_0) \end{array} \right\}, \qquad (5.41)$$

with the boundary conditions of Eq. (5.40). The state transition matrix has the following properties [170]:

1. *Identity:*

$$\Phi(t_0, t_0) = \mathbf{I}$$

2. *Inversion:*

$$\Phi(t_0, t) = \Phi^{-1}(t, t_0)$$

3. *Time Derivative:*

$$\dot{\Phi}(t, t_0) = \begin{bmatrix} A - BR^{-1}S^T & -BR^{-1}B^T \\ SR^{-1}S^T - Q & -A^T + SR^{-1}B^T \end{bmatrix} \Phi(t, t_0)$$

4. *Symplectic Nature:*

$$\Phi^T(t, t_0) \begin{pmatrix} 0 & I \\ -I & 0 \end{pmatrix} \Phi(t, t_0) = \begin{pmatrix} 0 & I \\ -I & 0 \end{pmatrix}$$

Here, I is the identity matrix, and 0 the null matrix. The solution of linear optimal regulation problem stated above for a quadratic performance is called the linear-quadratic regulator (LQR).

5.4 Kalman Filter

Some mathematical modeling errors are always present in a plant model. These are classified as the *process noise* vector, which represents random variables applied as inputs to the plant, and the *measurement noise* vector representing the random errors due to imprecise measurement and feedback of the output variables. A linear, deterministic plant behaves like a stochastic system in the presence of process and measurement noise. Consider such a plant with the state vector $\mathbf{x}(t)$, output vector $\mathbf{y}(t)$, having the following linear, time-varying state-space representation:

$$\dot{\mathbf{x}}(t) = A(t)\mathbf{x}(t) + B(t)\mathbf{u}(t) + F(t)\mathbf{v}(t)$$
$$\mathbf{y}(t) = C(t)\mathbf{x}(t) + D(t)\mathbf{u}(t) + \mathbf{w}(t), \tag{5.42}$$

where $\mathbf{v}(t)$ is the process noise vector which arises due to modeling errors such as neglecting nonlinear, high-frequency, or stochastic dynamics, and $\mathbf{w}(t)$ is the measurement noise vector. By assuming $\mathbf{v}(t)$ and $\mathbf{w}(t)$ to be white noises with a zero mean, we can greatly simplify the model of the stochastic plant. Since the time-varying stochastic system is a nonstationary process, the random noises, $\mathbf{v}(t)$ and $\mathbf{w}(t)$, are also assumed to be nonstationary white noises for generality. A nonstationary white noise can be simply derived by passing a stationary white noise signal through an amplifier with a time-varying gain [168]. The correlation matrices of nonstationary white noises, $\mathbf{v}(t)$ and $\mathbf{w}(t)$, are expressed as follows [170]:

$$R_v(t, \tau) = S_v(t)\delta(t - \tau)$$
$$R_w(t, \tau) = S_w(t)\delta(t - \tau), \tag{5.43}$$

where $S_v(t)$ and $S_w(t)$ are the time-varying, power spectral density matrices of $v(t)$ and $\mathbf{w}(t)$, respectively, and $R_v(t, t)$ and $R_w(t, t)$, are the corresponding infinite covariance matrices, respectively.

A full-state feedback control system cannot be designed for a stochastic plant because its state vector, $\mathbf{x}(t)$, is unknown at any given time. Instead, one has to rely on an *estimated state vector*, $\hat{\mathbf{x}}(t)$, that is derived from the measurement of the output vector, $\mathbf{y}(\tau)$, over a previous, finite time interval, $t_0 \leq \tau \leq t$. Thus, a subsystem of the controller must be dedicated to form an accurate estimate of the state vector from a finite record of the outputs. Such a subsystem is called an *observer* (or state estimator). The performance of the control system depends upon the accuracy and efficiency with which a state estimate can be supplied by the observer to the control system, despite the presence of process and measurement noise.

In order to take into account the fact that the measured outputs as well as the state variables are random variables, we need an observer that estimates the state vector based upon a statistical description of the output and state. Such an observer is the *Kalman filter*, which can be regarded as an optimal observer that minimizes the covariance matrix of the *estimation error*,

$$\hat{\mathbf{e}}(t) = \mathbf{x}(t) - \hat{\mathbf{x}}(t). \tag{5.44}$$

In order to understand why it may be useful to minimize the covariance of estimation error, we note that the estimated state, $\hat{\mathbf{x}}(t)$, is based on the measurement of the output, $\mathbf{y}(\tau)$, while knowing the applied input vector, $\mathbf{u}(\tau)$, for a finite time, $t_0 \leq \tau \leq t$. However, being a nonstationary signal, an accurate average of $\mathbf{x}(t)$ would require measuring the output for an infinite time, i.e., gathering infinite number of samples from which the expected value of $\mathbf{x}(t)$ could be derived. Therefore, the best estimate that the Kalman filter could obtain for $\mathbf{x}(t)$ is not the true mean, but a *conditional mean*, $\mathbf{x}_m(t)$, based on only a finite time record of the output expressed as follows by the expected value operator, $E(.)$:

$$\mathbf{x}_m(t) = E[\mathbf{x}(t)|\mathbf{y}(\tau), t_0 \leq \tau \leq t]. \tag{5.45}$$

The deviation of the estimated state vector, $\hat{\mathbf{x}}(t)$, from the conditional mean, $\mathbf{x}_m(t)$, is given by

$$\Delta\mathbf{x}(t) = \hat{\mathbf{x}}(t) - \mathbf{x}_m(t). \tag{5.46}$$

The conditional covariance matrix is defined as the covariance matrix based on a finite record of the output and can be written as follows:

$$R_e(t, t) = E[\mathbf{x}(t)\mathbf{x}^T(t)] - \hat{\mathbf{x}}(t)\mathbf{x}_m^T(t) - \hat{\mathbf{x}}^T(t)\mathbf{x}_m(t) + \hat{\mathbf{x}}(t)\hat{\mathbf{x}}^T(t)$$

$$= E[\mathbf{x}(t)\mathbf{x}^T(t)] - \mathbf{x}_m(t)\mathbf{x}_m^T(t) + \Delta\mathbf{x}(t)\Delta\mathbf{x}^T(t). \tag{5.47}$$

It is evident from Eq. (5.47) that the best estimate of the state vector, i.e., $\Delta\mathbf{x}(t) = \mathbf{0}$, would result in a minimization of the conditional covariance matrix, $R_e(t, t)$. In other words, the minimization of $R_e(t, t)$ yields the optimal observer (Kalman filter). The state equation of the full-order Kalman filter is that of a time-varying observer, and is expressed as follows:

$$\dot{\hat{\mathbf{x}}} = \hat{A}(t)\hat{\mathbf{x}}(t) + \hat{B}(t)\mathbf{u}(t) + L(t)\mathbf{y}(t), \tag{5.48}$$

where $L(t)$ is the gain matrix and $\hat{A}(t), \hat{B}(t)$ the coefficient matrices of the Kalman filter. Thus, we have the following state equation for the estimation error dynamics:

$$\dot{\hat{\mathbf{e}}}(t) = \hat{A}(t)\hat{\mathbf{e}}(t) + [A(t) - L(t)C(t) - \hat{A}(t)]\mathbf{x}(t)$$
$$+ [B(t) - L(t)D(t) - \mathbf{B}(t)]\mathbf{u}(t) + F(t)v(t) - L(t)\mathbf{w}(t). \tag{5.49}$$

In order that the estimation error dynamics be independent of the state and control variables, we require

$$\hat{A}(t) = A(t) - L(t)C(t)$$
$$\hat{B}(t) = B(t) - L(t)D(t), \tag{5.50}$$

which yields

$$\dot{\hat{\mathbf{e}}}(t) = [A(t) - L(t)C(t)]\hat{\mathbf{e}}(t) + F(t)v(t) - L(t)\mathbf{w}(t). \tag{5.51}$$

Given that $v(t)$ and $\mathbf{w}(t)$ are white, nonstationary signals with zero mean values, we have their linear combination also a white and nonstationary, random signal with a zero mean:

$$\mathbf{z}(t) = F(t)v(t) - L(t)\mathbf{w}(t), \tag{5.52}$$

that drives the estimation error dynamics of the Kalman filter:

$$\dot{\hat{\mathbf{e}}}(t) = [A(t) - L(t)C(t)]\hat{\mathbf{e}}(t) + \mathbf{z}(t). \tag{5.53}$$

A solution to the error state equation for a given initial error, $\hat{\mathbf{e}}(t_0)$, and a corresponding initial error covariance, $R_{\mathbf{e}}(t_0, t_0) = E[\hat{\mathbf{e}}(t_0)\hat{\mathbf{e}}^T(t_0)] = \mathbf{R}_{\mathbf{e}0}$, can be expressed as follows:

$$\hat{\mathbf{e}}(t) = \Phi(t, t_0)\hat{\mathbf{e}}(t_0) + \int_{t_0}^{t} \Phi(t, \lambda)\mathbf{z}(\lambda)d\lambda, \tag{5.54}$$

where $\Phi(t, t_0)$ is the error transition matrix corresponding to the homogeneous error state equation,

$$\dot{\hat{\mathbf{e}}}(t) = [A(t) - L(t)C(t)]\hat{\mathbf{e}}(t) = \hat{A}(t)\hat{\mathbf{e}}(t). \tag{5.55}$$

The conditional error covariance is then derived (dropping the conditional notation for simplicity) by noting that $\mathbf{z}(t)$ is a zero mean, white noise, i.e., $E[\mathbf{z}(t)] = \mathbf{0}$ and

$$R_{\mathbf{z}}(t, \tau) = S_{\mathbf{z}}(t)\delta(t - \tau), \tag{5.56}$$

where $\mathbf{S}_z(t)$ is the time-varying, power spectral density matrix of $\mathbf{z}(t)$, resulting in the following [168]:

$$\mathbf{R}_e(t,t) = E[\hat{\mathbf{e}}(t)\hat{\mathbf{e}}^T(t)] = \Phi(t,t_0)\mathbf{R}_e(t_0,t_0)\Phi^T(t,t_0) + \int_{t_0}^{t} \Phi(t,\lambda)\mathbf{S}_z(\lambda)\Phi^T(t,\lambda)d\lambda.$$

(5.57)

Differentiation of Eq. (5.57) with time, along with the fact that the error transition matrix satisfies its own homogeneous state equation,

$$\frac{\mathrm{d}}{\mathrm{d}t}\{\Phi(t,t_0)\} = \hat{\mathbf{A}}(t)\Phi(t,t_0),$$

(5.58)

yields the following MRE for the optimal error covariance matrix:

$$\frac{\mathrm{d}}{\mathrm{d}t}\{\mathbf{R}_e(t,t)\} = \hat{\mathbf{A}}(t)\mathbf{R}_e(t,t) + \mathbf{R}_e(t,t)\hat{\mathbf{A}}^T(t) + \mathbf{S}_z(t).$$

(5.59)

The MRE for the Kalman filter must be integrated forward in time, subject to the initial condition,

$$\mathbf{R}_e(t_0,t_0) = \mathbf{R}_{e0}.$$

(5.60)

When the process and measurement noise are cross-correlated, we have

$$\mathbf{R}_{vw}(t,\tau) = E[v(t)\mathbf{w}^T(\tau)] = \mathbf{S}_{vw}(t)\delta(t-\tau),$$

(5.61)

where $\mathbf{S}_{vw}(t)$ is the *cross-spectral density matrix* of $v(t)$ and $\mathbf{w}(t)$. This leads to the following expression for the measurement noise power spectral density:

$$\mathbf{S}_w = \mathbf{F}\mathbf{S}_v\mathbf{F}^T - 2\mathbf{F}\mathbf{S}_{vw}\mathbf{L}^T + \mathbf{L}\mathbf{S}_w\mathbf{L}^T.$$

(5.62)

Finally, the following optimal Kalman filter gain matrix is derived:

$$\mathbf{L} = \left[\mathbf{R}_e\mathbf{C}^T + \mathbf{F}\mathbf{S}_{vw}\right]\mathbf{S}_w^{-1},$$

(5.63)

where the optimal error covariance matrix is the positive definite solution of the following MRE:

$$\frac{\mathrm{d}}{\mathrm{d}t}\{\mathbf{R}_e(t,t)\} = \mathbf{A}_G(t)\mathbf{R}_e(t,t) + \mathbf{R}_e(t,t)\mathbf{A}_G^T(t)$$

$$- \mathbf{R}_e(t,t)\mathbf{C}^T(t)\mathbf{S}_w^{-1}(t)\mathbf{C}(t)\mathbf{R}_e(t,t) + \mathbf{F}(t)\mathbf{S}_G(t)\mathbf{F}^T(t),$$

(5.64)

and

$$\mathbf{A}_G(t) = \mathbf{A}(t) - \mathbf{F}(t)\mathbf{S}_{vw}(t)\mathbf{S}_w^{-1}(t)\mathbf{C}(t)$$

$$\mathbf{S}_G(t) = \mathbf{S}_v(t) - \mathbf{S}_{vw}(t)\mathbf{S}_w^{-1}(t)\mathbf{S}_{vw}^T(t).$$

(5.65)

It is interesting to note the following one-to-one equivalence between the Kalman filter and the LQR of the previous section, in that both the LQR gain and the Kalman filter gain are based upon the solution to an MRE:

$$\mathsf{A}^T(t) \Leftrightarrow \mathsf{A}(t)$$

$$\mathsf{C}^T(t) \Leftrightarrow \mathsf{B}(t)$$

$$\mathsf{S_w}(t) \Leftrightarrow \mathsf{R}(t)$$

$$\mathsf{F}(t)\mathsf{S}_v(t)\mathsf{F}^T(t) \Leftrightarrow \mathsf{Q}(t)$$

$$\mathsf{F}(t)\mathsf{S}_{vw}(t) \Leftrightarrow \mathsf{S}(t).$$

Hence, Kalman filter is termed the *dual* of the LQR regulator.

A major simplification in Kalman filter design occurs when the process and measurement noise are uncorrelated with each other, i.e., $\mathsf{S}_{vw}(t) = \mathbf{0}$. This is a common situation in many applications (especially ASE) for which we have

$$\mathsf{L} = \mathsf{R_e}\mathsf{C}^T\mathsf{S_w^{-1}}, \tag{5.66}$$

where

$$\frac{\mathrm{d}}{\mathrm{d}t}\{\mathsf{R_e}(t,t)\} = \mathsf{A}(t)\mathsf{R_e}(t,t) + \mathsf{R_e}(t,t)\mathsf{A}^T(t)$$

$$- \mathsf{R_e}(t,t)\mathsf{C}^T(t)\mathsf{S_w^{-1}}(t)\mathsf{C}(t)\mathsf{R_e}(t,t) + \mathsf{F}(t)\mathsf{S}_v(t)\mathsf{F}^T(t). \tag{5.67}$$

It is now clear that a practical ASE problem must be solved such that the feedback law of LQR problem employs the states estimated by the Kalman filter as its feedback variables. A controller thus obtained is called the linear-quadratic-Gaussian (LQG) compensator, and has the following state equations:

$$\dot{\mathbf{x}} = \mathsf{A}(t)\mathbf{x} - \mathsf{B}(t)\mathsf{R}^{-1}(t)\left[\mathsf{S}^T(t) + \mathsf{B}^T(t)\tilde{\mathsf{P}}(t)\right]\hat{\mathbf{x}}(t), \tag{5.68}$$

$$\dot{\hat{\mathbf{x}}} = \left\{\mathsf{A}(t) - \mathsf{L}(t)\mathsf{C}(t) - \mathsf{B}\mathsf{R}^{-1}(t)[\mathsf{B}^T(t)\tilde{\mathsf{P}}(t) + \mathsf{S}^T(t)]\right\}\hat{\mathbf{x}}(t) + \mathsf{L}(t)\mathbf{y}(t), \tag{5.69}$$

where $\tilde{\mathsf{P}}(t)$ and $\mathsf{L}(t)$ are determined from the solutions to the respective MREs, Eqs. (5.32) and (5.59) subject to their respective boundary conditions, Eqs. (5.33) and (5.60).

5.5 Infinite-Horizon Linear Optimal Control

A practical regulator problem requires dissipation of all errors to zero when the time becomes large compared with the timescale of plant dynamics. Hence, the much slower plant dynamics can often be essentially approximated by linear, time-invariant (LTI) systems where the plant coefficient matrices, $\mathsf{A},\mathsf{B},\mathsf{C},\mathsf{D}$, are constants.

Such a regulation problem is referred to as *infinite-horizon control*, because the control interval can be taken to be infinite in comparison with the plant dynamics. In such cases, both LQR and Kalman filter designs are greatly simplified by having an infinite-control interval. The MRE solutions, $P(t)$, $R_e(t, t)$, approach their respective steady-state values given by $P(\infty)$, $R_e(\infty, \infty)$, which can then be expressed simply as the constants P, R_e. The governing equation for a time-invariant LQR problem is derived simply by putting $\dot{P} = 0$ in the MRE, Eq. (5.32), resulting in the following *algebraic Riccati equation* (ARE):

$$0 = \left(A - BR^{-1}S^T\right)^T P + P\left(A - BR^{-1}S^T\right)$$

$$- PBR^{-1}B^TP + Q - SR^{-1}S^T. \tag{5.70}$$

The optimal feedback control law is obtained from the algebraic Riccati solution,

$$\mathbf{u}(t) = -R^{-1}\left(B^TP + S^T\right)\mathbf{x}(t), \tag{5.71}$$

where the cost coefficient matrices, Q, R, S, are constants. For asymptotic stability of the regulated system, all the eigenvalues of the closed-loop dynamics matrix,

$$A - BR^{-1}\left(B^TP + S^T\right),$$

must be in the left-half s-plane, which requires that the ARE solution, P must be a symmetric, positive semidefinite matrix [170], i.e., a matrix with all eigenvalues being either greater than, or equal to zero. There may not always be a positive semidefinite solution; on the other hand, there could be several such solutions of which it cannot be determined which one is to be regarded as the best one. However, it can be proved [62] that if the following sufficient conditions are satisfied, there exists a unique, symmetric, positive semidefinite solution to the ARE:

- The control cost coefficient matrix, R, is symmetric and positive definite, the matrix $(Q - SR^{-1}S^T)$ is symmetric and positive semidefinite, and the pair $(A - BR^{-1}S^T, Q - SR^{-1}S^T)$ is *detectable* (i.e., its unobservable subsystem is stable).
- The pair (A, B) is either controllable, or at least *stabilizable* (i.e., its uncontrollable subsystem is stable).

The infinite-horizon control can be extended to the Kalman filter, because the latter is the dual of the LQR problem. In the stochastic sense, a constant error covariance matrix, R_e, implies a stationary white noise process. If the estimation error of a linear system is stationary, the system must be driven by stationary processes. Therefore, an LTI Kalman filter essentially involves the assumption of stationary, zero mean, Gaussian white (ZMGWN) models for both process noise, $v(t)$, and measurement noise, $w(t)$. Thus, the error covariance matrix must now satisfy the following ARE:

$$0 = A_GR_e + R_eA_G^T - R_eC^TS_w^{-1}CR_e + F\left(S_v - S_{vw}S_w^{-1}S_{vw}^T\right)F^T, \tag{5.72}$$

where $\mathsf{S_w}, \mathsf{S}_\nu, \mathsf{S}_{\nu\mathsf{w}}$ are constant spectral density matrices and

$$\mathsf{A_G} = \mathsf{A} - \mathsf{FS}_{\nu w}\mathsf{S_w^{-1}C}. \tag{5.73}$$

The constant Kalman filter gain matrix is the following:

$$\mathsf{L} = \left(\mathsf{R_e C}^T + \mathsf{FS}_{\nu w}\right)\mathsf{S_w^{-1}}. \tag{5.74}$$

Clearly, the ARE for the Kalman filter must also have a symmetric, positive semidefinite solution, $\mathsf{R_e}$, for an asymptotically stable Kalman filter dynamics. Furthermore, by satisfying sufficient conditions that are dual to those stated above for the LQR problem, a unique, positive semidefinite solution to the ARE can be found.

The ARE is at the heart of both LQR and Kalman filter design for LTI systems. Being a nonlinear algebraic equation, it must be solved numerically, such as by iteration of the following *Lyapunov equation* for a symmetric matrix, X:

$$\mathsf{AX} + \mathsf{XA}^T + \mathsf{Q} = 0. \tag{5.75}$$

There are several efficient algorithms for iteratively solving the ARE, such as the one programmed in the MATLAB function, *are* [110].

Next, we will consider how the infinite horizon LQR and Kalman filter can be applied to an ASE problem of Type (a).

5.6   Adverse Aereoservoelastic Interaction

Consider a realistic illustration of ASE coupling encountered by a tailless delta winged fighter-type aircraft (Fig. 5.5). Neglecting long-period (phugoid) dynamics involving airspeed variation, the aircraft has a rigid longitudinal flight dynamics model consisting of the state vector $\mathbf{x}_r = (\alpha, \theta, q)^T$, where α is the angle of attack, θ the pitch angle, and q the pitch rate as shown in Fig. 5.5. The rigid longitudinal state equations are given by

$$\dot{\mathbf{x}}_r = \mathsf{A}_r\mathbf{x}_r + \mathsf{B}_r\delta_c, \tag{5.76}$$

where

$$\mathsf{A}_r = \begin{bmatrix} \dfrac{Z_\alpha}{mU - Z_{\dot{\alpha}}} & -\dfrac{mg\sin\Theta_e}{mU - Z_{\dot{\alpha}}} & \dfrac{mU + Z_q}{mU - Z_{\dot{\alpha}}} \\ 0 & 0 & 1 \\ \dfrac{M_\alpha}{J_{yy}} + \dfrac{M_{\dot{\alpha}}Z_\alpha}{J_{yy}(mU - Z_{\dot{\alpha}})} & -\dfrac{M_{\dot{\alpha}}(mg\sin\Theta_e)}{J_{yy}(mU - Z_{\dot{\alpha}})} & \dfrac{M_q}{J_{yy}} + \dfrac{M_{\dot{\alpha}}(mU + Z_q)}{J_{yy}(mU - Z_{\dot{\alpha}})} \end{bmatrix}$$

$$\tag{5.77}$$

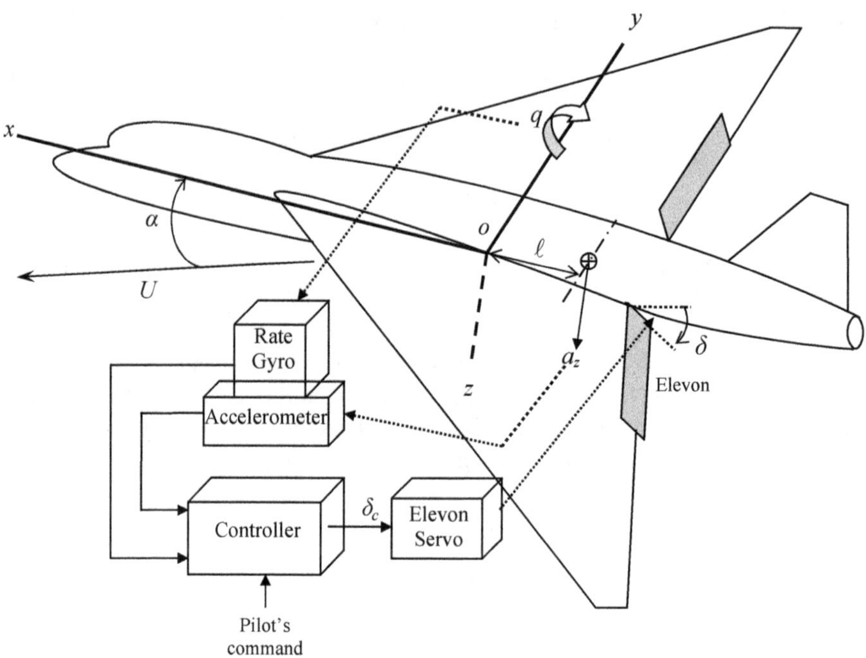

Fig. 5.5 Longitudinal control system for a tailless delta fighter aircraft

is the state dynamics matrix for longitudinal, short-period, rigid body dynamics [171], and

$$
\mathsf{B}_r = \begin{pmatrix} \dfrac{Z_\delta}{mU - Z_{\dot\alpha}} \\[2ex] 0 \\[2ex] \dfrac{M_\delta}{J_{yy}} + \dfrac{M_{\dot\alpha} Z_\delta}{J_{yy}(mU - Z_{\dot\alpha})} \end{pmatrix}. \tag{5.78}
$$

The aircraft is intentionally designed to be statically unstable in open loop ($M_\alpha > 0$) for better maneuverability. This feature requires a closed-loop flight control system for stability augmentation. The longitudinal control is provided by a pair of elevons, whose deflection, δ, is driven by a second-order actuator called the *elevon servo*, with the transfer function

$$
\frac{\delta}{\delta_c} = \frac{\omega_a^2}{s^2 + 2\zeta_a \omega_a s + \omega_a^2}, \tag{5.79}
$$

where δ_c is the commanded elevon deflection, ω_a the actuator's natural frequency, and ζ_a, its damping ratio. An accelerometer located on the fuselage at a distance ℓ aft of the center of mass senses the normal acceleration, $a_z = U(\dot\alpha - q) + \ell\dot q$, while a rate gyro measures the pitch rate, q. Thus, the rigid body dynamics has three vehicle states—$\mathbf{x}_r$, two actuator states (a fifth order system), a single input, δ, and two outputs, a_z, q. If one adds to the model the structural dynamic states, $\mathbf{x}_s$, with structural state equations,

$$\dot{\mathbf{x}}_s = \mathsf{A}_s\mathbf{x}_s + \mathsf{B}_s\delta_c, \tag{5.80}$$

the following state equations of the aeroelastic plant are obtained:

$$\dot{\mathbf{x}} = \mathsf{A}\mathbf{x} + \mathsf{B}\delta_c, \tag{5.81}$$

where the augmented state model is given by

$$\mathbf{x} = \begin{Bmatrix} \mathbf{x}_r \\ \mathbf{x}_s \\ \delta \\ \dot{\delta} \end{Bmatrix}, \tag{5.82}$$

$$\mathsf{A} = \begin{pmatrix} \mathsf{A}_r & 0 & \mathsf{B}_r & 0 \\ 0 & \mathsf{A}_s & \mathsf{B}_s & 0 \\ 0 & 0 & 0 & 1 \\ 0 & 0 & -\omega_a^2 & -2\zeta_a\omega_a \end{pmatrix}, \tag{5.83}$$

$$\bar{\mathsf{B}}_r = \begin{pmatrix} 0 \\ 0 \\ 0 \\ 0 \\ \omega_a^2 \end{pmatrix}. \tag{5.84}$$

In level flight equilibrium ($\Theta_e = 0$), the pitch angle θ can be discarded as the state variable, and the state-space model of the rigid aircraft's short-period dynamics is given by

$$\mathsf{A}_r = \begin{pmatrix} -0.3078 & 1.0 \\ 11.3 & -3.17 \end{pmatrix}, \qquad \mathsf{B}_r = \begin{pmatrix} -0.0114 \\ -8.25 \end{pmatrix}$$

$$\mathsf{C}_r = \begin{pmatrix} 0 & 1.0 \\ -71.587 & -6.34 \end{pmatrix}, \qquad \mathsf{D}_r = \begin{pmatrix} 0 \\ 20 \end{pmatrix}. \tag{5.85}$$

The rigid aircraft is an unstable system with eigenvalues $s = 1.9146, -5.3924$.

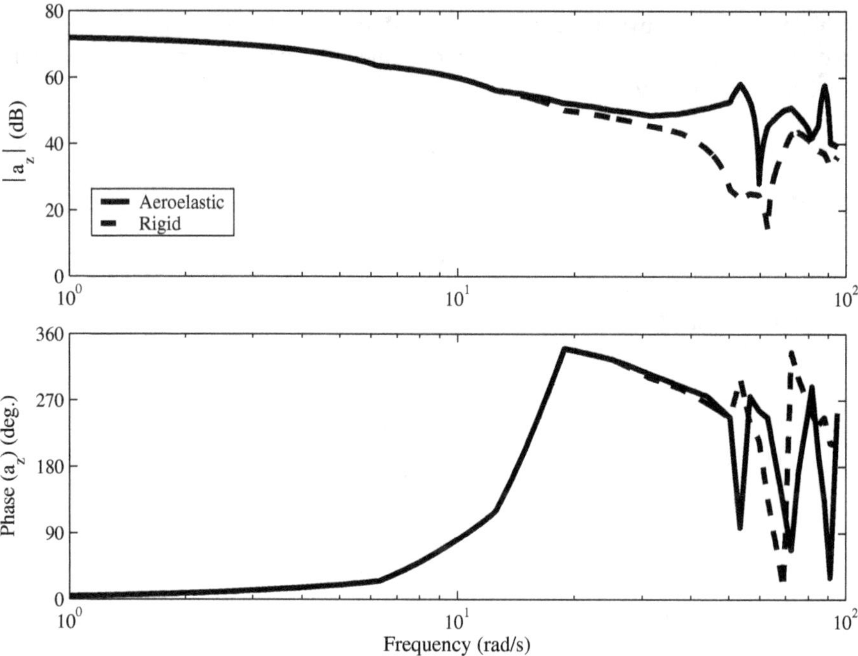

Fig. 5.6 Open-loop rigid and aeroelastic longitudinal frequency response of a tailless delta fighter aircraft at $M = 0.9$ level flight condition

The structural analysis of the fighter aircraft reveals 20 in vacuo modes below 50 Hz natural frequency. Of these, two wing-bending modes, and two combined wing-torsion/ fuselage-bending modes, along with the actuator mode of $\omega_a = 9$ Hz, $\zeta = 0.41$, are expected to fall within the closed-loop control system bandwidth of 10 Hz. The aeroelastic frequency response of the normal acceleration output with subsonic aerodynamic model of doublet-lattice method (Chap. 3) based on the first four flexible modes at freestream Mach number of 0.9 (but an equivalent Mach number normal to the leading edge of 0.52) and standard sea level is compared with the rigid body response in Fig. 5.6 for frequencies below 100 rad/s. The much higher acceleration magnitudes (about 40 dB higher at the first mode) for the aeroelastic system at peaks corresponding to the approximate structural natural frequencies can be clearly observed, while the phase is slightly lagged behind the rigid response due to the unsteady aerodynamic effects associated with the aeroelastic model. The pitch rate response is shown in Fig. 5.7. The aeroelastic effect on the pitch rate magnitude is seen to be much smaller than that on the normal acceleration response at the selected accelerometer location, but is still 10 dB more than the rigid response at the first peak. This example shows the importance of aeroelastic modes in the outputs of the system, which are fed to the controller.

In order to stabilize the statically unstable aircraft, an active stability augmentation system is designed that takes normal acceleration output, $y = a_z$, and feeds it to an

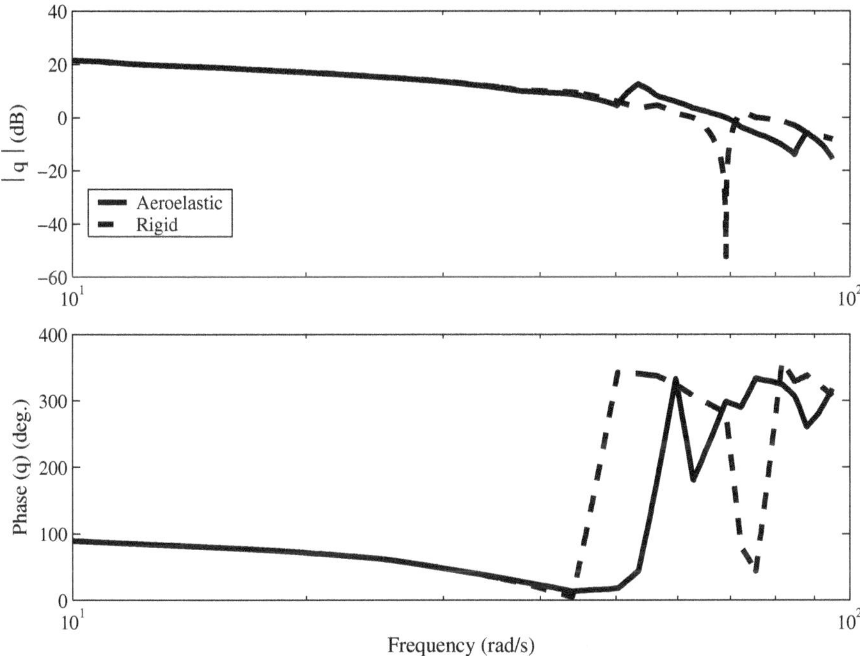

Fig. 5.7 Open-loop rigid and aeroelastic longitudinal frequency response of a tailless delta fighter aircraft at $M = 0.9$ level flight condition

observer-based controller for determining the control input, $u = \delta_c$. The pilot's commands for the desired angle of attack, α_c and the desired pitch rate, q_c, are the inputs of the closed-loop system, of which the block diagram is depicted in Fig. 5.8. For the purpose of demonstrating inadvertent aeroservoelastic interaction, assume that the design procedure neglects the vehicle's structural dynamics, and is entirely based upon the rigid, short-period model given by Eq. (5.76). First, a linear, full-order observer is designed with the gain matrix L, for the rigid aircraft's output equation,

$$\mathbf{y} = \mathbf{C}_r\mathbf{x}_r + \mathbf{D}_r\mathbf{u}, \qquad\qquad (5.86)$$

and the following state equation:

$$\dot{\hat{\mathbf{x}}} = (\mathbf{A}_r - \mathbf{L}\mathbf{C}_r)\,\hat{\mathbf{x}} + (\mathbf{B}_r - \mathbf{L}\mathbf{D}_r)\,\mathbf{u} + \mathbf{L}\mathbf{y}. \qquad\qquad (5.87)$$

Since the plant is observable with the normal acceleration output, the observer gain matrix, L, is selected by either eigenstructure assignment [168] (pole placement) for the observer dynamics matrix, $\mathbf{A}_r - \mathbf{L}\mathbf{C}_r$, or by the Kalman filter approach. Taking the latter approach with the following parameters:

$$\mathbf{S}_{v\mathbf{w}}(t) = \mathbf{0}, \mathbf{S}_v(t) = 10^{-2}\mathbf{I}, \mathbf{S}_\mathbf{w}(t) = 1, \mathbf{F} = \mathbf{I},$$

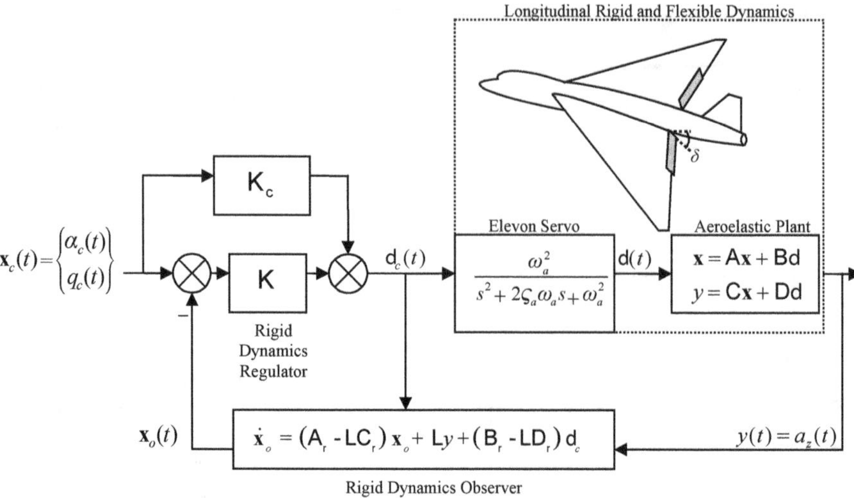

Fig. 5.8 Block diagram of the closed-loop aeroservoelastic system of the fighter aircraft based on feedback stabilization of unstable rigid modes

the observer gain matrix is determined to be the following:

$$\mathsf{L} = \begin{pmatrix} -0.1102 \\ -0.1340 \end{pmatrix}. \tag{5.88}$$

The rigid observer's poles are located at $s = -3.8997, -8.3177$. After the observer is designed, the loop is closed with a linear control law,

$$u = \mathsf{K_d x}_d + \mathsf{K}\left(\mathbf{x}_c - \hat{\mathbf{x}}\right), \tag{5.89}$$

where K is the regulator gain matrix, and the feedforward gain matrix, $\mathsf{K_d}$, is added to ensure the pilot's commands, $\mathbf{x}_d = (\alpha_c, q_c)^T$, are followed without any steady-state error. We will not design the feedforward gains here, since it does not serve our present purpose, and the reader can refer to controls references [168], [171] for its details. Next, the infinite horizon LQR problem is solved with

$$\mathsf{S} = \mathbf{0}, \ \mathsf{Q} = 10^{-6}\mathsf{I}, \ \mathsf{R} = 1,$$

in order to determine the regulator gains to be the following:

$$\mathsf{K} = (-2.3435, \quad -0.4609). \tag{5.90}$$

The regulated plant has stable eigenvalues of $s = -1.9146, -5.3924$, implying that the unstable pole of the plant has been turned around into its mirror image about the imaginary axis, while the stable pole is unchanged.

However, while the rigid plant has been stabilized by the controller designed without giving any regard to the aircraft's structural dynamics, the same is not true

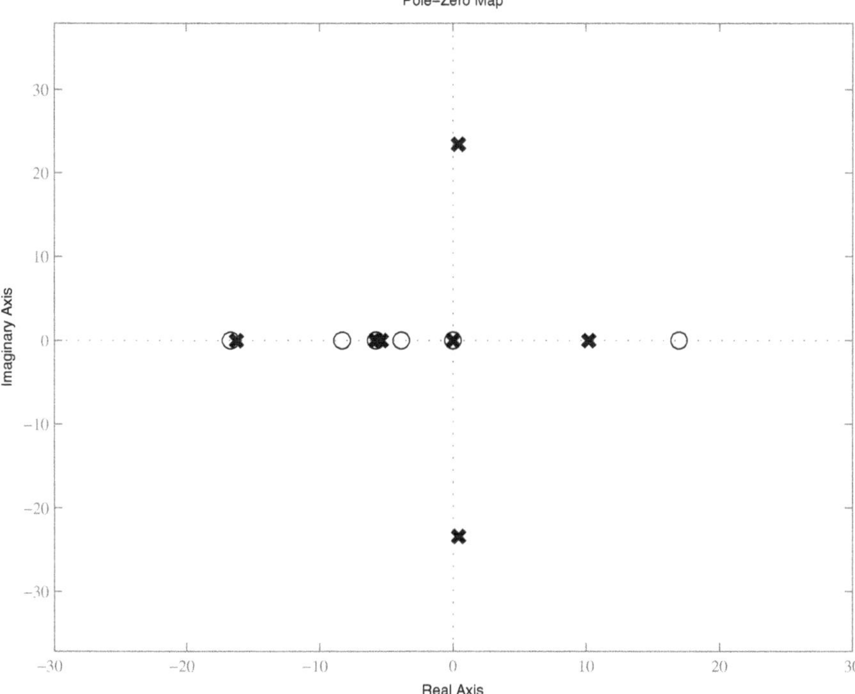

Fig. 5.9 Pole–zero map of the closed-loop aeroservoelastic system of the fighter aircraft for the transfer function $a_z(s)/\alpha_c(s)$ at $M = 0.9$ level flight condition. (x: pole, o: zero)

for the total aeroservoelastic system. Here, it is to be noted that the original aeroelastic plant was stable, but after the feedback control is applied to stabilize the rigid plant the aeroservoelastic system has become unstable! This can be seen in Figs. 5.9–5.12, which show the stability analysis of the overall ASE system (Fig. 5.8) with state equations

$$
\begin{Bmatrix} \dot{\mathbf{x}} \\ \dot{\hat{\mathbf{x}}} \end{Bmatrix} = \begin{pmatrix} \mathsf{A} & -\mathsf{BK} \\ \mathsf{LC} & \mathsf{A_r} - \mathsf{B_r K} - \mathsf{LC_r} \end{pmatrix} \begin{Bmatrix} \mathbf{x} \\ \hat{\mathbf{x}} \end{Bmatrix} + \begin{pmatrix} \mathsf{B(K + K_c)} \\ \mathsf{B_r(K + K_c)} \end{pmatrix} \mathbf{x}_c. \quad (5.91)
$$

Figure 5.9 is the pole–zero map of the aeroservoelastic system. It can be seen that the LQR-based rigid dynamics controller—while attempting to cancel the unstable rigid pole by placing a closed-loop zero at the same location—moves three of the neglected aeroelastic modes into the right-half s-plane.

The Nyquist plot of $a_z(s)/\alpha_c(s)$ for the complete aeroservoelastic system is shown in Fig. 5.10, where several positive (clockwise) encirclements of the point $s = -1$ can be observed. Since there is only one right-half s-plane pole of the open-loop

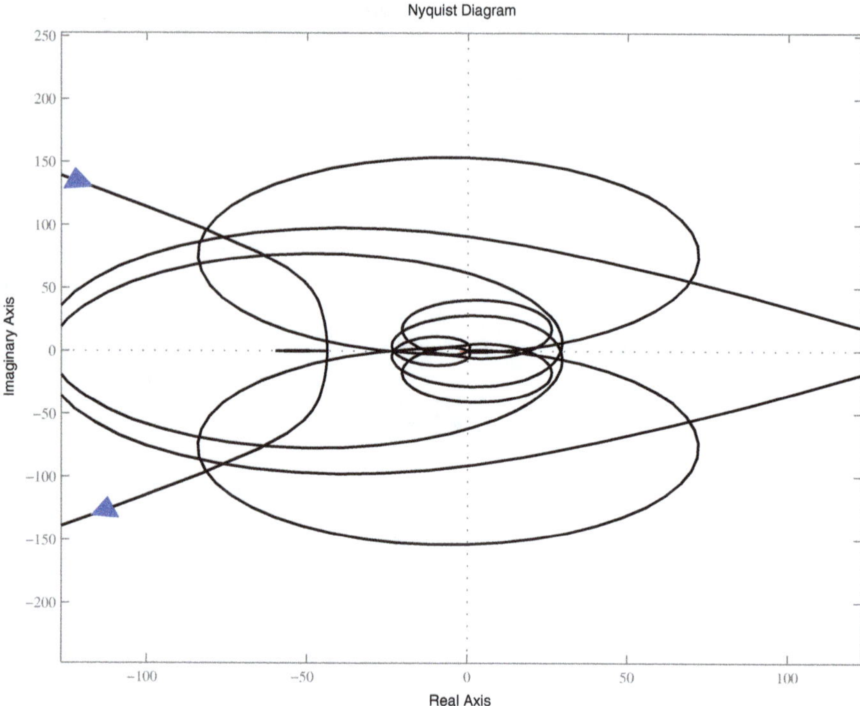

Fig. 5.10 Nyquist locus of the aeroservoelastic system of the fighter aircraft for the transfer function $a_z(s)/\alpha_c(s)$ at $M = 0.9$ level flight condition

transfer function, the Nyquist stability theorem [168][3] implies there are several unstable poles of the ASE system. Figure 5.11 compares the step responses to a commanded change in the angle of attack ($\alpha_c = 1.9°$) of the rigid aircraft and the flexible aircraft with the controller designed to stabilize the rigid mode. Here the feedforward gain matrix, $\mathbf{K}_c$, is taken to be zero. While the rigid aircraft has a stable response settling to a steady state in 2 s, the response of the ASE system is unstable, crossing 15 g normal acceleration in less than 0.5 s.

The associated commanded elevon deflections for the rigid and ASE systems are compared in Fig. 5.12, showing that the maximum allowable elevon deflection is quickly crossed for the flexible aircraft. Clearly, the inadvertent ASE instability will destroy the aircraft if any step changes are commanded. Such unstable ASE interactions between the flight control system and the structural modes are thought

---

[3] Nyquist stability theorem states that a feedback system has Z unstable poles, if and only if, the locus of open-loop transfer function for $s = i\omega, -\infty < \omega < \omega$ encircles the point $(-1, 0)$ in the clockwise direction exactly $N = Z - P$ times, where P is the number of poles of the open-loop transfer function in the right-half s-plane, provided no pole–zero cancellations have occurred in the open-loop system.

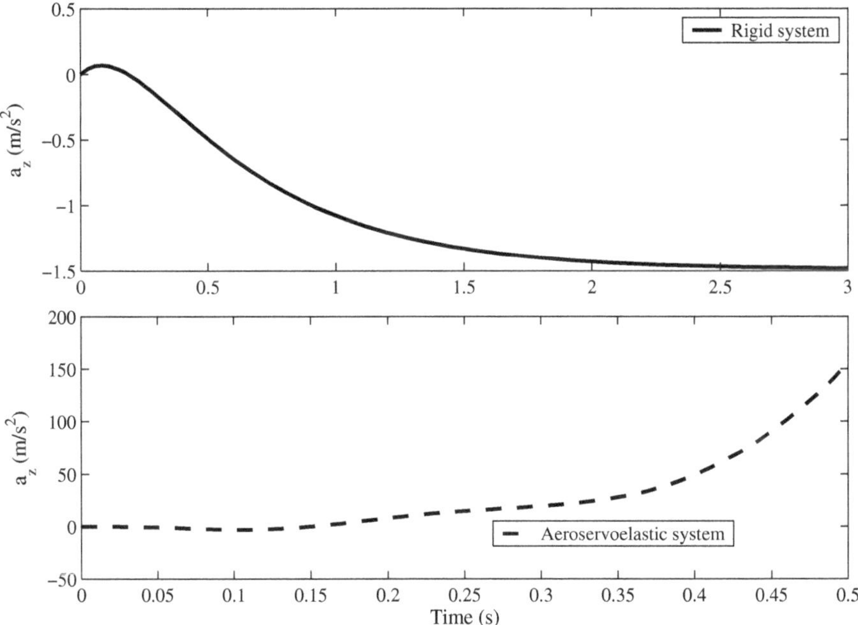

Fig. 5.11 Normal acceleration step response for commanded 1.9° step change in the angle of attack of the fighter aircraft, showing a stabilized rigid aircraft, but an unstable aeroservoelastic system

to have been responsible for crashes of several fighter prototypes (e.g., Lockheed F-22 and Taiwan IDF fighter). The unstable ASE system is also shown by the negative gain margin in the Bodé plot shown in Fig. 5.13.

5.6.1   Closed-Loop Stabilization of the ASE System

The problem of unstable aeroelastic modes is solved by a reprogramming the flight control computer with the LQR and Kalman filter gains based upon the aeroelastic (rather than the rigid) plant. The Kalman filter is designed with the following parameters:

$$S_{vw}(t) = \mathbf{0}, S_v(t) = 10^{-4}\mathsf{I}, S_w(t) = 1, \mathsf{F} = \mathsf{I},$$

while the LQR parameters are taken to be

$$\mathsf{S} = \mathbf{0}, \mathsf{Q} = 10^{-2}\mathsf{C}^T\mathsf{C}, \mathsf{R} = 1.$$

The dimensions of the aeroelastic plant are (38×38), and the resulting regulator and observer gains are listed in Table 5.1.

The regulator (LQR) and observer (Kalman filter) poles are listed in Tables 5.2 and 5.3, respectively. Note that the first two are rigid poles, while the rest are aeroelastic

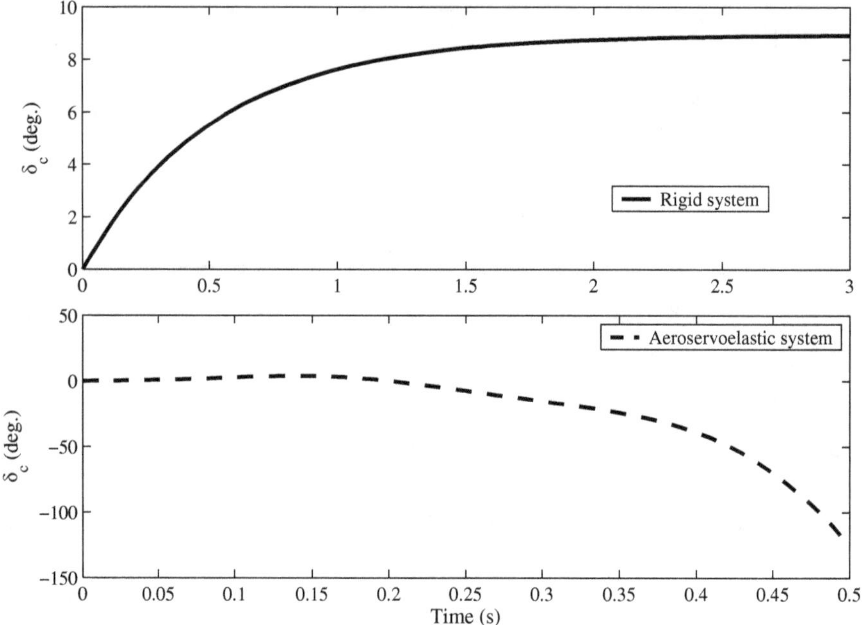

Fig. 5.12 Commanded elevon deflection response for commanded 1.9° step change in the angle of attack of the fighter aircraft, showing a stabilized rigid aircraft, but an unstable aeroservoelastic system

poles. As expected all the closed-loop poles are in the left-half s-plane. The observer has 8 real poles and 15 complex conjugate pairs of poles, whereas the regulator has 10 real poles and 14 complex pair of poles.

Figure 5.14 shows the pole–zero map of the aeroservoelastic system, wherein all the poles are now in the left-half plane. The right-half plane zeros of the open-loop plant, as well as of the rigid stabilized plant in Fig. 5.9—which indicated the non-minimum phase [168] "tail wag the dog" behavior of the rigid pitch dynamics controlled by the elevons—have also been moved to the left-half s-plane, thereby improving the closed-loop transient response. The Nyquist diagram of the closed-loop transfer function $a_z(s)/\alpha_c(s)$ is now seen to be stable for the complete ASE system in Fig. 5.15. The response to an initial pitch rate perturbation of 1°/s is shown in Fig. 5.16 for the normal acceleration output, the commanded and actual elevon deflections, the angle of attack, and the pitch rate, which confirms the well behaved closed-loop response settling in less than 3 s. The waviness in the normal acceleration response is due to the aeroelastic modes, the first four of these contribute individually to the normal acceleration of the closed-loop ASE system as shown in Fig. 5.17. The well-damped structural response is evident, and the magnitude of a particular mode decreases as its natural frequency increases. This fact is also confirmed by Fig. 5.18, which is the Bodé gain and phase plot of the closed-loop transfer function $a_z(s)/\alpha_c(s)$ with aeroelastic modes included in the plant model. The high robustness

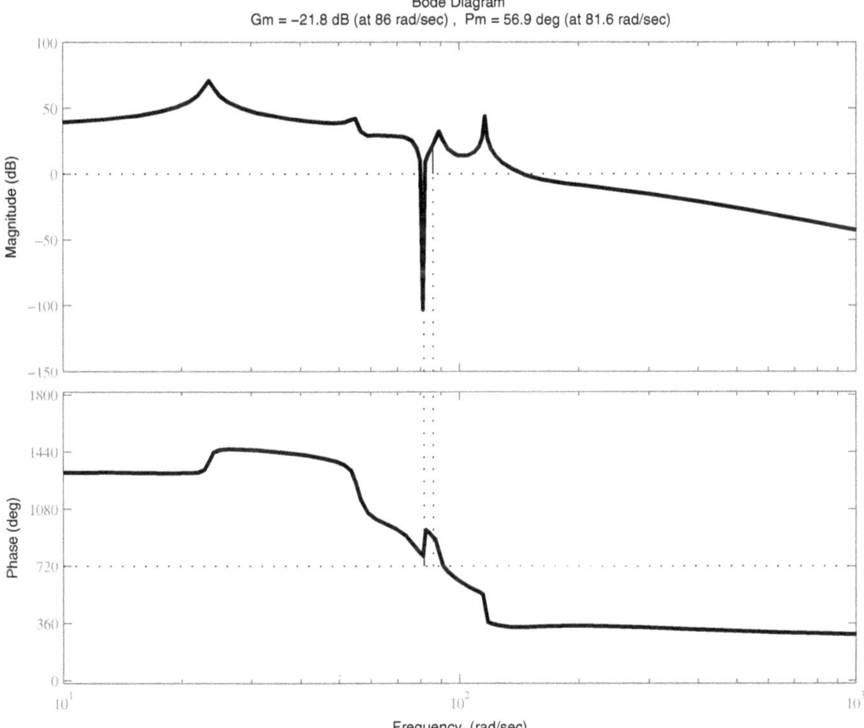

Fig. 5.13 Bodé plot for the transfer function $a_z(s)/\alpha_c(s)$ of the fighter aircraft, showing a negative gain margin hence an unstable aeroservoelastic system

of the complete ASE system is evident in Fig. 5.18 showing an infinite gain margin and a phase margin of 90°. The raising of the DC gain of the open-loop plant in order to achieve the positive gain margin (thus closed-loop stabilization) can be also seen in the Bodé plot.

5.6.2   *Active Maneuver Load Alleviation*

A clever ASE controller design can alleviate the peak normal acceleration experienced by a maneuvering aircraft for a given pitch rate, thereby reducing the structural fatigue and increasing the aircraft's service life. In a fighter-type aircraft, such an active alleviation can have the additional benefit of increasing the load factor for maneuvering, thereby increasing the maneuverability and making the aircraft more competitive vis-a-vis its contemporaries. Let us briefly consider a typical load alleviation example for our simple, tailless aircraft with a pair of trailing-edge elevons.

Table 5.1 Regulator and observer gain matrices for the tailless fighter aircraft at $U = 306$ m/s and standard sea level

K^T	L
-103.66	-2.8736
-20.505	-6.8318
-36.649	-0.20525
109.24	-0.04129
58.78	-0.042994
387.66	0.0011039
-7.6391	-0.027634
0.06913	-0.00028134
6.0975	-0.0014706
-83.762	-0.018054
-45.864	0.0094441
10.086	0.002567
130.45	-0.00089379
4.2596	-7.5657×10^{-5}
13.72	0.0010321
2.9304	-0.010766
-0.39514	0.0079591
-2.7375	-0.0025357
-1.3979	-0.20684
8.5541	-0.023023
1.5883	-0.051001
1.7031	-0.0025858
-3.3272	-0.037805
-2.3168	-0.0038365
-1.014	-0.00082518
-4.4959	-0.0055202
11.472	-0.014006
8.6004	0.005582
-4.8648	2.4076×10^{-5}
-5.0704	-6.8233×10^{-8}
-0.9564	6.6789×10^{-7}
-6.503	1.8302×10^{-6}
5.6262	-1.0659×10^{-6}
-0.43071	4.0438×10^{-7}
0.76822	-1.1487×10^{-7}
-0.90782	1.9441×10^{-7}
0.021961	-1.5219×10^{-9}
-0.0043702	2.3804×10^{-10}

Table 5.2 LQR poles of the tailless fighter aircraft at $U = 306$ m/s and standard sea level

Mode No.	Pole
1	$-2.2512 + 0i$
2	$-5.4649 + 0i$
3	$-14.61 + 0i$
4	$-16.974 + 0i$
5	$-51.947 + 0i$
6	$-34.522 \pm 40.753i$
7	$-53.386 \pm 5.2487i$
8	$-55.965 + 0i$
9	$-2.0083 \pm 56.442i$
10	$-56.876 \pm 4.773i$
11	$-23.855 \pm 65.835i$
12	$-63.495 \pm 39.034i$
13	$-4.6203 \pm 77.837i$
14	$-86.402 + 0i$
15	$-2.1758 \pm 89.017i$
16	$-98.633 + 0i$
17	$-107.78 \pm 14.109i$
18	$-1.6801 \pm 116.12i$
19	$-125.53 \pm 21.428i$
20	$-157.15 \pm 16.057i$
21	$-189.47 \pm 22.9i$
22	$-208.55 \pm 15.317i$
23	$-215.87 + 0i$
24	$-275.1 + 0i$

The actual modern fighter-type aircraft also has leading-edge flaps, which provide a better controllability and offer a greater reduction in the maneuvering loads.

Let us consider the response shown in Fig. 5.19 for a commanded unit step change in the angle of attack of the fighter aircraft with the original design presented above (solid line). This is compared with the response of the controller redesigned with a maneuver load alleviation objective (dashed line), using the following parameters for the regulator (the observer is unchanged):

$$\mathsf{S} = \mathbf{0}, \mathsf{Q} = 10^{-9}\mathsf{B}\mathsf{B}^{T}, \mathsf{R} = 1.$$

While the original system behaves in a jagged manner with a high peak acceleration and pitch rate, and requires very large control deflections (which are likely to be saturated in the actual implementation), the much smoother acceleration response

Table 5.3 Kalman filter poles of the tailless fighter aircraft at $U = 306$ m/s and standard sea level

Mode No.	Pole
1	$-2.0744 + 0i$
2	$-5.4649 + 0i$
3	$-15.89 + 0i$
4	$-18.173 + 0i$
5	$-52.073 + 0i$
6	$-51.866 \pm 5.1181i$
7	$-13.552 \pm 52.182i$
8	$-56.175 \pm 7.8233i$
9	$-33.57 \pm 45.861i$
10	$-21.539 \pm 53.441i$
11	$-65.785 + 0i$
12	$-67.055 \pm 13.021i$
13	$-4.6089 \pm 76.479i$
14	$-88.803 + 0i$
15	$-1.0911 \pm 89.019i$
16	$-100.29 + 0i$
17	$-108.14 \pm 14.51i$
18	$-0.40361 \pm 116.08i$
19	$-125.54 \pm 21.569i$
20	$-157.16 \pm 16.078i$
21	$-189.51 \pm 22.876i$
22	$-208.63 \pm 15.294i$
23	$-215.98 + 0i$
24	$-274.8 + 0i$

of the maneuver alleviation design is striking in comparison. It also requires only a third of the peak acceleration, and a fourth of the control deflection of the original design. The shaping of the aeroelastic modes in order to achieve a smaller initial peak acceleration for load alleviation is well evident in Fig. 5.20, which plots the normal acceleration contributions of the first four aeroelastic modes. The achievement of smooth aeroelastic behavior translates into a smaller peak acceleration for all the modes. The cost of load alleviation is a reduction in the gain margin from infinity to about 60 dB, which is quite adequate for practical implementation.

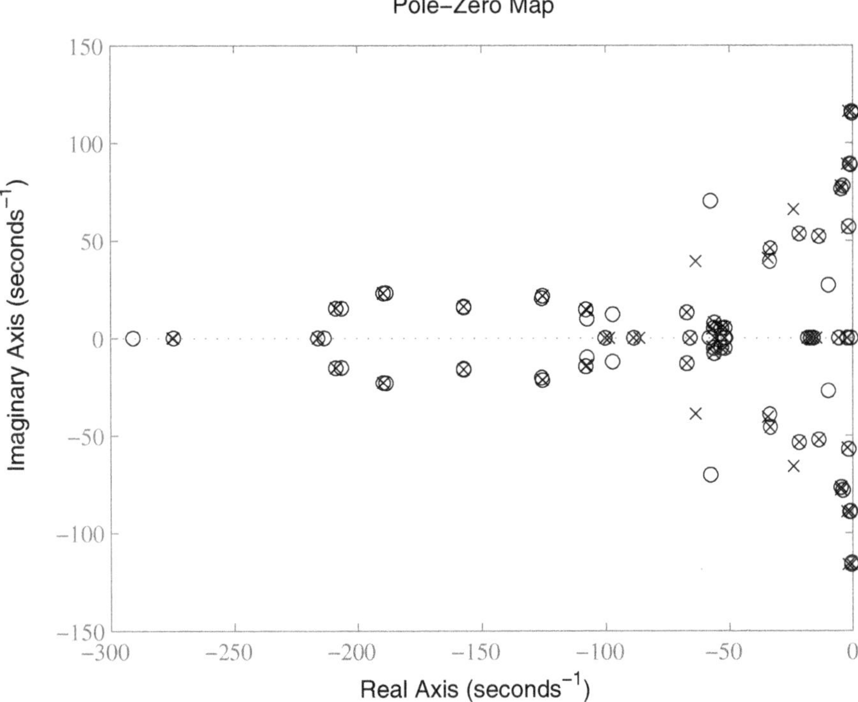

Fig. 5.14 Pole–zero map of the aeroservoelastic system for the fighter aircraft for the transfer function $a_z(s)/\alpha_c(s)$ with aeroelastic modes included in the plant model (x: pole; o: zero)

5.7   Robust Control of Linear Time-Invariant Systems

Although the example in the previous section could achieve a robust controller design for the fighter aircraft, there is no guarantee that separately designed regulator and observer can produce an acceptable behavior in the overall closed-loop system. For example, both LQR and Kalman filter have good stability and performance properties by themselves, but when combined into an LQG compensator can result in an appreciable degradation of such properties. This is because in a combination, neither can be regarded as being the optimal solution of the overall problem. In order to recover the good stability and performance characteristics displayed by LQR and Kalman filter, design of a *robust* controller is necessary. There are two main robust design methods, namely the LQG/LTR and H_2/H_∞ design, and we will briefly discuss them here.

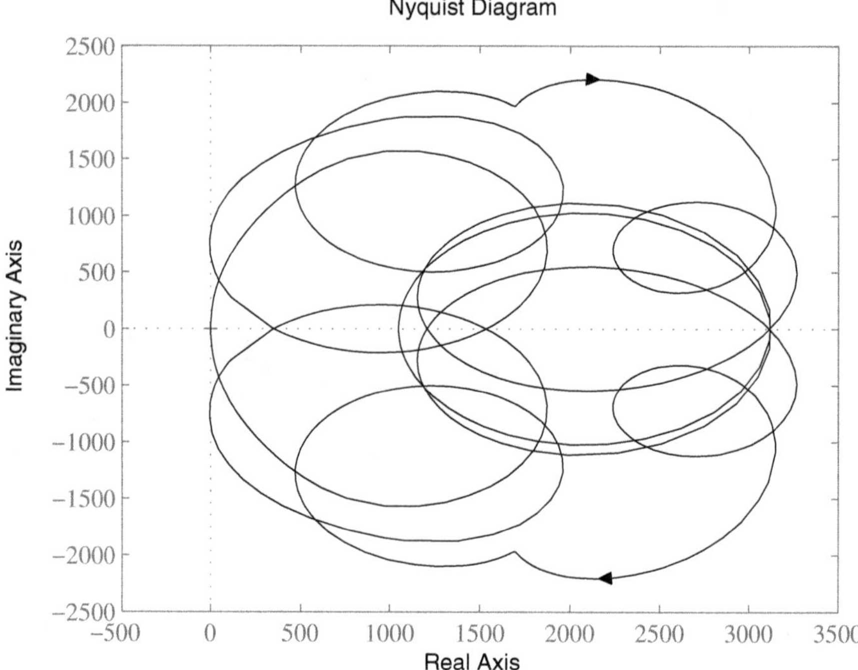

Fig. 5.15 Nyquist locus of the aeroservoelastic system for the fighter aircraft for the transfer function $a_z(s)/\alpha_c(s)$ with aeroelastic modes included in the plant model

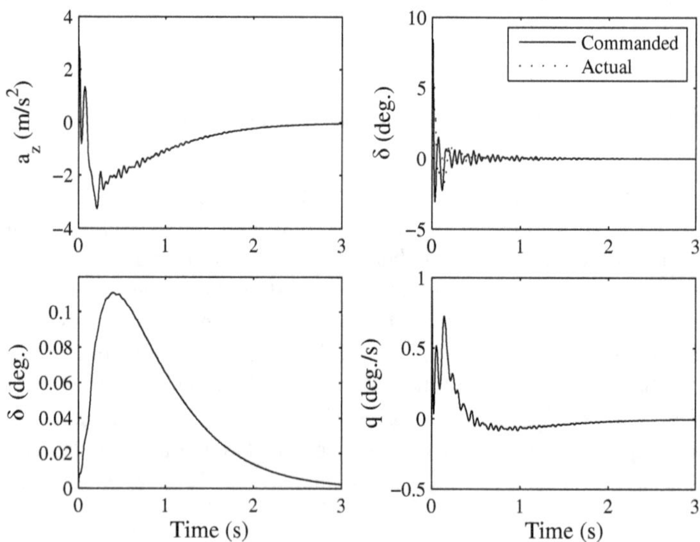

Fig. 5.16 Closed-loop response for an initial perturbation in the pitch rate of 1°/s of the fighter aircraft with aeroelastic modes stabilization, showing a stable and well-behaved aeroservoelastic system

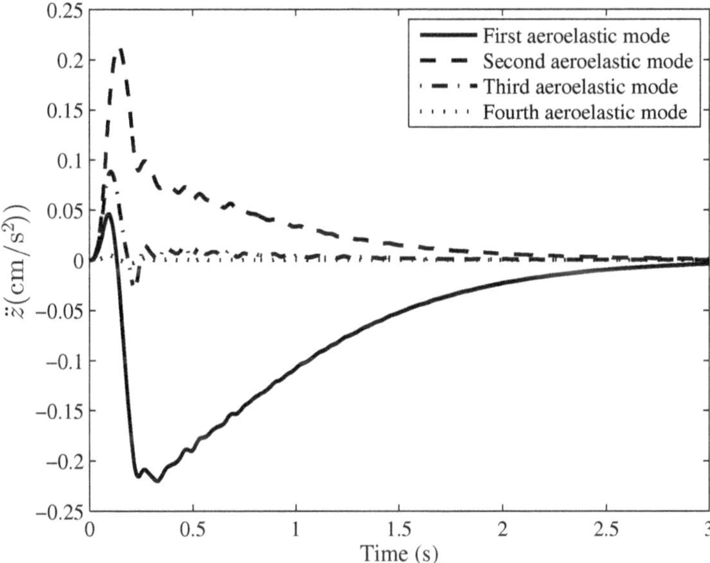

Fig. 5.17 Normal acceleration response of the first four aeroelastic modes for an initial perturbation in the pitch rate of 1°/s of the fighter aircraft with aeroelastic modes stabilization, confirming a stable and well-behaved aeroservoelastic system

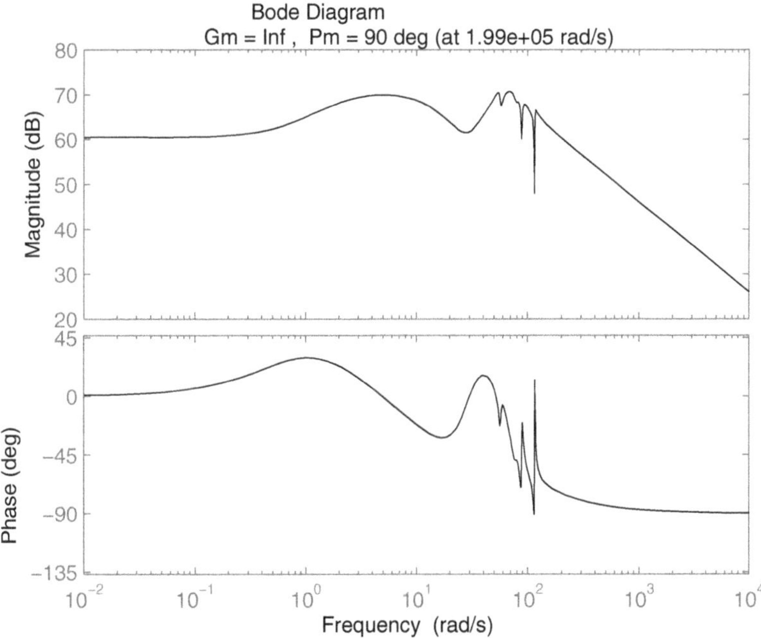

Fig. 5.18 Bodé plot for the transfer function $a_z(s)/\alpha_c(s)$ of the fighter aircraft with aeroelastic modes stabilization, showing an infinite gain margin and a 90° phase margin, hence a robust aeroservoelastic system.

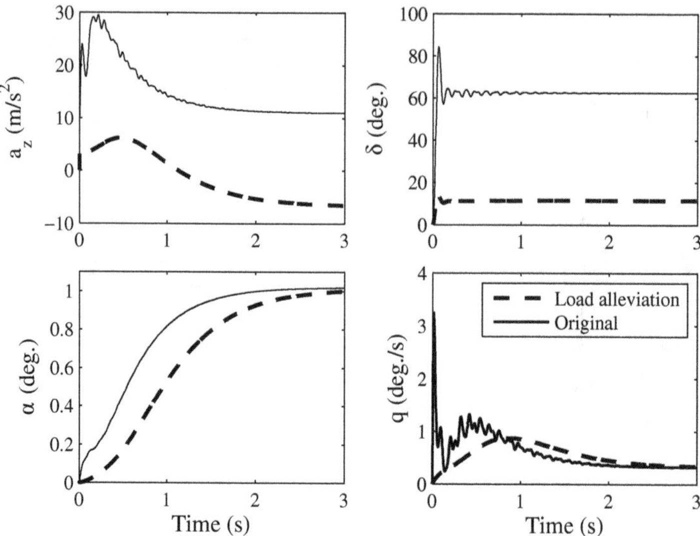

Fig. 5.19 Closed-loop response for a commanded step change in the angle of attack of the fighter aircraft with and without maneuver load alleviation

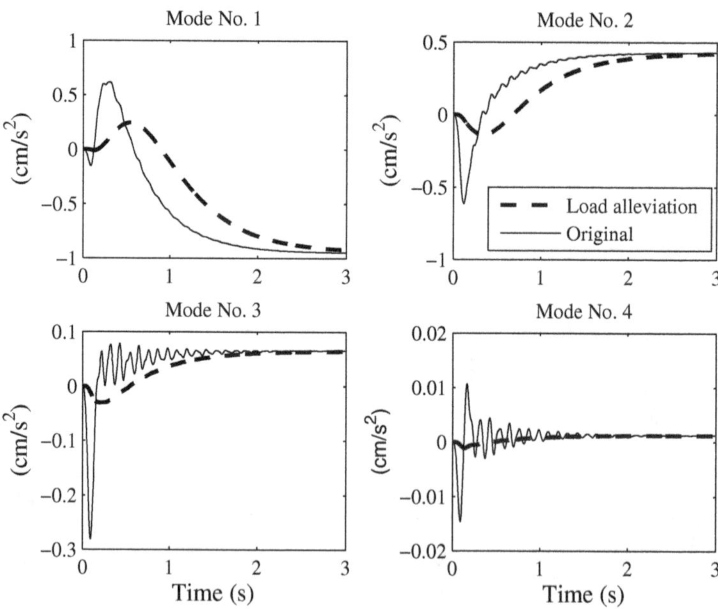

Fig. 5.20 Normal acceleration response of the first four aeroelastic modes for a commanded step change in the angle of attack of the fighter aircraft with and without maneuver load alleviation

5.7.1   LQG/LTR Method

The procedure by which an LQR and a Kalman filter are designed separately for a linear, time-invariant plant, and then put together to form an LQG compensator can lead to an unacceptable design in the presence of parametric disturbances. Consider a general case where the output vector, $\mathbf{y}(t)$, is to match a desired output (also called a reference, or commanded output), $\mathbf{y_d}(t)$, in the steady state. Such a reference output is usually commanded by the terminal controller (not shown in the figure). Clearly, the measured signal given to Kalman filter is $[\mathbf{y}(t) - \mathbf{y_d}(t)]$, based upon which (as well as the known input vector, $\mathbf{u}(t)$) it supplies the estimated state for feedback to the LQR regulator. Since the design of the LQG compensator—specified by the gain matrices, $(\mathbf{K},\mathbf{L})$—depends upon the chosen LQR cost parameters, $\mathbf{Q},\mathbf{R},\mathbf{S}$, and the selected Gaussian white noise spectral densities, $\mathbf{S_w},\mathbf{S}_v,\mathbf{S_{vw}}$, it is possible to design infinitely many compensators for a given plant. Usually, there are certain performance and robustness requirements specified for the closed-loop system that indirectly restrict the choice of the cost parameters to a given range. Being based upon optimal control, an LQG compensator has excellent performance features for a given set of cost parameters, but its robustness is subject to the extent the performance is degraded by state estimation through the Kalman filter. If the filter gains are too small, the estimation error does not tend to zero fast enough for the feedback to be accurate. On the other hand, if the Kalman filter has very large gains, there is an amplification of process and measurement noise by feedback, thereby reducing the overall robustness of the control system. Clearly, a balance must be struck in selecting the Kalman filter design parameters, $\mathbf{S_w},\mathbf{S}_v,\mathbf{S_{vw}}$, such that a good robustness is obtained without unduly sacrificing performance.

In order to study the robustness of an LQG compensated system, refer to the block diagram of the control system transformed to Laplace domain in a negative feedback configuration. For simplicity, consider a *strictly proper* plant [i.e, $\mathbf{D} = \mathbf{0}$] of order n, represented by the following transfer matrix:

$$\mathbf{G}(s) = \mathbf{C}\,(s\mathbf{I} - \mathbf{A})^{-1}\,\mathbf{B}$$

of dimension $\ell \times m$, where ℓ is the number of outputs, and m the number of inputs. An LQG compensator of dimension $m \times \ell$ has transfer matrix,

$$\mathbf{H}(s) = -\mathbf{K}\,(s\mathbf{I} - \mathbf{A} + \mathbf{BK} + \mathbf{LC})^{-1}\,\mathbf{L}.$$

The process noise is represented by a ZMGWN disturbance, $v(s)$, appearing at the plant's output, while the ZMGWN measurement noise, $\mathbf{w}(s)$, affects the feedback loop as shown. The overall system's transfer matrix, $\mathbf{T}(s)$, from the desired output to the output, is called the *transmission matrix*. On the other hand, the effect of the process noise on the output is given by the transfer matrix, $\mathbf{S}(s)$, called the *sensitivity matrix*. Both $\mathbf{T}(s)$ and $\mathbf{S}(s)$ are derived as follows:

$$\mathbf{y} = v + \mathbf{Gu} = v + \mathbf{G}\left[\mathbf{H}\left(\mathbf{y_d} - \mathbf{y} - \mathbf{w}\right)\right], \tag{5.92}$$

or

$$(I + GH)\, \mathbf{y} = v + GH\left(\mathbf{y_d} - \mathbf{w}\right), \tag{5.93}$$

thereby implying

$$\mathbf{y} = (I + GH)^{-1}\, v + (I + GH)^{-1}\, GH\left(\mathbf{y_d} - \mathbf{w}\right), \tag{5.94}$$

or

$$\mathbf{y}(s) = S(s)v(s) + T(s)\left[\mathbf{y_d}(s) - \mathbf{w}(s)\right], \tag{5.95}$$

where

$$S(s) = \left[I + G(s)H(s)\right]^{-1} \tag{5.96}$$

$$T(s) = \left[I + G(s)H(s)\right]^{-1} G(s)H(s).$$

Because it can be easily shown that $T(s) = I - S(s)$, the transmission matrix is also called *complementary sensitivity*.

The sensitivity and transmission matrices give us criteria for defining robustness. For a good robustness with respect to the process noise, the sensitivity matrix, $S(s)$, should have a small "magnitude," while a good robustness with respect to the measurement noise requires that the transmission matrix, $T(s)$, must have a small "magnitude." However, it is unclear what we mean by the magnitude of a matrix. One can assign a scalar measure, such as the Euclidean norm [170] for vectors, to matrices. Since we are now dealing with square arrays of complex numbers, $M(s)$, where $s = i\omega$, suitable norms for such a matrix, $M(i\omega)$, at a given frequency, ω, are based upon its *singular values* (also called *principal gains*), defined as the positive square roots of the eigenvalues of the following real matrix,

$$M^{T}(-i\omega)M(i\omega).$$

The *Hilbert* (or spectral) norm is the largest singular value at a given frequency, ω, denoted as

$$\bar{\sigma}\left\{M(i\omega)\right\},$$

or simply as $\bar{\sigma}(M)$. The magnitude of a frequency dependent, complex matrix is thus represented by its largest singular value. Clearly, for robustness with respect to the process noise, we require $\bar{\sigma}(S)$ should be as small as possible, whereas robustness with respect to the measurement noise can be achieved by minimizing $\bar{\sigma}(T)$. However, there is a serious conflict between simultaneously minimizing both $\bar{\sigma}(S)$ and $\bar{\sigma}(T)$, because the minimization of one results in the maximization of the other (and vice versa). A design compromise is obtained by choosing different ranges of frequencies for the minimization of $\bar{\sigma}(S)$ and $\bar{\sigma}(T)$.

The process noise, $v(s)$, is generally experienced due to modeling errors, which—while being present at all frequencies—usually have their maximum effect on a physical model at the lowest frequencies. An example is the neglect of higher-order dynamics while deriving the plant model, such as structural flexibility of a flight vehicle. Of all the neglected modes, that with the smallest natural frequency typically has the largest contribution to the transfer matrix. As the natural frequency increases, the concerned modes have successively smaller magnitude contributions. Thus, it makes sense to minimize $\bar{\sigma}(\mathsf{S})$ in the low-frequency range. On the other hand, measurement noise has its largest contribution at high frequencies, and thus $\bar{\sigma}(\mathsf{T})$ should be minimized at high frequencies. However, we should also remember that $\mathsf{T}(s)$ is the overall system's transfer matrix. A minimization of $\bar{\sigma}(\mathsf{T})$ at higher frequencies has the additional effect of slowing down the closed-loop system's performance in tracking a reference signal, which implies a larger tracking error at a given time. Therefore, in the interest of maintaining a reasonable performance, the minimization of $\bar{\sigma}(\mathsf{T})$ must take place outside the desired *bandwidth* of the control system.

As discussed above, the gains of the regulator and the Kalman filter must be selected in such a way that there is as little loss as possible of both the performance and robustness by combining the two in a single compensator. The transfer matrix $\mathsf{H}(s)\mathsf{G}(s)$ denotes the transfer of the input, $\mathbf{u}(s)$, back to itself if the loop is broken at the input, and is thus called the *return ratio at input*. If all the states are available for measurement, there is no need for a Kalman filter and the ideal return ratio at input would be the following:

$$\mathsf{H}(s)\mathsf{G}(s) = -\mathsf{K}\,(s\mathsf{I} - \mathsf{A})^{-1}\,\mathsf{B}.$$

On the other hand, if the feedback loop is broken at the output, the transfer matrix, $\mathsf{G}(s)\mathsf{H}(s)$, represents the transfer of the output, $\mathbf{y}(s)$, back to itself and is called the *return ratio at output*. If there is no regulator in the system, then the ideal return ratio at output is given by

$$\mathsf{G}(s)\mathsf{H}(s) = -\mathsf{C}\,(s\mathsf{I} - \mathsf{A})^{-1}\,\mathsf{L}.$$

If one can recover the ideal return ratio at either the plant's input or the output by suitably designing the LQG compensator, the best possible combination of the LQR regulator and the Kalman filter is achieved and the design process is called *loop-transfer recovery* (LTR). For simplicity in the following discussion, we assume there is no cross-correlation between process and measurement noise, i.e., $\mathsf{S}_{vw} = 0$. It can be proved [108] that by selecting

$$\mathsf{F} = \mathsf{B}\,; \qquad \mathsf{S}_v = \rho\mathsf{S}_w$$

and making the positive scalar parameter ρ arbitrarily large, the LQG return ratio at input,

$$\mathsf{H}(s)\mathsf{G}(s) = -\mathsf{K}\,(s\mathsf{I} - \mathsf{A} + \mathsf{BK} + \mathsf{LC})^{-1}\,\mathsf{LC}\,(s\mathsf{I} - \mathsf{A})^{-1}\,\mathsf{B}$$

can be made to approach the ideal return ratio at input. The following procedure for LTR at the input is thus commonly applied:

- Select an LQR regulator by a suitable choice of the weighting matrices $\mathsf{Q},\mathsf{R},\mathsf{S}$ such that good performance and robustness with state feedback are obtained.
- Select $\mathsf{F} = \mathsf{B}$, $\mathsf{S}_v = \rho\mathsf{S}_\mathsf{w}$ and increase ρ until the desired state feedback properties are recovered in the closed-loop system.

A variation of this approach can be applied for LTR at the plant's output, beginning with the design of Kalman filter, and then iterating for the LQR gain until the ideal return ratio at output is recovered:

- Select a Kalman filter by a suitable choice of the weighting matrices, $\mathsf{S}_v, \mathsf{S}_\mathsf{w}, \mathsf{S}_{v\mathsf{w}}$, such that a satisfactory return ratio at output obtained.
- Select $\mathsf{Q} = \rho\mathsf{R}$ and increase ρ until the ideal return ratio at output is recovered in the closed-loop system.

Further details of LQG/LTR methods can be found in [108] and [62]. MATLAB's *Control Systems Toolbox* [110] provides the functions *lqr* and *lqe* for carrying out the LQR and Kalman filter designs, respectively. Alternatively, the *Robust Control Toolbox* [110] has specialized functions *lqg* for "hands-off" LQG compensator design, and *ltru* and *ltry* for LTR at plant input and output, respectively. However, for a beginner we recommend the use of a basic ARE solver, such as *are*, in order that the relevant design steps are clearly understood.

5.7.2 H_2/H_∞ *Control*

Rather than indirectly designing a feedback compensator via a regulator and a Kalman filter, an alternative design approach is to extend the frequency domain design methodology commonly used for single variable (SISO) systems, to the multivariable control system. The regulator and Kalman filter then result naturally from such a design, in which the singular value spectra replace the Bodé gain plot, and parameters analogous to the gain and phase margins of an SISO system [cf. Chap. 2 of [168]] are addressed for a robust, multivariable design directly in the frequency domain. In deriving such a compensator, one minimizes a combined, frequency weighted measure of sensitivity, complementary sensitivity, and transfer matrix from disturbance to plant output, integrated over a range of frequencies. There are two such frequency domain, optimal control methods that only differ by the matrix norm sought to be minimized, namely the H_2 and H_∞ synthesis.

Consider a strictly proper plant, $\mathsf{G}(s)$, with control inputs, $\mathbf{u}(s)$, and measured outputs, $\mathbf{y}(s)$. All other inputs to the control system, namely, the reference output $\mathbf{y_d}(s)$, process noise $v(s)$, and measurement noise $\mathbf{w}(s)$, are clubbed together in a single vector, $\mathbf{d}(s)$, called *external disturbances*. An output feedback compensator, $\mathsf{H}(s)$, is to be designed for simultaneously minimizing an integrated measure of sensitivity, $\mathsf{S}(s)$, and complementary sensitivity, $\mathsf{T}(s)$, of the output with respect to the disturbance, and transfer matrix, $\mathsf{G_u}(s) = -\mathsf{H}(s)\mathsf{S}(s)$, from disturbance to plant input. However, since these objectives are contradictory, different ranges of frequency are specified for the minimization of each objective. This is practically implemented through weighting the concerned matrix by *frequency weights* as follows:

$$\mathsf{W_S}(s)\mathsf{S}(s) = \mathsf{W_S}(s)\big[\mathsf{I} + \mathsf{G}(s)\mathsf{H}(s)\big]^{-1}$$

$$\mathsf{W_T}(s)\mathsf{T}(s) = \mathsf{W_T}(s)\big[\mathsf{I} + \mathsf{G}(s)\mathsf{H}(s)\big]^{-1}\mathsf{G}(s)\mathsf{H}(s) \qquad (5.97)$$

$$\mathsf{W_u}(s)\mathsf{G_u}(s) = -\mathsf{W_u}(s)\mathsf{H}(s)\mathsf{S}(s).$$

Here, $\mathsf{W_S}(s)$ is a strictly proper, square transfer matrix, and $\mathsf{W_T}(s), \mathsf{W_u}(s)$ are square transfer matrices. The plant is thus augmented by the frequency weighting matrices with the additional outputs, $\mathbf{z_1}(s), \mathbf{z_2}(s), \mathbf{z_3}(s)$, called *error signals* :

$$\mathbf{y}(s) = \mathsf{G}(s)\mathbf{u}(s) + \mathbf{d}(s)$$

$$\mathbf{z_1}(s) = \mathsf{W_S}(s)\mathbf{y}(s) = \mathsf{W_S}(s)\big[\mathsf{G}(s)\mathbf{u}(s) + \mathbf{d}(s)\big] \qquad (5.98)$$

$$\mathbf{z_2}(s) = \mathsf{W_T}(s)\mathsf{G}(s)\mathbf{u}(s)$$

$$\mathbf{z_3}(s) = \mathsf{W_u}(s)\mathbf{u}(s).$$

The output feedback control law is given by

$$\mathbf{u}(s) = -\mathsf{H}(s)\mathbf{y}(s). \qquad (5.99)$$

Note that the compensator design is entirely dependent on the chosen frequency weights (rather than on the LQR cost parameters and noise spectral densities of the LQG case). Furthermore, the controller design is based upon output feedback, which requires the following inherent observer dynamics that is part of the augmented plant:

$$\dot{\hat{\mathbf{x}}} = (\mathsf{A} - \mathsf{FC})\hat{\mathbf{x}} + \mathsf{B}\mathbf{u} + \mathsf{F}\mathbf{y}. \qquad (5.100)$$

Since the coefficients $\mathsf{A,F,C}$ depend upon the chosen frequency weights, a stable observer requires a judicious choice of the frequency weights.

The overall closed-loop transfer matrix, $\mathsf{G_c}(s)$, from the disturbance, $\mathbf{d}$, to the error vector, $\mathbf{z} = (\mathbf{z_1}, \mathbf{z_2}, \mathbf{z_3})^T$, is thus given by

$$\mathbf{z}(s) = \mathsf{G_c}(s)\mathbf{d}(s) = \begin{bmatrix} \mathsf{W_S}(s)\mathsf{S}(s) \\ -\mathsf{W_T}(s)\mathsf{T}(s) \\ -\mathsf{W_u}(s)\mathsf{G_u}(s) \end{bmatrix} \mathbf{d}(s). \qquad (5.101)$$

In order to see how frequency weights can be selected, let us define the following frequency integral of a proper and asymptotically stable transfer matrix, $\mathsf{Q}(i\omega)$, as its H_2-*norm* :

$$\parallel \mathsf{Q} \parallel_2 = \sqrt{\frac{1}{2\pi} \int_{-\infty}^{\infty} \mathrm{tr}\big[\mathsf{Q}(i\omega)\mathsf{Q}^T(-i\omega)\big]\, d\omega}, \qquad (5.102)$$

where tr(.) refers to the trace of a square matrix (i.e., sum of its diagonal elements). Another related operator norm for $\mathbf{Q}(i\omega)$, is its H_∞-*norm* defined by

$$\| \mathbf{Q} \|_\infty = \sup_\omega \left[\bar{\sigma} \{\mathbf{Q}(i\omega)\} \right], \tag{5.103}$$

where $\sup_\omega(.)$ is the supremum (the maximum value) with respect to the frequency. Both the operator norms, $\| \cdot \|_2$ and $\| \cdot \|_\infty$, provide a single, positive real number measuring the largest possible magnitude of a frequency dependent matrix. Therefore, one can employ either of them for minimizing the combined sensitivity reflected by the closed-loop transfer matrix, $\mathbf{G}_c(s)$.

Finally, the H_2/H_∞ design method can be summarized by considering the following H_2-norm of $\mathbf{G}_c(s)$:

$$\| \mathbf{G}_c \|_2 = \| \mathbf{W}_S \mathbf{S} \|_2 + \| \mathbf{W}_T \mathbf{T} \|_2 + \| \mathbf{W}_u \mathbf{G}_u \|_2 . \tag{5.104}$$

The minimization of $\| \mathbf{G}_c \|_2$ guarantees a simultaneous minimization of the H_2-norm of the weighted sensitivity, complementary sensitivity, and $\mathbf{G}_u(s)$. We also note that the power spectral density of the error vector, $\mathbf{z}(t)$, when the disturbance, $\mathbf{d}(t)$, is a white noise of unit intensity (i.e, has identity matrix as its power spectral density) is the following:

$$\mathbf{S}_z(\omega) = \mathbf{Z}(i\omega)\mathbf{Z}^T(-i\omega) = \mathbf{G}_c(i\omega)\mathbf{G}_c^T(-i\omega). \tag{5.105}$$

Thus, the minimization of $\| \mathbf{G}_c \|_2$ directly results in a minimization of the error power spectral density, which is an objective quite similar to that of the LQG compensator. Therefore, one can proceed to derive the compensator that minimizes the H_2-norm of the error in much the same manner as the LQG case.

The augmented plant can be represented in an LTI state-space form as follows:

$$\begin{aligned}
\dot{\mathbf{x}} &= \mathbf{A}\mathbf{x} + \mathbf{B}\mathbf{u} + \mathbf{F}\mathbf{d} \\
\mathbf{y} &= \mathbf{C}\mathbf{x} + \mathbf{d} \\
\mathbf{z} &= \mathbf{M}\mathbf{x} + \mathbf{N}\mathbf{u}.
\end{aligned} \tag{5.106}$$

The optimal H_2 synthesis then consists of deriving an LQR regulator with $\mathbf{Q} = \mathbf{M}^T\mathbf{M}, \mathbf{R} = \mathbf{I}$ and gain, $\mathbf{K} = \mathbf{B}^T\mathbf{P}$ such that

$$\mathbf{H}(s) = -\mathbf{K}\left(s\mathbf{I} - \mathbf{A} + \mathbf{B}\mathbf{B}^T\mathbf{P} + \mathbf{F}\mathbf{C}\right)^{-1}\mathbf{F}, \tag{5.107}$$

where $\mathbf{P}$ is a symmetric, positive semidefinite solution to the following algebraic Riccati equation :

$$0 = \mathbf{A}^T\mathbf{P} + \mathbf{P}\mathbf{A} - \mathbf{P}\mathbf{B}\mathbf{B}^T\mathbf{P} + \mathbf{M}^T\mathbf{M}. \tag{5.108}$$

For simplicity of discussion, let us assume

$$\mathbf{N}^T\mathbf{M} = 0 ; \qquad \mathbf{N}^T\mathbf{N} = \mathbf{I}. \tag{5.109}$$

A good way to ensure a stable observer dynamics $(\mathsf{A}\text{-}\mathsf{FC})$ is by replacing F with a Kalman filter gain, L, which requires a particular structure for the frequency weights. Evidently, a major part of H_∞ design involves selection of suitable frequency weights.

An optimization procedure can be adopted for minimizing the H_∞-norm of $\mathsf{G}_c(s)$ based upon the augmented plant. However, now there being no direct relationship between H_∞-norm of the closed-loop transfer matrix and the objective function of an LQG design, it is difficult to know in advance what is the minimum value of $\|\,\mathsf{G}_c\,\|_\infty$ that would lead to an acceptable design. A practical method [64] is to choose a positive number, γ, and aim at deriving a stabilizing compensator that achieves

$$\|\,\mathsf{G}_c\,\|_\infty = \sup_{\omega}\left[\bar{\sigma}\left\{\mathsf{G}_c(i\omega)\right\}\right] \leq \gamma. \tag{5.110}$$

By decreasing γ until the compensator fails to stabilize the system, one has found the limit on the minimum value of $\|\,\mathsf{G}_c\,\|_\infty$. It is to be noted that if γ is increased to a large value, the design approaches that of the optimal H_2 (or LQG) compensator. Thus, one can begin iterating for γ from a value corresponding to a baseline H_2 (or LQG) design.

There is a serious drawback of the H_∞ approach in that it does not automatically produce a stabilizing compensator. We have just now demonstrated the existence of a nonpositive Lyapunov function, $f(t)$, which shows that a compensator designed by the H_∞ method does not meet the sufficient conditions for stability in the sense of Lyapunov (Appendix-D). Therefore, stability of the H_∞ compensator must be separately ensured by requiring that the dynamics matrix, $(\mathsf{A} - \mathsf{B}\mathsf{B}^T\mathsf{P})$, must have all the eigenvalues in the left-half s-plane.

The following optimization procedure by [64] can be implemented for the solution of the H_∞ design problem:

(a) Select a set of frequency weights, $\mathsf{W}_\mathsf{S}(s), \mathsf{W}_\mathsf{T}(s), \mathsf{W}_\mathsf{u}(s)$, for augmenting the plant, $\mathsf{G}(s)$. These are typically diagonal matrices of proper transfer functions. Also, ensure that the chosen frequency weights yield a stable observer dynamics $(\mathsf{A}\text{-}\mathsf{FC})$.

(b) Select a value for γ (usually 1).

(c) Solve Eq. (5.108) for a symmetric, positive semidefinite matrix, P. If no such solution exists, go back to (b) and increase γ. If a symmetric, positive semidefinite P exists which yields a stable dynamics matrix, $(\mathsf{A} - \mathsf{B}\mathsf{B}^T\mathsf{P})$ try to find a better solution by going back to (b) and decreasing γ.

(d) If the closed-loop properties are satisfactory, stop. Otherwise, go back to (a) and modify the frequency weights.

MATLAB's Robust Control Toolbox (RCT) [110] function, *hinfopt*, automates the H_∞ design procedure of [64], even for those cases where Eq. (5.110) does not hold. However, it replaces γ by its reciprocal, and aims at maximizing the reciprocal for a stabilizing controller. MATLAB-RCT contains another function *h2LQG* for automated H_2 synthesis. Both *hinfopt* and *h2LQG* first require augmenting the model by adding frequency weights through a dedicated function *augtf*.

5.8   Active Flutter Suppression

Airplanes are designed to be light in weight, and hence have highly flexible structures compared to other vehicles. In the interests of fuel efficiency, airplanes also must cruise at much higher speeds, and hence encounter aerodynamic loading several orders of magnitude larger than that of surface transport vehicles. The combination of structural flexibility, inertia, and aerodynamic loads can lead to a destructive dynamic coupling called flutter, which is unique to airplanes. Here, the unsteady airloads build-up in magnitude in a resonance like manner with the structural vibrations, which in turn grow in amplitude, ultimately causing a catastrophic failure of the wing or the tail from which there can be no safe recovery. It is not surprising then that flutter be regarded as the Nemesis which must be avoided at all costs, or at least be pushed out of the operating envelope by either passive or active means.

Let us begin with the state equations of the aeroelastic plant, which were derived in Chap. 4:

$$\dot{\mathbf{x}} = \mathbf{A}\mathbf{x} + \mathbf{B}\mathbf{Q}_c, \tag{5.111}$$

where

$$\mathbf{x} = \left(\mathbf{q}^T, \dot{\mathbf{q}}^T, \mathbf{x}_a^T\right)^T,$$

$$\mathbf{A} = \begin{pmatrix} \mathbf{0} & \mathbf{I} & \mathbf{0} & \mathbf{0} \\ -\bar{\mathbf{M}}^{-1}\bar{\mathbf{K}} & -\bar{\mathbf{M}}^{-1}\bar{\mathbf{C}} & \bar{\mathbf{M}}^{-1}\mathbf{N}_a & \bar{\mathbf{M}}^{-1}\mathbf{N}_g \\ \boldsymbol{\Gamma}_a & \mathbf{A}_a & & \mathbf{0} \\ \boldsymbol{\Gamma}_g & \mathbf{0} & \mathbf{A}_g & \end{pmatrix},$$

and

$$\mathbf{B} = \begin{pmatrix} \mathbf{0} \\ \bar{\mathbf{M}}^{-1}\mathbf{I} \\ \mathbf{0} \end{pmatrix}.$$

To the aeroelastic plant, the state equations of m control surface actuators with deflections vector, $\boldsymbol{\delta} = (\delta_1, \delta_2, \dots, \delta_m)^T$, and control torques input vector, $\mathbf{u} = (u_1, u_2, \dots, u_m)^T$, are added as follows:

$$\dot{\mathbf{x}}_c = \mathbf{A}_c\mathbf{x}_c + \mathbf{B}_c\mathbf{u}, \tag{5.112}$$

where

$$\mathbf{x}_c = \left(\boldsymbol{\delta}, \dot{\boldsymbol{\delta}}, \boldsymbol{\xi}_c^T\right)^T,$$

and

$$\mathbf{Q}_c = \mathbf{C}_c\mathbf{x}_c + \mathbf{D}_c\mathbf{u}.$$

The actuator and the aeroelastic system are two subsystems of the open-loop aeroeservoelastic (ASE) plant, with the overall state vector

$$\mathbf{X} = \left(\mathbf{x}^T, \mathbf{x}_c^T\right)^T,$$

overall state equation,

$$\dot{\mathbf{X}} = \bar{\mathbf{A}}\mathbf{X} + \bar{\mathbf{B}}\mathbf{u}, \tag{5.113}$$

and output equation,

$$\mathbf{y} = \bar{\mathbf{C}}\mathbf{X} + \bar{\mathbf{D}}\mathbf{u}, \tag{5.114}$$

where

$$\bar{\mathbf{A}} = \begin{pmatrix} \mathbf{A} & \mathbf{BC}_c \\ \mathbf{0} & \mathbf{A}_c \end{pmatrix}, \quad \bar{\mathbf{B}} = \begin{pmatrix} \mathbf{BD}_c \\ \mathbf{B}_c \end{pmatrix},$$

$$\bar{\mathbf{C}} = \begin{pmatrix} \mathbf{C} & \mathbf{DC}_c \end{pmatrix}, \quad \bar{\mathbf{D}} = \mathbf{DD}_c.$$

The ASE plant must be stabilized with the following linear feedback control law [4]:

$$u = -\mathbf{K}\hat{\mathbf{X}}, \tag{5.115}$$

where $\hat{\mathbf{X}}$ is the estimated state vector generated by a linear observer with the following state equations:

$$\dot{\mathbf{X}} = \left(\bar{\mathbf{A}} - \mathbf{L}\bar{\mathbf{C}}\right)\mathbf{X} + \left(\bar{\mathbf{B}} - \mathbf{L}\bar{\mathbf{D}}\right)\mathbf{u} + \mathbf{L}\hat{\mathbf{y}}. \tag{5.116}$$

One could alternatively employ a reduced-order observer [171] for a reduction in the order of the ASE system.

Since the flutter suppression problem is a regulator problem, we can consider the desired output vector (i.e., normal accelerations at selected locations) to be zero.

---

[4] Note that here we are not designing separate control laws for the servo actuator and the aeroelastic system. Instead, we have clubbed the two as being the subsystems of the ASE plant. The controller and observer gains of the actuator and aeroelastic subsystems can be extracted from the respective gain matrices for an implementation on a flight control computer.

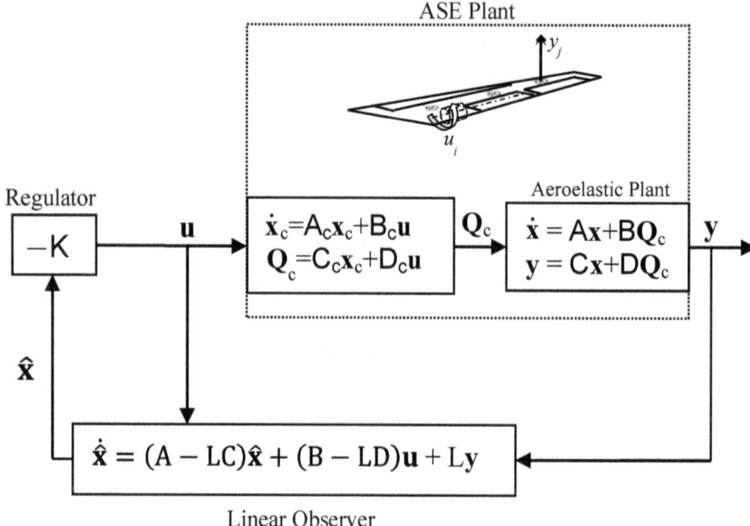

Fig. 5.21 Block diagram of a flutter suppression system

The state equations of the regulated ASE system (depicted by a block diagram in
Fig. 5.21) are, therefore, the following:

$$\left\{\begin{matrix} \dot{x} \\ \dot{\hat{x}} \end{matrix}\right\} = \begin{pmatrix} \bar{A} & -\bar{B}K \\ L\bar{C} & \bar{A} - \bar{B}K - L\bar{C} \end{pmatrix} \left\{\begin{matrix} x \\ \hat{x} \end{matrix}\right\}. \tag{5.117}$$

In order to consider an example of active flutter suppression, let us consider the
modified DAST-ARW1 wing whose open-loop flutter speed was obtained in Chap. 4
to be 284.7 m/s at a standard altitude of 7.6 km. The corresponding flutter Mach
number is 0.9192. This wing is equipped with a trailing-edge control surface and a
4th-order actuator (with 2 lag-parameters) for driving the control surface with actuator
torque input, u. An outboard accelerometer provides the output y, while six spanwise
locations are selected for the aeroelastic deflections. The resulting aeroelastic plant
(with 2 lag-parameters) is of order 24, therefore the overall ASE plant is of order 28.
To this plant is added a regulator and observer designed by the LQG/LTR procedure
in order to yield the best combination of maximum control torque magnitude and
robustness. The selected design parameters are as follows:

$$S_{vw}(t) = 0, \quad S_v(t) = 10^{-12}\bar{B}\bar{B}^T, \quad S_w(t) = 1, \quad F = I,$$

while the LQR parameters are taken to be

$$S = 0, \quad Q = 5000\bar{C}^T\bar{C}, \quad R = 1,$$

which results in the gain matrices of the Kalman filter and the regulator, listed in
Table 5.4.

Table 5.4 Regulator and Kalman filter gains for the active flutter suppression system

K^T	L
3.9766×10^6	-9.6977×10^{-5}
-7.0627×10^7	0.00044932
-1.6821×10^7	-0.00057742
-8.2738×10^7	-0.00033756
-1.2608×10^8	3.2843×10^{-5}
-9.0589×10^7	-0.00080993
-4.2405×10^9	2.7185×10^{-7}
-4.121×10^9	-2.9803×10^{-7}
-2.8641×10^9	-2.4542×10^{-6}
-6.318×10^9	1.2329×10^{-6}
2.1176×10^9	-3.6159×10^{-7}
1.2046×10^9	-1.0203×10^{-6}
-9853.7	-1.2291×10^{-7}
-52637	6.0445×10^{-8}
-94157	-2.4566×10^{-6}
-1.139×10^5	7.738×10^{-7}
-1.4867×10^5	-7.3826×10^{-7}
-3.8988×10^5	-1.019×10^{-6}
7.9125×10^{11}	-2.339×10^{-10}
4.0714×10^{12}	1.0837×10^{-9}
7.8012×10^{12}	-1.393×10^{-9}
1.1572×10^{13}	-8.1399×10^{-10}
1.7344×10^{13}	7.9229×10^{-11}
3.0728×10^{13}	$-1.9536e \times 10^{-9}$
1.793×10^{10}	2.6796×10^{-7}
-6.1962×10^6	5.7031×10^{-6}
-3.1483×10^9	2.6798×10^{-7}
4.9487×10^8	1.376×10^{-11}

The closed-loop response to an initial tip displacement is compared in Fig. 5.22 with that of the open-loop ASE plant for supercritical speed $M = 0.95$ at altitude 7.6 km. Note that the closed-loop acceleration response is stable at the given supercritical condition, has smaller initial peak magnitude than that of the open-loop syatem, and settles to zero in less than 1 s. The required control torque is plotted in Fig. 5.23, showing a peak magnitude of 59.25 N m per m of initial tip displacement. The Bodé plot of the closed-loop ASE system is shown in Fig. 5.24, indicating a gain margin of 37 dB and an infinite phase margin. Therefore, the design is quite robust to

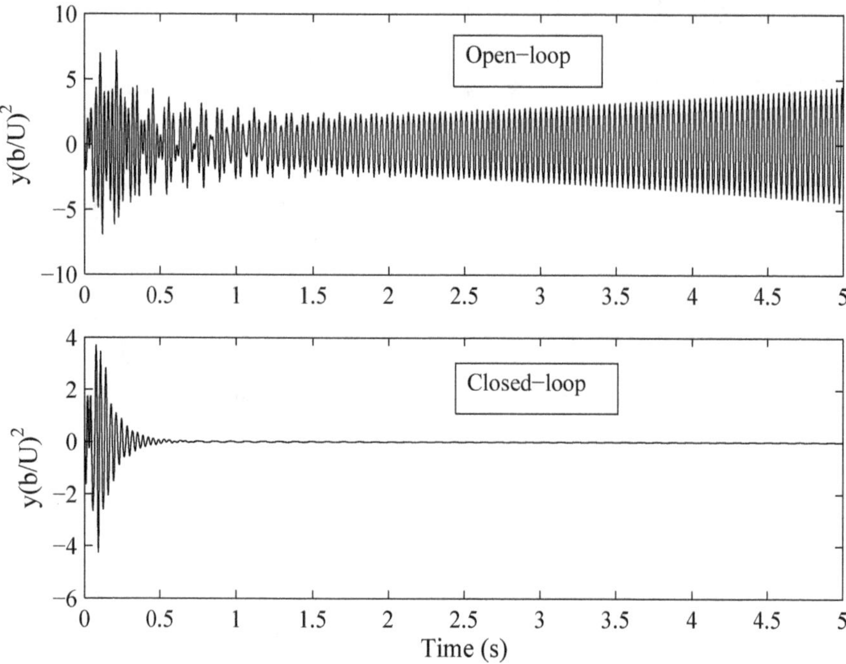

Fig. 5.22 Open-loop and closed-loop response of the flutter suppression system for an initial unit tip displacement at supercritical speed $M = 0.95$ and altitude 7.6 km

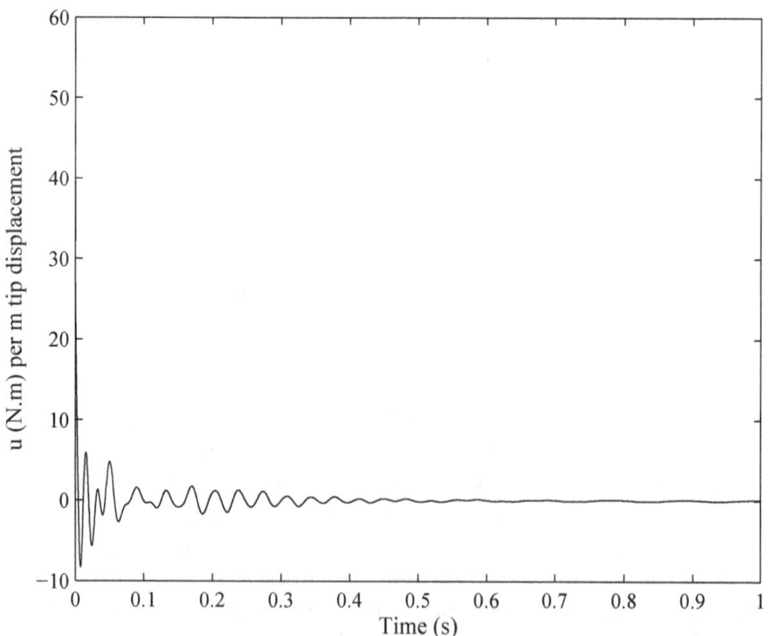

Fig. 5.23 Closed-loop control-surface torque input of the flutter suppression system for an initial tip displacement at supercritical speed $M = 0.95$ and altitude 7.6 km

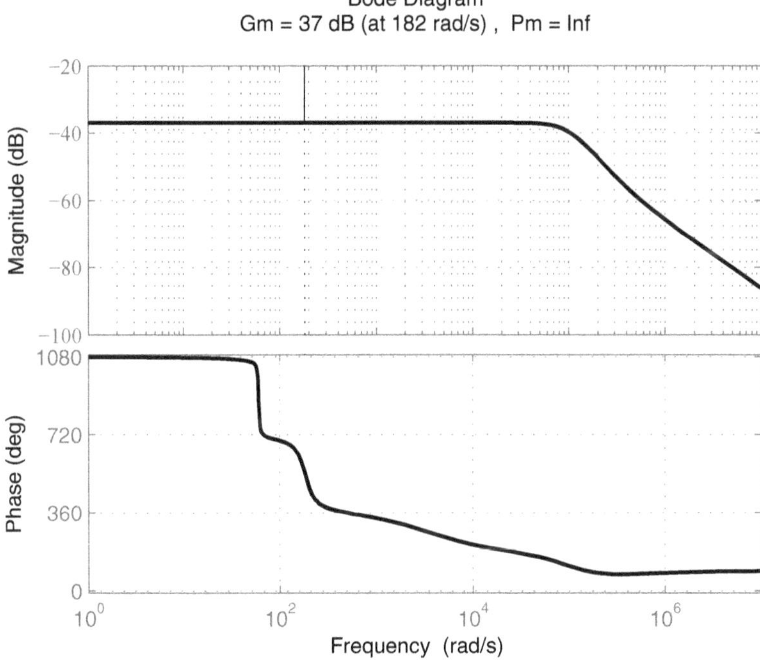

Fig. 5.24 Bodé plot of the closed-loop flutter suppression system at supercritical speed $M = 0.95$ and altitude 7.6 km

parametric variations in the ASE plant. Adequate robustness to high-frequency measurement noise is also evident in the magnitude roll-off of 20 dB per decade. This design is ready to be implemented under actual flight conditions, where the Nyquist bandwidth provided by digital controller will cut-off all high-frequency signals, such as that due to shock-induced turbulence at transonic speeds.

With a slight modification involving the addition of gust states, $\mathbf{x}_g(t)$, in the ASE plant (Chap. 4), the concept of active flutter suppression is readily extended to gust load alleviation for better ride comfort/ targeting accuracy while penetrating an atmospheric gust [22]. Here, the design need not necessarily be carried out for a supercritical case, and the open-loop plant need not be unstable. The optimal control strategy, especially LQG/LTR design, is ideally suited for such an application. The author has devised an optimization strategy [172] especially for the gust alleviation problem, where the regulator design is based upon a quadratic cost function for minimization including the rate of change of normal acceleration, in order to produce a much smoother closed-loop response (hence smaller aerodynamic loading) than possible by a traditional LQR method.

Chapter 6
Nonlinear Aeroservoelastic Applications

6.1 Nonlinear Aeroservoelasticity

Aeroservoelastic models can be nonlinear in nature due to the presence of structural, aerodynamic, or control nonlinearities. For example, a wing undergoing large amplitude vibrations may encounter significantly inelastic behavior. Similarly, the unsteady transonic flow and the flow at large angles of attack are nonlinear due to mixed subsonic–supersonic regions and separation, respectively. Periodically, separating and reattaching flows due to shock-wave/boundary-layer interactions are also inherently nonlinear. Lastly, nonlinear control elements such as adaptive controllers, and saturated and rate-limited actuators, also result in a nonlinear ASE system.

6.2 Describing Functions for Nonlinear ASE

Since ASE systems operate in a closed loop, the effects of nonlinearities are hardly confined to the nonlinear elements themselves, but can be felt throughout the system. At the outset, the treatment of nonlinearities might appear to be a daunting task, because there is no systematic procedure available for designing a nonlinear control system. Fortunately, nonlinear ASE systems can be analyzed by a useful approximation, called the *describing function method* [158].

Briefly stated, the describing function approach takes advantage of the fact that many nonlinear control systems of practical interest can be represented by a nonlinear function block in negative series feedback with a linear, stable transfer function (Fig. 6.1). Thus a simple harmonic signal,

$$u = u_0 \sin{(\omega t)}, \tag{6.1}$$

passing through the nonlinear block is effectively amplified and distorted into several harmonics, represented by the Fourier series,

© Springer Science+Business Media, LLC 2015
A. Tewari, *Aeroservoelasticity*, Control Engineering,
DOI 10.1007/978-1-4939-2368-7_6

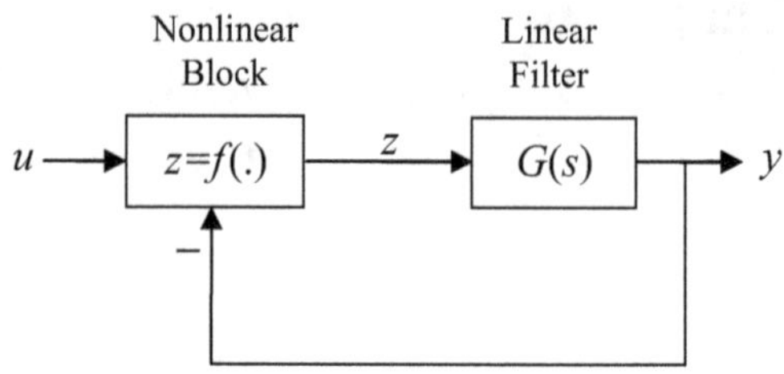

Fig. 6.1 Nonlinear system representation for describing function approximation

$$z = \sum_{k=1}^{\infty} a_k \cos{(k\omega t)} + b_k \sin{(k\omega t)}, \tag{6.2}$$

where

$$a_k = \frac{1}{\pi} \int_{-\pi}^{\pi} z(t) \cos{(k\omega t)} \mathrm{d}(\omega t), \tag{6.3}$$

$$b_k = \frac{1}{\pi} \int_{-\pi}^{\pi} z(t) \sin{(k\omega t)} \mathrm{d}(\omega t), \tag{6.4}$$

of which all the higher harmonics except the fundamental one are suppressed by the linear transfer function acting as a low-pass filter in a feedback loop[1]. Consequently, the output signal can be approximated by only the fundamental harmonic, which is amplified and phase-lagged compared to the input signal, and given by

$$y = a_1 \cos{(\omega t)} + b_1 \sin{(\omega t)} = y_0 \sin{(\omega t + \phi)}. \tag{6.5}$$

Due to the presence of the nonlinear block, the amplification ratio, y_0/u_0, and the phase angle, ϕ, are functions of both the forcing frequency, ω, and input magnitude, u_0. Therefore, the nonlinear element is effectively modeled as the following describing function:

$$N(u_0, \omega) = \frac{y_0 e^{j(\omega t + \phi)}}{u_0 e^{j\omega t}} = \frac{y_0}{u_0} e^{j\phi}. \tag{6.6}$$

In the multivariable case of m inputs and n outputs, the describing functions matrix, $\mathbf{N}$, can be defined as follows:

$$\mathbf{y}_0 = \mathbf{N}(\mathbf{u}_0, \omega) e^{j\Phi} \mathbf{u}_0, \tag{6.7}$$

---

[1] Here, the nonlinearity is assumed to be odd, i.e.,

$$0 = \int_{-\pi}^{\pi} z(t) \mathrm{d}(\omega t),$$

which is a good assumption for most systems.

where

$$
e^{j\phi} = \begin{pmatrix}
e^{j\phi_{11}} & e^{j\phi_{12}} & \cdots & e^{j\phi_{1m}} \\
e^{j\phi_{21}} & e^{j\phi_{22}} & \cdots & e^{j\phi_{2m}} \\
\cdots & \cdots & \cdots & \cdots \\
e^{j\phi_{n1}} & e^{j\phi_{n2}} & \cdots & e^{j\phi_{nm}}
\end{pmatrix}. \tag{6.8}
$$

The typical aeroservoelastic application involves a regulation problem in which an aeroelastic system must be stabilized. In such a case, if the aeroelastic behavior is nonlinear due to structural, aerodynamic, or control nonlinearities, the regulated system can be generally represented[2] as the block diagram of Fig. 6.2. Here, a linear aeroelastic system with a state-space representation given by

$$
\dot{\mathbf{x}} = \mathbf{A}\mathbf{x} + \mathbf{B}\mathbf{u}, \tag{6.9}
$$

is to be stabilized in the presence of structural nonlinearities, aerodynamic nonlinearities created by either flow separation, or shock waves, and controller nonlinearities, all of which are collectively represented in a nonlinear functional form by the block

$$
\mathbf{z} = \mathbf{f}(\mathbf{x}, \mathbf{u}), \tag{6.10}
$$

while a linear state-feedback regulator,

$$
\mathbf{u} = -\mathbf{K}\mathbf{x}, \tag{6.11}
$$

closes the loop. Clearly, the linear plant plus the regulator act as a stable, linear filter in closed loop with the nonlinear block. Therefore, the requirements of the describing function approximation are satisfied by such ASE systems. When the nonlinearities are large, the controller parameters, $\mathbf{K}$, can be suitably adjusted with respect to $\mathbf{x}$, by an adaptation mechanism, based upon an online estimation of nonlinear behavior of the plant by suitable describing functions.

In the present chapter, three examples of aeroservoelastic systems with nonlinear unsteady aerodynamic behavior are considered. These are: (a) flapping-wing flight for the incompressible case, (b) shock-induced buffett in the transonic regime, and (c) transonic flutter. Since both applications contain a linear aeroelastic subsystem, this can be regarded as the linear postfilter, $G(s)$, of Fig. 6.1 in a negative feedback loop with the unsteady aerodynamic nonlinear block, and a linear, stabilizing feedback controller.

---

[2] The state-feedback, linear regulator can be easily replaced by a linear observer-based compensator, if the states are not directly measurable (Chap. 5).

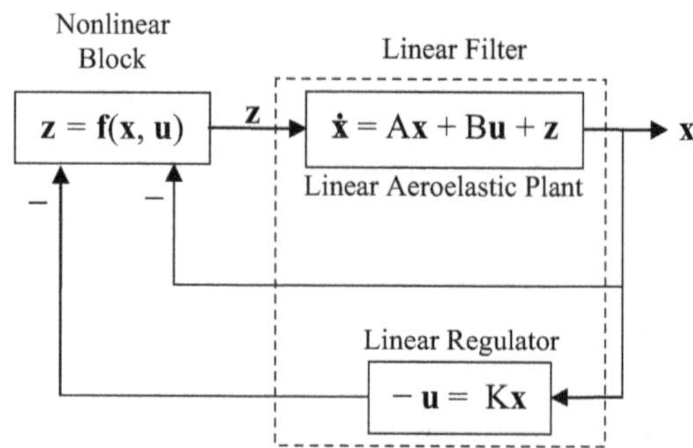

Fig. 6.2 Nonlinear aeroservoelastic system representation for describing function approximation

6.3 Flapping-Wing Flight

The design of modern airplanes became possible when people finally abandoned the concept of bird-like flapping-wing flight in favor of separate mechanisms for propulsion and lift generation. George Cayley proposed this conceptual breakthrough in a coin engraving c. 1799, which showed lift and drag on a fixed wing. Such a wing must be propelled forward by an engine in order to overcome the drag. These ideas, almost a century later, led to the first successful controlled flight of a heavier than air machine (Wright Flyer 1903). While modern aviation was thus founded on fixed wings, it never really approached the seemingly effortless and graceful flight of the birds. Airplanes need large airports to operate from and create massive noise pollution in their wake. Thus, the desire of taking-off and landing from one's own backyard and soaring majestically into the skies like birds has always been a dream, as well as an area of active research.

Emulating the bird flight through flapping, pitching, spanning, and morphing motions requires highly flexible wings over which the local flow can be controlled very effectively by relative motion of structural parts. While such a control is provided by several powerful muscles in a bird, it can be artificially produced by motors (linear and rotary) attached to the structure at discrete points. The structure should be capable of many different degrees of freedom through a set of articulated joints actuated by the motors. Such a highly actuated structure can be modeled by a lumped parameter approach using rigid elements connected by flexible joints.

The simplest structural model for flapping-wing flight is a two-dimensional airfoil suspended by linear and torsional springs representing lumped stiffnesses in the heave, $h(t)$, and pitch, $\theta(t)$, degrees of freedom. A vertical force, $F(t)$, excites the flapping (heaving) motion, while the rotation about the pitch axis can either be excited by a motor, or by a trailing-edge control surface with deflection, $\beta(t)$, as shown in Fig. 6.3; in this case, an actuator of torque output $H(t)$ is required about the control surface hinge line. The coupled equations of linear structural vibration (Chap. 2) can be written as follows:

$$\mathbf{M\ddot{q}} + \mathbf{Kq} = \mathbf{Q}_a + (F, 0, H)^T, \tag{6.12}$$

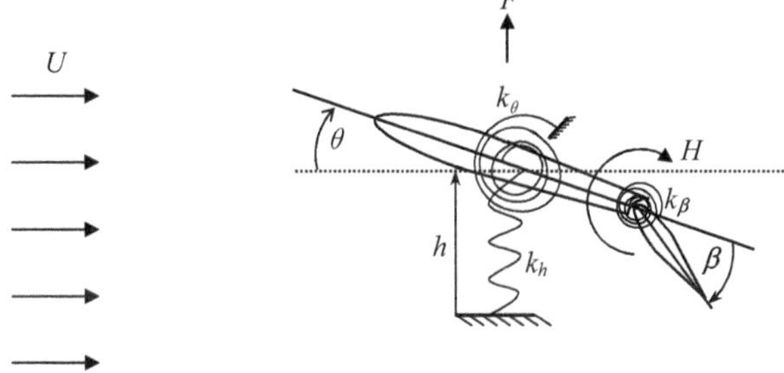

Fig. 6.3 Structural model for a simple flapping-wing system with a control surface

where $\mathbf{q} = (h, \theta, \beta)^T$ is the generalized coordinates vector, and M, K are the following generalized mass and stiffness matrices of the structure:

$$\mathsf{M} = \begin{pmatrix} m & mx_\theta & m_c x_\beta \\ mx_\theta & I_\theta & m_c x_c x_\beta \\ m_c x_\beta & m_c x_c x_\beta & I_\beta \end{pmatrix} \tag{6.13}$$

$$\mathsf{K} = \begin{pmatrix} k_h & 0 & 0 \\ 0 & k_\theta & 0 \\ 0 & 0 & k_\beta \end{pmatrix},$$

with m being the wing's mass per unit span, I_θ its moment of inertia per unit span about the pitch axis, x_θ the distance of wing's center of mass aft of the pitch axis, m_c the mass per unit span of the control surface, x_c the distance of control surface's hinge line aft of the pitch axis, I_β the control surface's moment of inertia per unit span about its hinge line, and x_β the distance of control surface's center of mass aft of the hinge line. Here, $\mathbf{Q}_a(t)$ is the generalized aerodynamic force vector, assumed to be linearly related to the generalized coordinates and their time derivatives, as well as to additional state variables called the *aerodynamic lag state* vector, $\mathbf{x}_a(t)$, required for modeling the influence of a circulatory wake (Chap. 3). Thus, we write

$$\mathbf{Q}_a = \mathsf{M}_a \ddot{\mathbf{q}} + \mathsf{C}_a \dot{\mathbf{q}} + \mathsf{K}_a \mathbf{q} + \mathsf{N}_a \mathbf{x}_a , \tag{6.14}$$

where $\mathsf{M}_a, \mathsf{C}_a$, and K_a are the generalized aerodynamic inertia, aerodynamic damping, and aerodynamic stiffness matrices, respectively, containing both circulatory and noncirculatory effects, and N_a is the *aerodynamic lag* coefficient matrix associated with the circulatory wake. In continuation with the linear aeroelastic model, the aerodynamic lag states are assumed to be governed by the following linear state equations driven by the generalized coordinates and their rates:

$$\dot{\mathbf{x}}_a = \mathsf{F}_a \mathbf{x}_a + \Gamma_a \begin{Bmatrix} \mathbf{q} \\ \dot{\mathbf{q}} \end{Bmatrix}, \tag{6.15}$$

where F_a and Γ_a are the aerodynamic coefficient matrices corresponding to circulatory wake effects. Furthermore, the aeroelastic equations of motion, Eqs. (6.12) and (6.14) can be combined to yield

$$(M - M_a)\ddot{q} - C_a\dot{q} + (K - K_a)q = N_a x_a + (F, 0, H)^T. \tag{6.16}$$

Hence, the linear aeroelastic system has the following state-space representation:

$$\dot{x} = Ax + Bu, \tag{6.17}$$

where

$$A = \begin{pmatrix} 0 & I & 0 \\ -\bar{M}^{-1}\bar{K} & \bar{M}^{-1}C_a & -\bar{M}^{-1}N_a \\ \Gamma_a & & F_a \end{pmatrix}, \tag{6.18}$$

and

$$B = \begin{pmatrix} 0 \\ I \\ 0 \end{pmatrix}, \tag{6.19}$$

where $\bar{M} = M - M_a$, $\bar{K} = K - K_a$,

$$x = \begin{Bmatrix} q \\ \dot{q} \\ x_a \end{Bmatrix}, \tag{6.20}$$

is the *augmented state vector*, and $u = (F, 0, H)^T$, the control input vector.

The aerodynamic damping term occurring on the left-hand side of Eq. (6.16), $-C_a\dot{q}$, offers an important insight into the aeroelastic motion. For an unforced motion ($u = 0$), the energy removed from the system by aerodynamic damping results in a stable motion where both $q(t)$ and $x_a(t)$ tend to zeros in the steady limit, $t \to \infty$. On the other hand, the aerodynamic energy input to the system by the circulatory wake due the term on the right-hand side of Eq. (6.16), $N_a x_a$, can exceed the damping term under certain conditions, resulting in aeroelastic instability. At the limiting case of flutter, this linearized inviscid model produces a self-sustained oscillation, where the energy dissipated by damping is exactly compensated by that fed into the system by a circulatory wake, in which case no external inputs are necessary for maintaining a constant amplitude. However, if the motion amplitude is large, there is a breakdown of aerodynamic linearity, and the nonlinear viscous aerodynamics can again generate a limit-cycle oscillation at a certain flight condition, which does not need an external forcing. Thus the aerodynamic forces and moments due to the flapping motion require either an addition, or removal of energy by the force, $F(t)$,

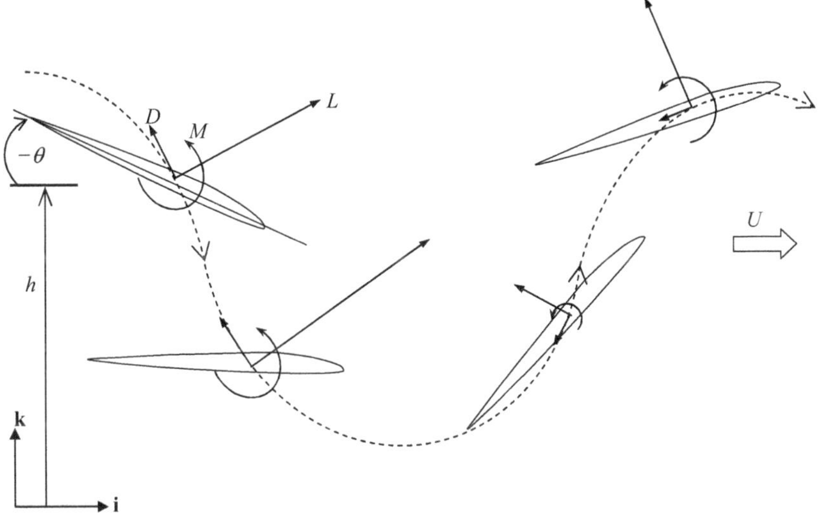

Fig. 6.4 Schematic diagram showing the flapping and pitching of an airfoil

and torque, $H(t)$, applied as external (control) inputs, in order to maintain a constant amplitude of vibration. Consequently, flapping-wing flight control systems should be designed to not only maximize the net lift and thrust per cycle, but also the smallest input magnitudes for maintaining a nearly self-sustained oscillation.

6.3.1 Lift and Thrust for Flapping Flight

The main objective of flapping-wing flight is to combine both lift and thrust generation into a single mechanism. The success of bird-like flight depends upon the efficiency of such a mechanism. In order to analyze a flapping wing, consider the simple harmonic pitch and heave motions of a thin airfoil shown in Fig. 6.4. The sectional lift, drag, and pitching moment of the airfoil per unit span are taken at the center of mass of the wing (Fig. 6.4), and defined with respect to the wing's trajectory. The aircraft is moving forward at an instantaneous speed, $U(t)$, while the wing is forced to pitch, $\theta(t)$, and heave, $h(t)$, relative to the flight direction. With reference to a right-handed body frame, $(\mathbf{i}, \mathbf{j}, \mathbf{k})$, fixed to the aircraft with $\mathbf{i}, \mathbf{k}$ denoting the plane of flight, the net velocity of the wing's center of mass is expressed as follows:

$$\mathbf{v} = U\mathbf{i} + \dot{h}\mathbf{k}, \tag{6.21}$$

and the angular velocity of the wing relative to the aircraft is given by

$$\omega = -\dot{\theta}\mathbf{j}. \tag{6.22}$$

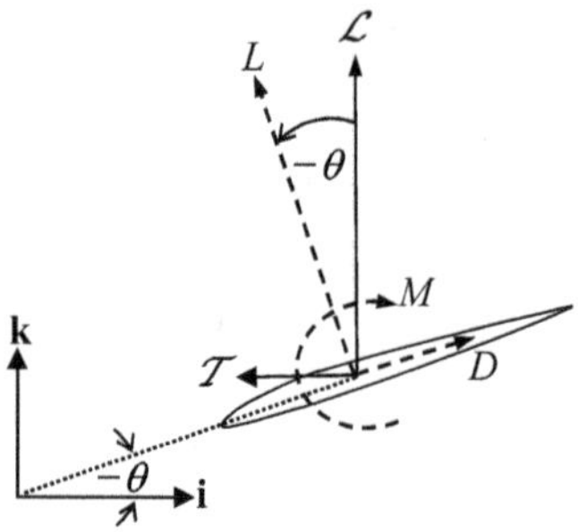

Fig. 6.5 Lift and thrust created by flapping and pitching of an airfoil

The instantaneous angle of attack of the wing is the following:

$$\alpha = \theta - \tan^{-1}\left(\frac{\dot{h}}{U}\right), \tag{6.23}$$

and the net upwash (relative flow component normal to the airfoil) due to the unsteady motion is given by

$$w = U\sin\theta - \dot{h}\cos\theta. \tag{6.24}$$

Aerodynamic relationships determine the local lift and drag as functions of the relative flow speed,

$$v = \sqrt{U^2 + \dot{h}^2}, \tag{6.25}$$

and either the angle of attack, α, or the upwash, w. An efficient fixed-wing operation requires that the angle of attack must be always close to the optimum value, which yields the largest lift-to-drag ratio. For a thin airfoil, this optimum angle of attack is quite small, therefore linearized aerodynamic theory is applicable for fixed wings. However, in case of flapping-wing flight, the instantaneous direction of flight of the aircraft, $\mathbf{i}$, is not the same as that of the wing. Therefore, the local instantaneous lift and drag per unit span of the wing must be resolved in the net flight direction, as shown in Fig. 6.5. Consequently, the net lift per unit span normal to the flight direction is given by

$$\mathcal{L} = L\cos\theta - D\sin\theta, \tag{6.26}$$

and the net thrust per unit span is the following:

$$\mathcal{T} = -L\sin\theta - D\cos\theta. \tag{6.27}$$

From Eqs. (6.26) and (6.27), it is clear that even if the airfoil always operatedat its optimum, local angle of attack, the net thrust over a flapping cycle would be either zero or negative, and the net lift-to-drag ratio would be far from its optimum value. If the pitch angle were constant during a flapping cycle, the downstroke ($\dot{h} < 0$) would increase the local angle of attack, α, thereby increasing L and creating a larger net lift for a thin wing ($L >> D$), but a reduction in net thrust. However, if the angle of attack is increased beyond a certain limit, it leads to the stalling of the airfoil, in which case $L \simeq 0$, and aerodynamic efficiency becomes zero. Similarly, with a constant value of θ, the upstroke ($\dot{h} > 0$) creates a reduction in the angle of attack and local lift magnitude L, thereby producing a smaller net lift and thrust. If the constant value of θ is either too large, or too small, the net lift and thrust for a complete flapping cycle would be either zero or negative. Therefore, a loss of aerodynamic efficiency (L/D) results whenever the pitch angle is maintained constant during a flapping cycle. As indicated in Fig. 6.4, the best case scenario is obtained if θ is negative during a downstroke, with a continuous reduction in its magnitude until reaching the lowest point, after which it must be continuously increased to a positive value corresponding to zero (or slightly negative) local lift angle of attack, $L \simeq 0$, during the upstroke. From this discussion, it is clear that the best aerodynamic efficiency can be achieved by suitably adjusting the amplitudes and phase angles of the heaving and pitching motions in a flapping oscillation.

In order to derive a functional aerodynamic model, consider, for example, a simple harmonic oscillation in heave, $h(t)$, driven by a motor about the flapping axis[3] such that

$$h = h_0 \cos{(\omega t)}, \tag{6.28}$$

where ω is the forcing frequency. Since the flapping could generate large angles of attack at which flow separation can cause nonlinear aerodynamic behavior, a nonlinear analysis is necessary. The forced pitching motion, $\theta(t)$, is assumed to involve only the fundamental harmonic of the driving frequency by the describing function approximation, and is given by:

$$\theta = \theta_0 \cos{(\omega t + \phi)}, \tag{6.29}$$

where $\phi(\omega, h_0)$, the phase angle of pitching motion relative to the heaving motion, is a function of the forcing frequency and amplitude, and the amplitudes $h_0, \theta_0(\omega, h_0)$ need not be small. The lift and thrust per unit span, averaged over a complete flapping cycle are given by

$$\bar{\mathcal{L}} = \frac{\omega}{2\pi} \int_0^{2\pi/\omega} (L \cos\theta - D \sin\theta)\, d\tau, \tag{6.30}$$

and

$$\bar{\mathcal{T}} = -\int_0^{2\pi/\omega} (L \sin\theta + D \cos\theta)\, d\tau. \tag{6.31}$$

The flapping frequency, ω, and amplitudes, h_0, dictate the values of local lift, drag, and pitching moment per unit span of the wing, and hence the average lift

---

[3] Due to the aerodynamic and inertial coupling, either of the two degrees of freedom can be considered to be the input for driving the other degree of freedom.

and thrust per cycle. Thus an unsteady aerodynamic model is necessary for relating the two-dimensional (sectional) lift, drag, and pitching moment to the flight speed, U, airfoil chord, c, viscosity coefficient, μ, atmospheric density, ρ (assumed constant), oscillation frequency, ω, and the heave amplitude by the following nonlinear functions:

$$
\begin{aligned}
L &= L(U, c, \mu, \rho, \omega, h_0), \\
D &= D(U, c, \mu, \rho, \omega, h_0), \\
M &= M(U, c, \mu, \rho, \omega, h_0).
\end{aligned}
\tag{6.32}
$$

By a dimensional analysis, it can be shown that the total number of nondimensional parameters upon which the sectional aerodynamic coefficients depend are the number of dependent variables (6), minus the number of independent (mass, length, time) dimensions, which leads to $6 - 3 = 3$ dimensionless variables. The resulting nondimensional relationships can thus be expressed as follows:

$$
\begin{aligned}
C_L &= \frac{L}{\frac{1}{2}\rho U^2 c} = C_L(Re, k, \hat{h}_0), \\
C_D &= \frac{D}{\frac{1}{2}\rho U^2 c} = C_D(Re, k, \hat{h}_0), \\
C_m &= \frac{M}{\frac{1}{2}\rho U^2 c^2} = C_m(Re, k, \hat{h}_0).
\end{aligned}
\tag{6.33}
$$

where the three aerodynamic parameters are the nondimensional heave amplitude, $\hat{h}_0 = h_0/c$, the Reynolds number,

$$
Re = \frac{\rho U c}{\mu},
\tag{6.34}
$$

and the reduced frequency (or Strouhal number),

$$
k = \frac{\omega c}{U}.
\tag{6.35}
$$

An accurate aerodynamic model for flapping wing—being viscous and nonlinear in nature due to separated flow at both leading and trailing edges—would require the solution of unsteady Navier-Stokes equations (Chap. 3). Such computations have been carried out by several researchers, such as Wang [190] and Zhu and Peng [198]. Less rigorous computational fluid dynamics (CFD) solutions of potential flows with leading-edge separation are also available in literature [125]. However, being time consuming and resource intensive in nature, it is not envisaged that a CFD model can be adopted for online aeroservoelastic computations. Instead, online systems identification of describing functions appears to be more suitable. For example, measurement of the normal acceleration as an output variable can produce an empirical model expressed by Eq. (6.33). To this end, we write the total acceleration of the wing's center of mass as follows:

$$\mathbf{a} = \frac{d\mathbf{v}}{dt} = \dot{U}\mathbf{i} + \ddot{h}\mathbf{k} - \left(\dot{\theta}\mathbf{j}\right) \times \left(U\mathbf{i} + \dot{h}\mathbf{k}\right)$$

$$= \left(\dot{U} - \dot{h}\dot{\theta}\right)\mathbf{i} + \left(\ddot{h} + U\dot{\theta}\right)\mathbf{k}, \tag{6.36}$$

and substitute Eq. (6.28) in order to yield the normal acceleration output given by

$$y = \ddot{h} + U\dot{\theta}$$

$$= -\omega^2 h_0 \cos\left(\omega t\right) - U\omega\theta_0 \sin\left(\omega t + \phi\right), \tag{6.37}$$

which can be rendered nondimensional as follows:

$$\hat{y} = \frac{yc}{U^2} = -k^2 \hat{h}_0 \cos\left(k\tau\right) - k\theta_0 \sin\left(k\tau + \phi\right), \tag{6.38}$$

where $\tau = tU/c$ is the nondimensional time.

The unknown pitching amplitude, θ_0, is a function of both reduced frequency and the heave amplitude, $\hat{h}_0$, and can be determined from the normal acceleration as follows:

$$\theta_0(\tau) = -\frac{k^2 \hat{h}_0 \cos\left(k\tau\right) + \hat{y}_0(\tau)}{k \sin\left(k\tau + \phi\right)}, \tag{6.39}$$

where the phase angle, $\phi(\tau)$, describes the time lag between the heaving and pitching degrees of freedom in a forced oscillation, and is therefore an important aeroelastic parameter. Since it is a function of the forcing frequency and motion amplitudes, the unknown phase angle, $\phi(\tau)$, can be determined from the measurement of the output, given a forcing condition (input), $k, \theta_0(\tau), \hat{h}_0$. A finite record of the output taken at specific time instants, $\tau_k, k = 1, 2, \ldots, n$ can be expressed in the following vector form:

$$\hat{\mathbf{y}} = \left\{\begin{array}{c} \hat{y}(\tau_1) \\ \hat{y}(\tau_2) \\ \vdots \\ \hat{y}(\tau_n) \end{array}\right\} = -k^2 \hat{h}_0 \left\{\begin{array}{c} \cos\left(k\tau_1\right) \\ \cos\left(k\tau_2\right) \\ \vdots \\ \cos\left(k\tau_n\right) \end{array}\right\} - k \left\{\begin{array}{c} \theta_0(\tau_1) \sin\left[k\tau_1 + \phi(\tau_1)\right] \\ \theta_0(\tau_2) \sin\left[k\tau_2 + \phi(\tau_2)\right] \\ \vdots \\ \theta_0(\tau_n) \sin\left[k\tau_n + \phi(\tau_n)\right] \end{array}\right\}, \tag{6.40}$$

which is solved for the vector

$$\mathbf{x} = \left\{\begin{array}{c} \theta_0(\tau_1) \sin\left[k\tau_1 + \phi(\tau_1)\right] \\ \theta_0(\tau_2) \sin\left[k\tau_2 + \phi(\tau_2)\right] \\ \vdots \\ \theta_0(\tau_n) \sin\left[k\tau_n + \phi(\tau_n)\right] \end{array}\right\} = -\frac{1}{k}\hat{\mathbf{y}} - k\hat{h}_0 \left\{\begin{array}{c} \cos\left(k\tau_1\right) \\ \cos\left(k\tau_2\right) \\ \vdots \\ \cos\left(k\tau_n\right) \end{array}\right\}. \tag{6.41}$$

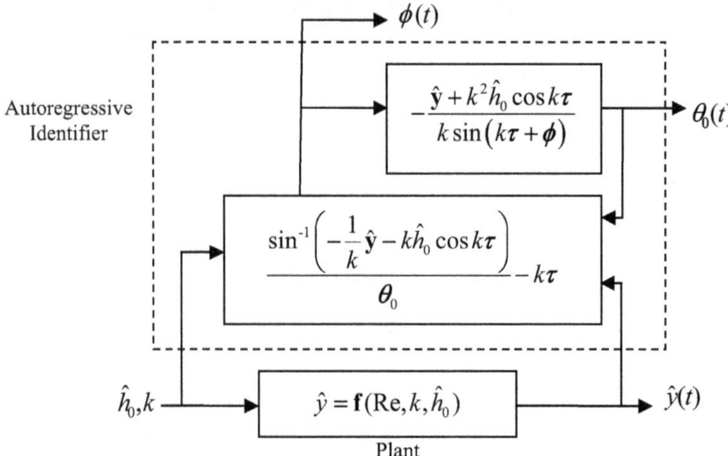

Fig. 6.6 Schematic diagram for determining the pitch amplitude and heave-pitch phase lag from measurement of the normal acceleration output at discrete time instants

Finally, the unknown phase angle is determined as the following vector at the given discrete time instants:

$$\phi = \begin{Bmatrix} \phi(\tau_1) \\ \phi(\tau_2) \\ \vdots \\ \phi(\tau_n) \end{Bmatrix} = \sin^{-1} \mathbf{x}/\theta_0 - k \begin{Bmatrix} \tau_1 \\ \tau_2 \\ \vdots \\ \tau_n \end{Bmatrix} = \sin^{-1} \mathbf{x}/\theta_0 - k\tau, \qquad (6.42)$$

where

$$\theta_0 = \begin{Bmatrix} \theta_0(\tau_1) \\ \theta_0(\tau_2) \\ \vdots \\ \theta_0(\tau_n) \end{Bmatrix}, \qquad (6.43)$$

and $(.)/(.)$ denotes element-wise division of two vectors. Since neither the phase nor the amplitude of pitching motion is known in advance, a recursive algorithm must be used for their determination, employing Eqs. (6.39)–(6.42) until convergence within a specified tolerance is obtained. Figure 6.6 depicts the recursive identification scheme for the amplitude and the phase lag based upon the normal acceleration output.

In order to develop an unsteady aerodynamic model, consider the net lift developed by the forced heaving and pitching oscillations as the force normal to the flight path, and described by the fundamental harmonic of the flapping frequency:

$$\mathcal{L}(t) = my = m\left(\ddot{h} + U\dot{\theta}\right)$$

$$= -m\omega^2 h_0 \cos(\omega t) - mU\omega\theta_0 \sin(\omega t + \phi) \tag{6.44}$$

$$= \mathcal{L}_0 \cos(\omega t + \psi), \tag{6.45}$$

where $\mathcal{L}_0(\omega, h_0)$ is the lift magnitude and $\psi(\omega, h_0)$ the lift phase angle. It is to be noted that the nonlinear aerodynamic behavior requires that the lift magnitude and phase be functions of the forcing amplitudes as well as the frequency. Since the pitching magnitude, θ_0, and phase, ϕ, profiles have already been identified from the normal acceleration data, it is now a simple matter to evaluate the lift amplitude and phase by the following recursive formulae:

$$\mathcal{L}_0(t) = -m\omega\frac{[\omega h_0 \cos(\omega t) + U\theta_0(t)\sin(\omega t + \phi(t))]}{\cos(\omega t + \psi(t))}, \tag{6.46}$$

$$\mathbf{x} = \left\{\begin{array}{c} \mathcal{L}_0(t_1)\cos[\omega t_1 + \psi(t_1)] \\[4pt] \mathcal{L}_0(t_2)\cos[\omega t_2 + \psi(t_2)] \\ \vdots \\ \mathcal{L}_0(t_n)\cos[\omega t_n + \psi(t_n)] \end{array}\right\}$$

$$= -m\omega^2 h_0 \left\{\begin{array}{c} \cos(\omega t_1) \\ \cos(\omega t_2) \\ \vdots \\ \cos(\omega t_n) \end{array}\right\} - mU\omega \left\{\begin{array}{c} \theta_0(t_1)\sin[\omega t_1 + \phi(t_1)] \\ \theta_0(t_2)\sin[\omega t_2 + \phi(t_2)] \\ \vdots \\ \theta_0(t_n)\sin[\omega t_n + \phi(t_n)] \end{array}\right\} \tag{6.47}$$

$$\psi = \left\{\begin{array}{c} \psi(t_1) \\ \psi(t_2) \\ \vdots \\ \psi(t_n) \end{array}\right\} = \cos^{-1}\mathbf{x}/\mathcal{L}_0 - \omega\left\{\begin{array}{c} t_1 \\ t_2 \\ \vdots \\ t_n \end{array}\right\} = \cos^{-1}\mathbf{x}/\mathcal{L}_0 - \omega t, \tag{6.48}$$

Fig. 6.7 Block diagram of an autoregressive nonlinear aerodynamic model for flapping flight by describing functions for lift and thrust

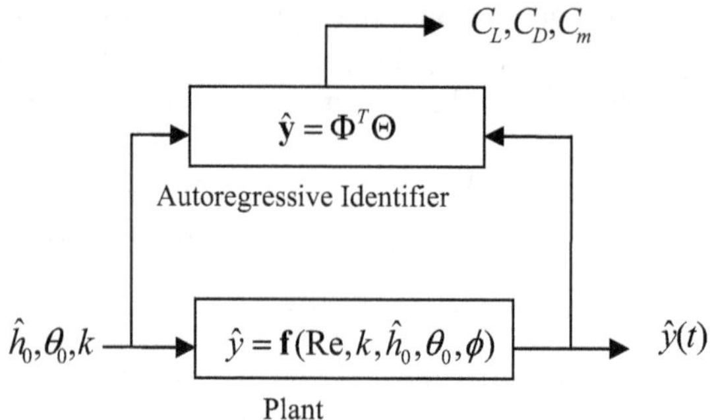

where

$$\mathcal{L}_0 = \left\{ \begin{array}{c} \mathcal{L}_0(t_1) \\ \mathcal{L}_0(t_2) \\ \vdots \\ \mathcal{L}_0(t_n) \end{array} \right\}. \tag{6.49}$$

The determination of the thrust magnitude and phase can be carried out in a similar manner by an autoregressive scheme. Figure 6.7 shows the recursive identification scheme for the lift and thrust based upon the normal acceleration output.

The net acceleration of the wing's center of mass can be derived in an inertial reference frame [169], and substituted into the aircraft's translational dynamics equation (Newton's second law). Similarly, the pitching dynamics of the aircraft (which is inherently coupled with the translational motion in the nonlinear case) can also be represented [169]. Finally, optimal feedback control laws can be derived [170] based upon the lift and pitching moment identified by the autoregressive scheme given above, in order to generate the maximum possible lift and thrust per cycle. This appears to be a promising research topic for the future.

The flapping-wing flight model can be extended to other nonlinear systems, for which controller is to be designed. A general adaptation scheme based upon the autoregressive identification of plant characteristics in a closed loop is depicted by a schematic diagram in Fig. 6.8. Note the presence of an outer adaptation loop which can adjust the controller gains based upon a changing plant model. Design of such adaptation laws can be carried out to ensure the stability of the overall adaptive control system [10]. Such adaptive techniques are very promising for a wide range of ASE applications, including those briefly discussed below.

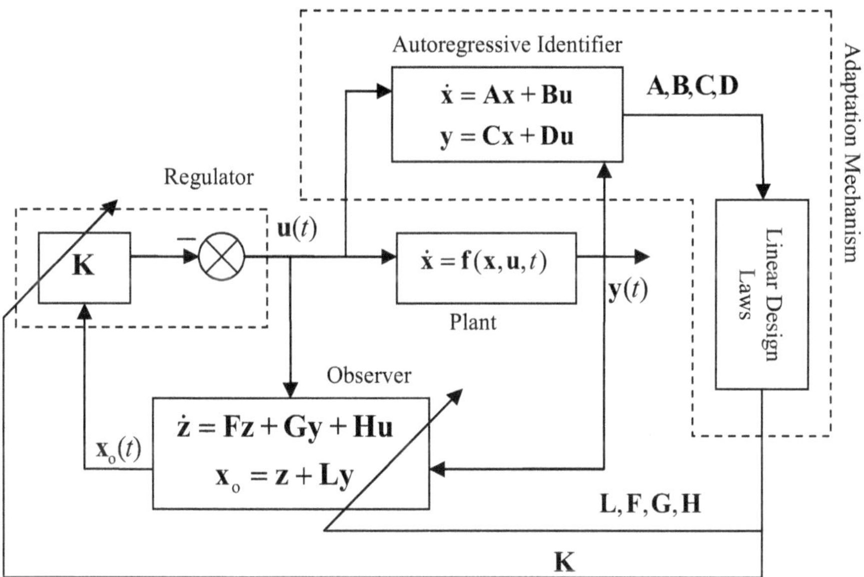

Fig. 6.8 Schematic block diagram of an adaptive control system based upon autoregressive identification of an uncertain plant

6.4 Shock-Induced Buffet

The transonic flight regime is home to several hazardous phenomena due to the presence of nearly normal shock waves on a wing's upper and lower surfaces. The shock waves are prone to self-induced oscillations in their locations, whereby unsteady pressure fluctuations can be amplified by interaction with structural vibration and rigid-body modes. When such oscillations assume larger amplitudes, they cause periodic strengthening of the shocks, resulting in cyclic separation and reattachment of boundary layers. The aeroelastic coupling in such a case can produce limit-cycle buffeting motions that are potentially disastrous to fighter type aircraft maneuvering at transonic speeds. Even high-subsonic airline transports can encounter shock-induced buffeting of wings and tails while cruising in gusty conditions.

Experimental research into transonic buffet [107] reveals two basic types of buffeting of wing-like structures: (a) shock-induced buffet of moderately swept wings and tails, and (b) vortex-induced buffet of highly swept, low-aspect-ratio wings. Type (a) buffet has a rich frequency spectrum, capable of exciting several structural modes. However, Type (b) buffet is usually of a single frequency due to its production by a strong leading edge vortex, whose strength is not modified appreciably by the presence of shock waves. Both types of buffeting, however, require nonlinear, viscous aerodynamic modeling due to separated flows. It appears that the single frequency peak of Type (b) buffet can be filtered out by a linear controller, whose gain is suitably adjusted with flight parameters (Mach number, altitude, and "g"-loading). Thus the gain-scheduling approach would be applicable here. Alternatively, a describing function model can be applied. However, due to the multimodal response of Type (a) buffet, an online identification of the aerodynamic parameters would be necessary. This can be achieved either by an autoregressive scheme presented above for the flapping-wing flight, or by Lyapunov-based model-reference adaptive control, discussed below by an illustrative example.

6.5   Transonic Flutter

Presence of shock waves on an aircraft wing operating in the transonic regime causes a hazardous two-fold flutter: (a) a dip in the flutter dynamic pressure, and (b) stall flutter due to shock-induced flow separation. The earliest encounters with transonic flutter were encountered toward the end of World-War II by fighter type aircraft in steep dives. The phenomena of flutter dip and stall flutter are illustrated by the experimental data in Figs. 6.9 and 6.10, respectively, obtained by Rivera et al. [141] on a rectangular planform with NACA 0012 airfoil, suspended by a flexible pitch and plunge apparatus in NASA Langley's Transonic Dynamics Tunnel. Figure 6.9 is based upon the data of this experiment, and shows that the classical flutter mechanism of pitch and plunge at subsonic Mach numbers changes to pure plunge instability above $M = 0.90$ and zero angle-of-attack, causing a rapid dip of flutter dynamic pressure to less than half of its peak value at $M = 0.85$. The second hazard of a dip in the flutter boundary due to increased angle-of-attack is seen in Fig. 6.10, where separated flow due to shock waves for angle-of-attack above $\alpha = 3.5$ degree symbol results in a stall condition at transonic Mach number of 0.78. Such an increase in the angle-of-attack can happen either due to an inadvertent maneuver, or a sharp-edged atmospheric gust. For sweptback wings, the pitching moment contributions due to the tip are such that the initial rise of flutter boundary seen in Fig. 6.9 is absent [199], resulting in a steepening decline of the flutter speed until the sonic flight speed. Such a behavior is depicted graphically in Fig. 6.11 [176] for three types of wing planforms and an early attempt to predict the transonic dip by an empirical Mach correction method for torsional stiffness [33]. An aeroservoelastic design that addresses the transonic flutter problem requires a simple and adequate model of the associated flutter mechanisms for both the phenomena (a) and (b).

The classical subsonic flutter mechanism for a high-aspect-ratio wing involves the primary bending and primary torsion modes. For a typical section, these translate into the plunge (heave) and pitch modes, respectively. The coupling of the two modes is such that as the dynamic pressure is increased, the bending (plunging) natural frequency, which is initially smaller, increases, while that of the torsion (pitch) mode decreases, until they coalesce at the flutter condition. Thus at the flutter speed, the two primary modes are almost merged into a single mode, whose damping becomes negative, and energy can be extracted from the system at the flutter frequency.

At transonic speeds, the flutter mechanism changes drastically due to the following phenomena:

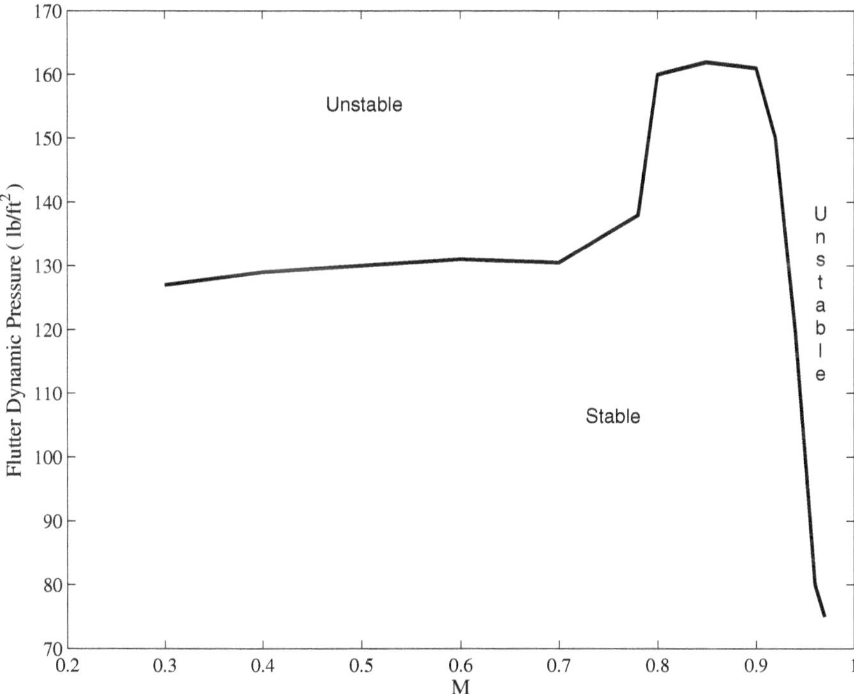

Fig. 6.9 Sudden dip in flutter boundary at transonic speeds on a NACA 0012 airfoil at zero angle-of-attack [141]

1. Increase in the quasi-steady lift-curve-slope, C_{L_α}, with Mach number until $M = 1$.
2. Aft movement of the aerodynamic center from nearly quarter-chord to mid-chord position as the speed is increased from subsonic to supersonic.

Although these aerodynamic effects partly explain why a transonic flutter dip could occur, the explanation of flutter mechanism responsible for the dip requires consideration of structural properties (stiffness and mass parameters). Zwaan [199] offers such an explanation with respect to a rectangular flat plate suspended by vertical and torsional springs.

6.5.1 Adaptive Suppression of Transonic LCO: Illustrative Example

An aeroservoelastic controller operating in transonic flight regime must necessarily be adaptive with respect to unknown parametric variations in the aerodynamic

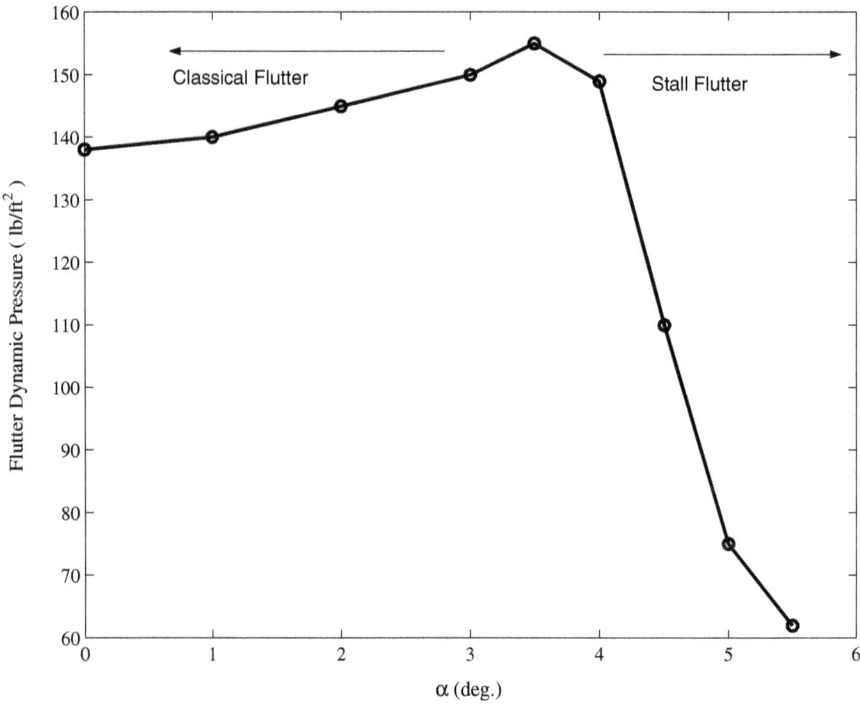

Fig. 6.10 Stall flutter at transonic speeds on a NACA 0012 airfoil at $M = 0.78$ [141]

characteristics of the plant. This renders the control law to be nonlinear, due to its dependence on operating conditions (hence state variables). There are several nonlinear control design techniques, which can be applied to such a system, such as Lyapunov-based controllers [60, 132], feedback linearization [78], sliding-mode (variable structure) control [53], model reference adaptation [10], integral back-stepping [89], etc. While a comprehensive discussion of such techniques is beyond our present approach, we will briefly consider an illustrative application of the Lyapunov-based back-stepping design to an interesting problem, where plant parameters are highly uncertain.

Consider an aircraft wing experiencing transonic limit-cycle oscillation (LCO) involving the primary torsion mode of natural frequency ω_p and damping ratio ζ_p. The unsteady aerodynamics feeding the LCO is caused by the shock-induced flow separation near the trailing edge, and can be represented by a nonlinear angular acceleration forcing term $f(\alpha, \dot{\alpha})$, where $\alpha(t)$ is the angle of attack, as follows:

$$f(\alpha) = \begin{cases} 0, & \alpha < \theta \\ \\ a\dot{\alpha}, & \alpha \geq \theta \end{cases}$$

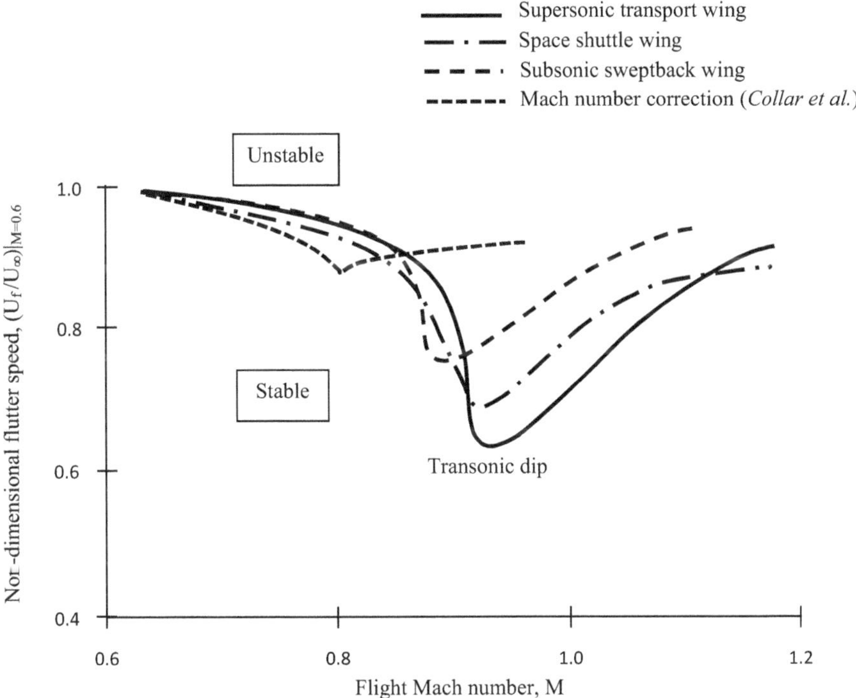

Fig. 6.11 Transonic flutter boundary dip for sweptback wings

with a, θ being unknown aerodynamic parameters.

The equation of motion of the single degree-of-freedom aeroelastic plant is the following:

$$\ddot{\alpha} + 2\omega_p \zeta_p \dot{\alpha} + \omega_p^2 \alpha = f(\alpha, \dot{\alpha}) + u,$$

where $u(t)$ is the angular acceleration control input provided by a trailing-edge control surface. The task of the ASE engineer is to suppress the LCO by designing a feedback control law,

$$u = g(\dot{\alpha}),$$

where $q = \dot{\alpha}$ is the pitch rate picked up by a rate gyro. However, the design task is complicated by the fact that the unsteady forcing parameters a, θ are unknown, and vary with the flight Mach number in an unpredictable manner.

In order to carry out an adaptive controller design, the plant state equations are expressed as follows:

$$\dot{\alpha} = q$$

$$\dot{q} = -2\omega_p \zeta_p q - \omega_p^2 \alpha + f(\alpha, \dot{\alpha}) + u,$$

and an output feedback controller is proposed,

$$u = -(k + x)q,$$

with a constant parameter $k > 0$, and an adaptive gain $x(t)$ with adaptation law

$$\dot{x} = q^2.$$

The resulting closed-loop system is of third order and is given by

$$\dot{\alpha} = q$$

$$\dot{q} = -\left[2\omega_p\zeta_p + (k + x)\right]q - \omega_p^2\alpha + f(\alpha, \dot{\alpha})$$

$$\dot{x} = q^2.$$

For stability analysis of this ASE system, let us construct the following Lyapunov function (Appendix-D):

$$V(\alpha, q, x) = \frac{1}{2}\omega_p^2\alpha^2 + \frac{1}{2}q^2 + \frac{1}{2}(x - a)^2$$

The time-derivative of V is then given by

$$\dot{V}(\alpha, q, x) = \omega_p^2\alpha\dot{\alpha} + q\dot{q} + \dot{x}(x - a),$$

or

$$\dot{V} = \omega_p^2\alpha q - \left[2\omega_p\zeta_p + (k + x)\right]q^2 - \omega_p^2\alpha q + f(\alpha, \dot{\alpha})q + q^2(x - a),$$

implying that

$$\dot{V} \leq -\left[2\omega_p\zeta_p + (k + x)\right]q^2 + aq^2 + q^2(x - a),$$

or that

$$\dot{V} \leq -\left[2\omega_p\zeta_p + k\right]q^2 < 0.$$

Since the time-derivative of the Lyapunov function is strictly negative-definite, by Lyapunov stability theorem (Appendix-D) the closed-loop adaptive ASE system is unconditionally and asymptotically stable, which implies that all initial perturbations decay to zero as $t \to \infty$. This fact is illustrated by the simulated response of the closed-loop system in Figs. 6.12, 6.13 , and 6.14, which is carried out by a Runge–Kutta algorithm [110, 111, 154] with $\omega_p = 50$ rad/s, $\zeta_p = 0.01$, $a = 2.9$, $\theta = 1.5°$, and $k = 0.1$, to initial condition $\alpha = 5°$, $q = 0.5$ rad/s.

It is important to highlight that the adaptation mechanism has regulated the ASE system without any knowledge of the unknown system parameters a, θ. This would not have been possible without such an adaptation law. Such an application could be also applied to multi-degree of freedom ASE systems, such as transonic flutter suppression and active control of aeroelastic buffeting at high angles-of-attack.

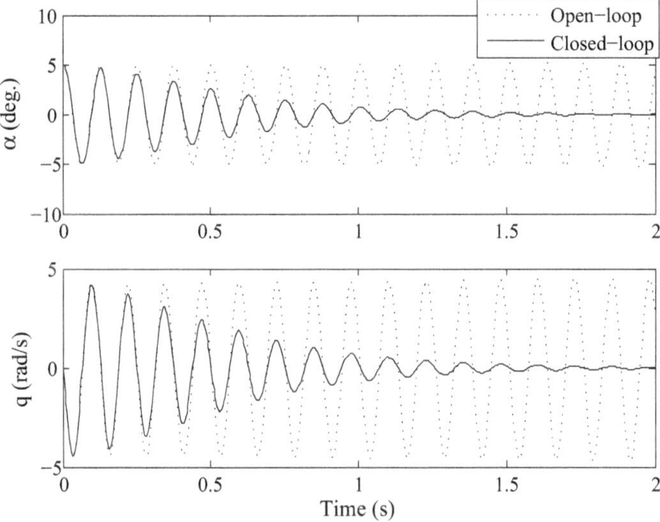

Fig. 6.12 Simulated angle of attack and pitch rate response of the adaptive aeroservoelastic system for transonic LCO suppression with initial angle of attack and pitch rate perturbation

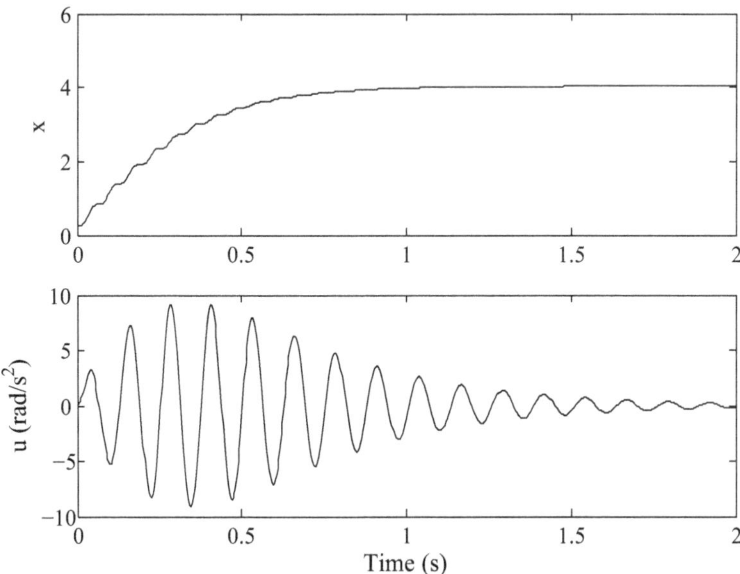

Fig. 6.13 Simulated control input of the adaptive aeroservoelastic system for transonic LCO suppression with initial angle of attack and pitch rate perturbation

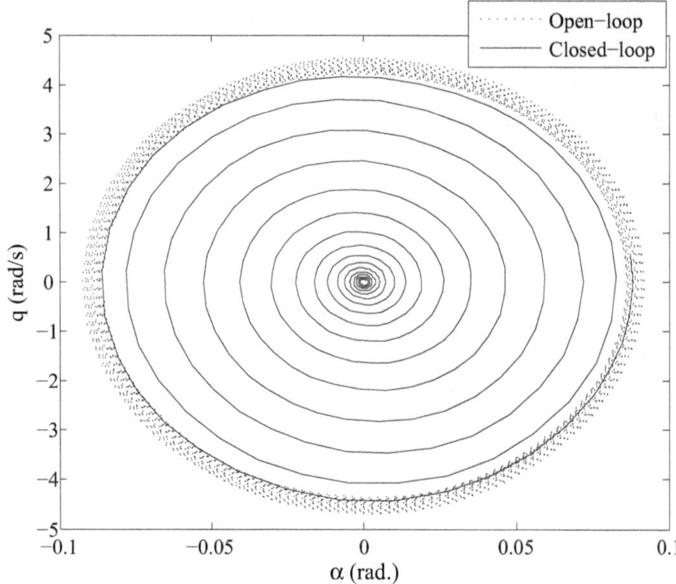

Fig. 6.14 Phase portrait of the adaptive aeroservoelastic system for transonic LCO suppression with initial angle of attack and pitch rate perturbation

Appendix-A

Listing of Mass and Stiffness Matrices

The global stiffness and mass matrices for the cantilevered, high aspect-ratio wing of Example 2.1, derived by the finite-element method with 20, two-noded, beam-shaft finite elements are listed below. The notation uses the identifier indices of the matrix element in parentheses (i, j) followed by its value immediately below.

Nonzero Elements of Global Mass Matrix

(1,1)	(1,2)	(1,3)	(1,4)	(1,41)	(1,42)	
27.3085	−0.0182	4.6858	−0.6203	18.9165	3.9498	

(2,2)	(2,3)	(2,4)	(2,41)	(2,42)		
0.2117	0.6203	−0.0787	−0.0380	0.4827		

(3,3)	(3,4)	(3,5)	(3,6)	(3,41)	(3,42)	(3,43)
26.8388	−0.0182	4.6045	−0.6096	−3.9498	17.9589	3.7469

(4,4)	(4,5)	(4,6)	(3,41)	(3,42)	(3,43)	
0.2081	0.6096	−0.0774	−0.4827	−0.0372	0.4579	

(5,5)	(5,6)	(5,7)	(5,8)	(5,42)	(5,43)	(5,44)
26.3690	4.5232	−0.5988	−0.0182	−3.7469	17.0234	3.5488

(6,6)	(6,7)	(6,8)	(6,42)	(6,43)	(6,44)	
0.2045	0.5988	−0.0760	−0.4579	−0.0363	0.4337	

(7,7)	(7,8)	(7,9)	(7,10)	(7,43)	(7,44)	(7,45)
25.8992	−0.0182	4.4419	−0.5880	−3.5488	16.1101	3.3555

© Springer Science+Business Media, LLC 2015
A. Tewari, *Aeroservoelasticity*, Control Engineering,
DOI 10.1007/978-1-4939-2368-7

(8,8)	(8,9)	(8,10)	(8,10)	(8,43)	(8,44)	(8,45)
0.2008	0.5880	−0.0746	−0.4337	−0.0354	0.4100	

(9,9)	(9,10)	(9,11)	(9,12)	(9,44)	(9,45)	(9,46)
25.4294	−0.0182	4.3606	−0.5773	−3.3555	15.2188	3.1669

(10,10)	(10,11)	(10,12)	(10,44)	(10,45)	(10,46)
0.1972	0.5773	−0.0733	−0.4100	−0.0346	0.3870

(11,11)	(11,12)	(11,13)	(11,14)	(11,45)	(11,46)	(11,47)
24.9596	−0.0182	4.2793	−0.5665	−3.1669	14.3498	2.9830

(12,12)	(12,13)	(12,14)	(12,45)	(12,46)	(12,47)
0.1935	0.5665	−0.0719	−0.3870	−0.0337	0.3645

(13,13)	(13,14)	(13,15)	(13,16)	(13,46)	(13,47)	(13,48)
24.4898	−0.0182	4.1980	−0.5558	−2.9830	13.5028	2.8039

(14,14)	(14,15)	(14,16)	(14,46)	(14,47)	(14,48)
0.1899	0.5558	−0.0705	−0.3645	−0.0328	0.3426

(15,15)	(15,16)	(15,17)	(15,18)	(15,47)	(15,48)	(15,49)
24.0201	−0.0182	4.1167	−0.5450	−2.8039	12.6780	2.6295

(16,16)	(16,17)	(16,18)	(16,47)	(16,48)	(16,49)
0.1862	0.5450	−0.0692	−0.3426	−0.0320	0.3213

(17,17)	(17,18)	(17,19)	(17,20)	(17,48)	(17,49)	(17,50)
23.5503	−0.0182	4.0354	−0.5342	−2.6295	11.8753	2.4599

(18,18)	(18,19)	(18,20)	(18,48)	(18,49)	(18,50)
0.1826	0.5342	−0.0678	−0.3213	−0.0311	0.3006

(19,19)	(19,20)	(19,21)	(19,22)	(19,49)	(19,50)	(19,51)
23.0805	−0.0182	3.9540	−0.5235	−2.4599	11.0947	2.2950

(20,20)	(20,21)	(20,22)	(20,49)	(20,50)	(20,51)
0.1790	0.5235	−0.0664	−0.3006	−0.0302	0.2805

(21,21)	(21,22)	(21,23)	(21,24)	(21,50)	(21,51)	(21,52)
22.6107	−0.0182	3.8727	−0.5127	−2.2950	10.3363	2.1348
(22,22)	(22,23)	(22,24)	(22,50)	(22,51)	(22,52)	
0.1753	0.5127	−0.0651	−0.2805	−0.0294	0.2609	
(23,23)	(23,24)	(23,25)	(23,26)	(23,51)	(23,52)	(23,53)
22.1409	−0.0182	3.7914	−0.5019	−2.1348	9.6000	1.9794
(24,24)	(24,25)	(24,26)	(24,51)	(24,52)	(24,53)	
0.1717	0.5019	−0.0637	−0.2609	−0.0285	0.2419	
(25,25)	(25,26)	(25,27)	(25,28)	(25,52)	(25,53)	(25,54)
21.6712	−0.0182	3.7101	−0.4912	−1.9794	8.8859	1.8288
(26,26)	(26,27)	(26,28)	(26,52)	(26,53)	(26,54)	
0.1680	0.4912	−0.0623	−0.2419	−0.0276	0.2235	
(27,27)	(27,28)	(27,29)	(27,30)	(27,53)	(27,54)	(27,55)
21.2014	−0.0182	3.6288	−0.4804	−1.8288	8.1938	1.6829
(28,28)	(28,29)	(28,30)	(28,53)	(28,54)	(28,55)	
0.1644	0.4804	−0.0610	−0.2235	−0.0267	0.2056	
(29,29)	(29,30)	(29,31)	(29,32)	(29,54)	(29,55)	(29,56)
20.7316	−0.0182	3.5475	−0.4696	−1.6829	7.5240	1.5417
(30,30)	(30,31)	(30,32)	(30,54)	(30,55)	(30,56)	
0.1608	0.4696	−0.0596	−0.2056	−0.0259	0.1884	
(31,31)	(31,32)	(31,33)	(31,34)	(31,55)	(31,56)	(31,57)
20.2618	−0.0182	3.4662	−0.4589	−1.5417	6.8762	1.4053
(32,32)	(32,33)	(32,34)	(32,55)	(32,56)	(32,57)	
0.1571	0.4589	−0.0582	−0.1884	−0.0250	0.1717	
(33,33)	(33,34)	(33,35)	(33,36)	(33,56)	(33,57)	(33,58)
19.7920	−0.0182	3.3849	−0.4481	−1.4053	6.2506	1.2736

(34,34)	(34,35)	(34,36)	(34,56)	(34,57)	(34,58)	
0.1535	0.4481	−0.0569	−0.1717	−0.0241	0.1556	

(35,35)	(35,36)	(35,37)	(35,38)	(35,57)	(35,58)	(35,59)
19.3223	−0.0182	3.3036	−0.4373	−1.2736	5.6471	1.1466

(36,36)	(36,37)	(36,38)	(36,57)	(36,58)	(36,59)	
0.1498	0.4373	−0.0555	−0.1556	−0.0233	0.1401	

(37,37)	(37,38)	(37,39)	(37,40)	(37,58)	(37,59)	(37,60)
18.8525	−0.0182	3.2223	−0.4266	−1.1466	5.0657	1.0244

(38,38)	(38,39)	(38,40)	(38,58)	(38,59)	(38,60)	
0.1462	0.4266	−0.0541	−0.1401	−0.0224	0.1252	

(39,39)	(39,40)	(39,59)	(39,60)	(40,40)	(40,59)	(40,60)
9.3088	−0.7219	−1.0244	2.3903	0.0722	−0.1252	−0.1878

(41,41)	(41,42)	(42,42)	(42,43)	(43,43)	(43,44)	(44,44)
57.7222	14.3137	56.7874	14.0800	55.8525	13.8463	54.9177

(44,45)	(45,45)	(45,46)	(46,46)	(46,47)	(47,47)	(47,48)
13.6126	53.9828	13.3789	53.0480	13.1451	52.1131	12.9114

(48,48)	(48,49)	(49,49)	(49,50)	(50,50)	(50,51)	(51,51)
51.1783	12.6777	50.2434	12.4440	49.3086	12.2103	48.3737

(51,52)	(52,52)	(52,53)	(53,53)	(53,54)	(54,54)	(54,55)
11.9766	47.4389	11.7429	46.5041	11.5092	45.5692	11.2754

(55,55)	(55,56)	(56,56)	(56,57)	(57,57)	(57,58)	(58,58)
44.6344	11.0417	43.6995	10.8080	42.7647	10.5743	41.8298

(58,59)	(59,59)	(59,60)	(60,60)			
10.3406	40.8950	10.1069	20.2138			

Nonzero Elements of Symmetric Global Stiffness Matrix ($\times 10^{-7}$)

(1,1)	(1,2)	(1,3)	(1,4)	(2,2)	(2,3)	(2,4)
93.0885	−0.4960	−45.6422	12.5496	9.3834	−12.5496	2.3004

(3,3)	(3,4)	(3,5)	(3,6)	(4,4)	(4,5)	(4,6)
89.4804	−0.4960	−43.8382	12.0535	9.0197	−12.0535	2.2095
(5,5)	(5,6)	(5,7)	(5,8)	(6,6)	(6,7)	(6,8)
85.8724	−0.4960	−42.0342	11.5575	8.6560	−11.5575	2.1185

(7,7)	(7,8)	(7,9)	(7,10)	(8,8)	(8,9)	(8,10)
82.2643	−0.4960	−40.2301	11.0615	8.2923	−11.0615	2.0276

(9,9)	(9,10)	(9,11)	(9,12)	(10,10)	(10,11)	(10,12)
78.6562	−0.4960	−38.4261	10.5654	7.9286	−10.5654	1.9367

(11,11)	(11,12)	(11,13)	(11,14)	(12,12)	(12,13)	(12,14)
75.0481	−0.4960	−36.6220	10.0694	7.5649	−10.0694	1.8458

(13,13)	(13,14)	(13,15)	(13,16)	(14,14)	(14,15)	(14,16)
71.4400	−0.4960	−34.8180	9.5734	7.2012	−9.5734	1.7548

(15,15)	(15,16)	(15,17)	(15,18)	(16,16)	(16,17)	(16,18)
67.8319	−0.4960	−33.0140	9.0774	6.8375	−9.0774	1.6639

(17,17)	(17,18)	(17,19)	(17,20)	(18,18)	(18,19)	(18,20)
64.2239	−0.4960	−31.2099	8.5813	6.4738	−8.5813	1.5730

(19,19)	(19,20)	(19,21)	(19,22)	(20,20)	(20,21)	(20,22)
60.6158	−0.4960	−29.4059	8.0853	6.1101	−8.0853	1.4821

(21,21)	(21,22)	(21,23)	(21,24)	(22,22)	(22,23)	(22,24)
57.0077	−0.4960	−27.6018	7.5893	5.7464	−7.5893	1.3911

(23,23)	(23,24)	(23,25)	(23,26)	(24,24)	(24,25)	(24,26)
53.3996	−0.4960	−25.7978	7.0932	5.3827	−7.0932	1.3002

(25,25)	(25,26)	(25,27)	(25,28)	(26,26)	(26,27)	(26,28)
49.7915	−0.4960	−23.9937	6.5972	5.0190	−6.5972	1.2093

(27,27)	(27,28)	(27,29)	(27,30)	(28,28)	(28,29)	(28,30)
46.1834	−0.4960	−22.1897	6.1012	4.6553	−6.1012	1.1184

(29,29)	(29,30)	(29,31)	(29,32)	(30,30)	(30,31)	(30,32)
42.5754	−0.4960	−20.3857	5.6051	4.2916	−5.6051	1.0274

(31,31)	(31,32)	(31,33)	(31,34)	(32,32)	(32,33)	(32,34)
38.9673	−0.4960	−18.5816	5.1091	3.9279	−5.1091	0.9365

(33,33)	(33,34)	(33,35)	(33,36)	(34,34)	(34,35)	(34,36)
35.3592	−0.4960	−16.7776	4.6131	3.5642	−4.6131	0.8456

(35,35)	(35,36)	(35,37)	(35,38)	(36,36)	(36,37)	(36,38)
31.7511	−0.4960	−14.9735	4.1170	3.2005	−4.1170	0.7547

(37,37)	(37,38)	(37,39)	(37,40)	(38,38)	(38,39)	(38,40)
28.1430	−0.4960	−13.1695	3.6210	2.8368	−3.6210	0.6637

(39,39)	(39,40)	(40,40)	(41,41)	(41,42)	(42,42)	(42,43)
13.1695	−3.6210	1.3275	3.2978	−1.6205	3.1842	−1.5637

(43,43)	(43,44)	(44,44)	(44,45)	(45,45)	(45,46)	(46,46)
3.0705	−1.5068	2.9568	−1.4500	2.8432	−1.3932	2.7295

(46,47)	(47,47)	(47,48)	(48,48)	(48,49)	(49,49)	(49,50)
−1.3364	2.6159	−1.2795	2.5022	−1.2227	2.3886	−1.1659

(50,50)	(50,51)	(51,51)	(51,52)	(52,52)	(52,53)	(53,53)
2.2749	−1.1090	2.1613	−1.0522	2.0476	−0.9954	1.9340

(53,54)	(54,54)	(54,55)	(55,55)	(55,56)	(56,56)	(56,57)
−0.9386	1.8203	−0.8817	1.7066	−0.8249	1.5930	−0.7681

(57,57)	(57,58)	(58,58)	(58,59)	(59,59)	(59,60)	(60,60)
1.4793	−0.7113	1.3657	−0.6544	1.2520	−0.5976	0.5976

Appendix-B

B.1 Aerodynamic Influence Coefficients for Vortex-Ring Elements

$$A_{ij} = \mathbf{v}_{ij} \cdot \mathbf{n}_j, \qquad (i = 1, \ldots, N; \ j = 1, \ldots, N) \tag{B.1}$$

$$\mathbf{v}_{ij} = \sum_{k=1}^{4} \mathbf{v}_{kij}, \qquad (i = 1, \ldots, N; \ j = 1, \ldots, N) \tag{B.2}$$

$$\mathbf{v}_{1ij} = -\frac{(\mathbf{r}_{2i} - \mathbf{r}_{1i}) \times [\, (\mathbf{r}_{1i} + \mathbf{r}_{2i})\,/2 - \mathbf{r}_j\,]}{4\pi \,|\,(\mathbf{r}_{1i} + \mathbf{r}_{2i})\,/2 - \mathbf{r}_j\,|^3} \tag{B.3}$$

$$\mathbf{v}_{2ij} = -\frac{(\mathbf{r}_{3i} - \mathbf{r}_{2i}) \times [\, (\mathbf{r}_{2i} + \mathbf{r}_{3i})\,/2 - \mathbf{r}_j\,]}{4\pi \,|\,(\mathbf{r}_{2i} + \mathbf{r}_{3i})\,/2 - \mathbf{r}_j\,|^3} \tag{B.4}$$

$$\mathbf{v}_{3ij} = -\frac{(\mathbf{r}_{4i} - \mathbf{r}_{3i}) \times [\, (\mathbf{r}_{3i} + \mathbf{r}_{4i})\,/2 - \mathbf{r}_j\,]}{4\pi \,|\,(\mathbf{r}_{3i} + \mathbf{r}_{4i})\,/2 - \mathbf{r}_j\,|^3} \tag{B.5}$$

$$\mathbf{v}_{4ij} = -\frac{(\mathbf{r}_{1i} - \mathbf{r}_{4i}) \times [\, (\mathbf{r}_{1i} + \mathbf{r}_{4i})\,/2 - \mathbf{r}_j\,]}{4\pi \,|\,(\mathbf{r}_{1i} + \mathbf{r}_{4i})\,/2 - \mathbf{r}_j\,|^3} \tag{B.6}$$

The subscript i identifies the vortex-ring element of which the influence is calculated, and subscript j stands for the panel at whose 3/4-chord, mid-span collocation point the induced velocity is specified. The local outward normal to the wing surface at the collocation point is denoted $\mathbf{n}_j$. The subscripts $1, 2, 3, 4$ refer to the corner points of the vortex-ring element, arranged in a clockwise direction.

The wake influence coefficients B_{ij}, C_{ij}, D_{ij} are similarly computed, with the sending (or receiving) panels being wake (rather than wing) panels. After the influence of panels on one side of the wing is computed, the one on the other (symmetrical) side is added by a mirror image of Biot-Savart law , with y_i replaced by $-y_i$, and the spanwise component of induced velocity $\mathbf{v}_{ij}$ also reversed in sign.

© Springer Science+Business Media, LLC 2015

A. Tewari, *Aeroservoelasticity,* Control Engineering,

DOI 10.1007/978-1-4939-2368-7

B.2　Aerodynamic Influence Coefficients for Quadrilateral Doublet Elements

Katz and Plotkin [86] (p. 287) present closed-form expressions for the velocity components induced by a doublet of spatially constant strength μ (or a vortex-ring of circulation $\Gamma = \mu$). The doublet (or vortex-ring) panel has corners arranged clockwise denoted by coordinates, (x_1, y_1, z_1), (x_2, y_2, z_2), (x_3, y_3, z_3), and (x_4, y_4, z_4), while the induced-velocity components (u, v, w) are evaluated at a point (x, y, z). For a quadrilateral doublet pointing along the z-axis, we have, by Green's integral:

$$\Phi(x, y, z) = \lim_{\epsilon \to 0} -\frac{\mu}{4\pi} \iiint \frac{z\,d\xi\,d\eta\,d\zeta}{[(x - \xi)^2 + (y - \eta)^2 + (z - \epsilon)^2]} \tag{B.7}$$

The velocity induced by the doublet is derived by differentiation, $(u, v, w)^T = \nabla\Phi$, and taking the limit $\epsilon \to 0$:

$$
\begin{aligned}
u = \frac{\mu}{4\pi} \Bigg[& \frac{z(y_1 - y_2)(r_1 + r_2)}{r_1 r_2 \left\{ r_1 r_2 - [(x - x_1)(x - x_2) + (y - y_1)(y - y_2) + z^2] \right\}} \\
& + \frac{z(y_2 - y_3)(r_2 + r_3)}{r_2 r_3 \left\{ r_2 r_3 - [(x - x_2)(x - x_3) + (y - y_2)(y - y_3) + z^2] \right\}} \\
& + \frac{z(y_3 - y_4)(r_3 + r_4)}{r_3 r_4 \left\{ r_3 r_4 - [(x - x_3)(x - x_4) + (y - y_3)(y - y_4) + z^2] \right\}} \\
& + \frac{z(y_4 - y_1)(r_4 + r_1)}{r_4 r_1 \left\{ r_4 r_1 - [(x - x_4)(x - x_1) + (y - y_4)(y - y_1) + z^2] \right\}} \Bigg]
\end{aligned} \tag{B.8}
$$

$$
\begin{aligned}
v = \frac{\mu}{4\pi} \Bigg[& \frac{z(x_2 - x_1)(r_1 + r_2)}{r_1 r_2 \left\{ r_1 r_2 - [(x - x_1)(x - x_2) + (y - y_1)(y - y_2) + z^2] \right\}} \\
& + \frac{z(x_3 - x_2)(r_2 + r_3)}{r_2 r_3 \left\{ r_2 r_3 - [(x - x_2)(x - x_3) + (y - y_2)(y - y_3) + z^2] \right\}} \\
& + \frac{z(x_4 - x_3)(r_3 + r_4)}{r_3 r_4 \left\{ r_3 r_4 - [(x - x_3)(x - x_4) + (y - y_3)(y - y_4) + z^2] \right\}} \\
& + \frac{z(x_1 - x_4)(r_4 + r_1)}{r_4 r_1 \left\{ r_4 r_1 - [(x - x_4)(x - x_1) + (y - y_4)(y - y_1) + z^2] \right\}} \Bigg]
\end{aligned} \tag{B.9}
$$

$$
\begin{aligned}
w = \frac{\mu}{4\pi} \Bigg[& \frac{[(x - x_2)(y - y_1) - (x - x_1)(y - y_2)](r_1 + r_2)}{r_1 r_2 \left\{ r_1 r_2 - [(x - x_1)(x - x_2) + (y - y_1)(y - y_2) + z^2] \right\}} \\
& + \frac{[(x - x_3)(y - y_2) - (x - x_2)(y - y_3)](r_2 + r_3)}{r_2 r_3 \left\{ r_2 r_3 - [(x - x_2)(x - x_3) + (y - y_2)(y - y_3) + z^2] \right\}}
\end{aligned}
$$

$$+ \frac{[(x - x_4)(y - y_3) - (x - x_3)(y - y_4)](r_3 + r_4)}{r_3 r_4 \left\{ r_3 r_4 - [(x - x_3)(x - x_4) + (y - y_3)(y - y_4) + z^2] \right\}}$$

$$\left. + \frac{[(x - x_1)(y - y_4) - (x - x_4)(y - y_1)](r_4 + r_1)}{r_4 r_1 \left\{ r_4 r_1 - [(x - x_4)(x - x_1) + (y - y_4)(y - y_1) + z^2] \right\}} \right] \qquad \text{(B.10)}$$

Here $r_i = \sqrt{(x - x_i)^2 + (y - y_i)^2 + (z - z_i)^2}$, $i = 1, 2, 3$.

Appendix-C

C.1 Listing of Generalized Aerodynamics Harmonic Data

The following is a listing of generalized aerodynamics transfer matrix , $D(s)$, for the harmonic case ($s = i\omega$), generated by Doublet-Lattice method for the DAST-ARW1 wing (Chap. 3) with one chordwise and six spanwise boxes at Mach number 0.8072 and standard altitude 7.6 km, for six structural modes (3 symmetric and 3 anti-symmetric) given in Chap. 4. This data is used for curve-fitting for the illustrative example presented in Chap. 4. Each column of the listing has values of elements of $D(i\omega)$ at 30 frequency points listed row-wise. For example, a term D(22,3,1) implies the element $(3, 1)$ of $D(i\omega)$ evaluated at the 22nd frequency point, i.e., for $\omega = 30$ Hz. The first six matrices correspond to the GAF data produced by a combination of the structural deformation modes, while the last matrix $D(:, :, 7)$ is the GAF data for the trailing-edge flap.

```
G(:,:,1) =

  Columns 1 through 3

       -1.0241 +            0i      -2.2079 +            0i      -3.7013 +            0i
       -1.0241 + 0.00028767i      -2.2079 -   0.0001723i      -3.7013 - 0.00053639i
       -1.0241 + 0.00057531i      -2.2078 - 0.00034463i      -3.7013 -   0.0010728i
       -1.0241 +   0.0011503i      -2.2078 - 0.00068962i      -3.7013 -   0.0021461i
       -1.0239 +   0.0028708i      -2.2076 -   0.0017301i       -3.701 -   0.0053724i
       -1.0233 +   0.0057066i      -2.2067 -   0.0035027i      -3.6999 -    0.010795i
       -1.0224 +   0.0084742i      -2.2054 -   0.0053575i      -3.6981 -    0.016314i
       -1.0211 +    0.011144i      -2.2035 -   0.0073288i      -3.6956 -    0.021968i
       -1.0194 +    0.013692i      -2.2011 -    0.009445i      -3.6924 -    0.027787i
       -1.0175 +    0.016098i      -2.1982 -    0.011729i      -3.6887 -    0.033795i
       -1.0126 +    0.020418i      -2.1911 -    0.016867i      -3.6793 -    0.046447i
       -1.0068 +    0.024008i      -2.1824 -    0.022849i      -3.6678 -    0.060034i
      -0.99238 +    0.028768i      -2.1605 -    0.037634i      -3.6385 -     0.09036i
      -0.96604 +    0.029495i      -2.1191 -    0.067423i      -3.5818 -     0.14454i
      -0.94642 +    0.025697i      -2.0871 -    0.092386i      -3.5371 -     0.18657i
      -0.91537 +    0.013834i      -2.0347 -     0.13701i      -3.4624 -     0.25795i
       -0.8943 +    0.002011i      -1.9976 -     0.17116i      -3.4085 -     0.31061i
      -0.86317 -    0.021208i      -1.9399 -     0.22835i       -3.323 -     0.39625i
      -0.84331 -    0.040058i      -1.9007 -     0.27002i      -3.2637 -     0.45716i
```

© Springer Science+Business Media, LLC 2015
A. Tewari, *Aeroservoelasticity,* Control Engineering,
DOI 10.1007/978-1-4939-2368-7

```
  -0.81577 -     0.07275i     -1.8419 -     0.33701i     -3.1721 -     0.55312i
  -0.79933 -    0.097027i     -1.8031 -      0.3841i     -3.1097 -     0.61944i
  -0.77816 -     0.13624i     -1.7462 -     0.45731i     -3.0148 -     0.72127i
  -0.76401 -     0.17061i     -1.7003 -     0.51974i     -2.9346 -      0.8073i
  -0.75312 -     0.20563i     -1.6558 -     0.58256i     -2.8534 -     0.89351i
   -0.7467 -     0.23356i     -1.6214 -     0.63254i     -2.7877 -     0.96213i
  -0.74054 -     0.27442i     -1.5713 -     0.70616i     -2.6879 -      1.0637i
  -0.73843 -     0.30046i     -1.5387 -     0.75385i     -2.6202 -      1.1302i
  -0.73748 -     0.33703i     -1.4908 -      0.8226i     -2.5169 -      1.2275i
  -0.73782 -     0.35948i     -1.4591 -     0.86626i     -2.4467 -      1.2905i
  -0.73883 -     0.39006i     -1.4115 -     0.92811i     -2.3389 -      1.3821i

Columns 4 through 6

   -5.4942 +           0i     -8.2153 +           0i     -12.367 +           0i
   -5.4942 - 0.00021782i     -8.2153 + 0.00074983i     -12.367 + 0.000018851i
   -5.4942 - 0.00043569i     -8.2153 +  0.0014996i     -12.367 + 0.00037694i
   -5.4941 - 0.00087188i     -8.2152 +  0.0029986i     -12.367 + 0.00075322i
   -5.4937 -  0.0021882i     -8.2148 +  0.0074864i     -12.367 +  0.0018715i
   -5.4924 -  0.0044359i     -8.2133 +   0.014902i     -12.365 +  0.0036632i
   -5.4903 -  0.0067962i     -8.2108 +   0.022186i     -12.362 +  0.0053067i
   -5.4873 -  0.0093123i     -8.2074 +   0.029291i     -12.358 +  0.0067507i
   -5.4835 -   0.012017i      -8.203 +    0.03618i     -12.353 +  0.0079587i
    -5.479 -   0.014933i     -8.1978 +   0.042829i     -12.347 +  0.0089035i
   -5.4678 -   0.021475i     -8.1848 +   0.055325i     -12.333 +  0.0099105i
   -5.4538 -    0.02906i     -8.1687 +    0.06663i     -12.314 +  0.0095896i
   -5.4183 -   0.047795i     -8.1276 +   0.085121i     -12.267 +  0.0043282i
   -5.3489 -   0.085857i     -8.0469 +    0.10132i     -12.175 -   0.016394i
   -5.2938 -    0.11809i     -7.9827 +    0.10414i     -12.101 -   0.039054i
   -5.2008 -    0.17633i     -7.8738 +   0.096661i     -11.974 -   0.086187i
   -5.1331 -    0.22129i     -7.7942 +   0.084238i     -11.882 -    0.12604i
   -5.0249 -    0.29686i     -7.6665 +   0.055452i     -11.732 -    0.19748i
   -4.9493 -    0.35193i     -7.5769 +   0.030239i     -11.626 -    0.25215i
   -4.8317 -    0.44016i     -7.4369 -   0.015227i      -11.46 -    0.34346i
    -4.751 -    0.50191i     -7.3405 -    0.04983i     -11.344 -    0.40979i
   -4.6268 -    0.59761i     -7.1916 -    0.10686i     -11.165 -    0.51628i
   -4.5205 -    0.67915i     -7.0639 -    0.15807i      -11.01 -    0.61049i
   -4.4118 -    0.76145i     -6.9331 -    0.21174i      -10.85 -    0.70886i
    -4.323 -    0.82735i     -6.8262 -    0.25598i      -10.72 -    0.79012i
   -4.1865 -    0.92566i     -6.6619 -    0.32383i     -10.518 -    0.91564i
   -4.0931 -    0.99049i     -6.5496 -     0.3697i      -10.38 -     1.0014i
   -3.9493 -     1.0863i     -6.3769 -    0.43904i     -10.167 -     1.1328i
   -3.8509 -     1.1489i     -6.2589 -    0.48543i     -10.022 -     1.2219i
   -3.6991 -      1.241i     -6.0775 -    0.55502i      -9.798 -     1.3578i


G(:,:,2) =

Columns 1 through 3

   -6.3597 +           0i     -14.749 +           0i     -22.439 +           0i
   -6.3597 +  0.0038241i     -14.749 +  0.0025984i     -22.439 +  0.0013866i
   -6.3597 +  0.0076481i     -14.749 +  0.0051966i     -22.439 +  0.0027729i
   -6.3596 +   0.015295i     -14.749 +   0.010391i     -22.439 +  0.0055436i
   -6.3588 +   0.038213i     -14.748 +   0.025948i     -22.438 +   0.013821i
   -6.3561 +   0.076254i     -14.745 +   0.051677i     -22.434 +   0.027376i
   -6.3516 +    0.11396i      -14.74 +   0.076987i     -22.429 +   0.040421i
   -6.3454 +     0.1512i     -14.734 +     0.1017i     -22.422 +   0.052747i
   -6.3376 +    0.18784i     -14.725 +    0.12566i     -22.413 +   0.064189i
   -6.3283 +    0.22378i     -14.715 +    0.14877i     -22.402 +   0.074616i
   -6.3055 +    0.29327i      -14.69 +    0.19202i     -22.376 +   0.092045i
```

```
   -6.2778 +     0.3592i      -14.66 +    0.23088i      -22.346 +    0.10443i
   -6.2101 +    0.47934i     -14.588 +    0.29404i      -22.272 +    0.11234i
   -6.0869 +    0.62871i     -14.458 +    0.34966i      -22.142 +   0.078138i
   -5.9954 +    0.70787i     -14.364 +    0.36029i      -22.051 +   0.023636i
   -5.8509 +    0.79726i     -14.221 +    0.33778i      -21.918 -    0.10454i
   -5.7528 +    0.83823i     -14.127 +    0.29814i      -21.837 -    0.21956i
   -5.6076 +    0.87359i     -13.994 +    0.20366i      -21.731 -    0.43351i
   -5.5144 +    0.88112i     -13.914 +    0.11865i      -21.675 -    0.60188i
   -5.3837 +    0.87125i     -13.808 -   0.039143i      -21.614 -    0.88971i
    -5.304 +    0.85249i      -13.75 -    0.16272i      -21.591 -     1.1032i
   -5.1977 +    0.80974i     -13.683 -    0.37238i      -21.585 -     1.4526i
    -5.122 +    0.76404i     -13.646 -    0.56651i      -21.605 -     1.7681i
    -5.058 +    0.71237i     -13.626 -    0.77561i      -21.648 -     2.1036i
   -5.0148 +    0.66868i     -13.622 -    0.95208i      -21.699 -     2.3852i
   -4.9616 +    0.60231i     -13.637 -     1.2295i      -21.801 -     2.8279i
   -4.9324 +    0.55913i     -13.658 -     1.4216i      -21.886 -     3.1358i
   -4.8949 +    0.49789i     -13.706 -     1.7186i      -22.037 -     3.6161i
   -4.8719 +    0.45998i     -13.747 -     1.9216i      -22.153 -     3.9482i
   -4.8368 +    0.40732i      -13.82 -     2.2328i      -22.346 -      4.464i

Columns 4 through 6

   -22.676 +          0i     -19.305 +          0i      -57.719 +          0i
   -22.676 - 0.00046291i     -19.305 -  0.0036554i      -57.719 +   0.0057027i
   -22.676 - 0.00092615i     -19.305 -  0.0073112i      -57.719 +    0.011405i
   -22.675 -  0.0018549i     -19.305 -   0.014625i      -57.718 +    0.022806i
   -22.674 -  0.0046827i     -19.303 -   0.036617i      -57.717 +    0.056955i
   -22.671 -  0.0096834i     -19.298 -   0.073608i      -57.711 +     0.11349i
   -22.665 -   0.015288i     -19.291 -     0.1113i      -57.701 +     0.16924i
   -22.656 -   0.021732i      -19.28 -    0.14995i      -57.688 +     0.22393i
   -22.646 -   0.029193i     -19.267 -    0.18975i      -57.671 +     0.27738i
   -22.634 -   0.037806i     -19.252 -    0.23083i      -57.651 +     0.32946i
   -22.605 -    0.05887i     -19.214 -     0.3172i      -57.603 +      0.4291i
   -22.569 -   0.085551i     -19.168 -    0.40974i      -57.544 +     0.52213i
   -22.485 -    0.15773i     -19.057 -    0.61547i      -57.399 +     0.68596i
   -22.338 -     0.3178i     -18.858 -    0.98041i      -57.131 +      0.8725i
   -22.234 -    0.46026i     -18.712 -     1.2617i      -56.929 +     0.95819i
   -22.084 -      0.726i      -18.49 -     1.7375i      -56.607 +      1.0329i
   -21.992 -    0.93591i     -18.346 -      2.088i      -56.384 +      1.0498i
   -21.876 -     1.2956i     -18.142 -     2.6584i      -56.048 +      1.0317i
   -21.816 -     1.5625i     -18.019 -     3.0649i      -55.825 +     0.99467i
   -21.754 -     1.9986i     -17.857 -      3.708i      -55.499 +     0.90861i
   -21.734 -     2.3106i     -17.765 -     4.1552i      -55.288 +     0.83477i
   -21.736 -     2.8061i     -17.654 -     4.8484i      -54.981 +      0.7052i
   -21.766 -     3.2412i     -17.585 -     5.4423i      -54.735 +     0.58424i
   -21.822 -     3.6938i     -17.537 -     6.0476i      -54.496 +     0.45493i
   -21.884 -     4.0675i     -17.512 -     6.5384i       -54.31 +     0.34724i
   -22.006 -     4.6456i     -17.497 -     7.2833i      -54.038 +     0.18122i
   -22.104 -     5.0423i      -17.5 -     7.7846i      -53.861 +    0.068808i
   -22.277 -     5.6538i     -17.523 -     8.5424i       -53.6 -     0.10067i
   -22.407 -     6.0724i     -17.549 -     9.0511i      -53.428 -     0.21344i
   -22.624 -     6.7173i     -17.604 -     9.8189i      -53.173 -     0.38121i



G(:,:,3) =

Columns 1 through 3

   -10.381 +          0i     -23.535 +          0i      -37.068 +          0i
   -10.381 +  0.0082468i     -23.535 +  0.0083001i      -37.068 +   0.0008031i
   -10.381 +   0.016493i     -23.535 +     0.0166i      -37.068 +    0.016061i
```

-10.381 +	0.032984i	-23.535 +	0.033196i	-37.067 +	0.032119i
-10.38 +	0.082411i	-23.533 +	0.082929i	-37.066 +	0.080221i
-10.374 +	0.16448i	-23.527 +	0.16542i	-37.059 +	0.15992i
-10.365 +	0.24588i	-23.518 +	0.24708i	-37.05 +	0.2386i
-10.353 +	0.32633i	-23.505 +	0.32755i	-37.036 +	0.31587i
-10.338 +	0.40559i	-23.488 +	0.40654i	-37.019 +	0.3914i
-10.319 +	0.48346i	-23.469 +	0.48382i	-36.999 +	0.46492i
-10.274 +	0.63443i	-23.421 +	0.63253i	-36.951 +	0.60525i
-10.22 +	0.77831i	-23.364 +	0.77258i	-36.894 +	0.73567i
-10.087 +	1.0427i	-23.227 +	1.0238i	-36.758 +	0.96331i
-9.8448 +	1.3775i	-22.983 +	1.3228i	-36.522 +	1.2135i
-9.6655 +	1.5597i	-22.808 +	1.4692i	-36.359 +	1.3173i
-9.3835 +	1.7739i	-22.545 +	1.6118i	-36.128 +	1.3805i
-9.1929 +	1.8792i	-22.376 +	1.6575i	-35.993 +	1.3637i
-8.9126 +	1.9847i	-22.143 +	1.6563i	-35.828 +	1.2559i
-8.734 +	2.0228i	-22.006 +	1.6117i	-35.75 +	1.1326i
-8.4855 +	2.037i	-21.837 +	1.4843i	-35.687 +	0.87745i
-8.3359 +	2.0217i	-21.75 +	1.3627i	-35.684 +	0.66438i
-8.1397 +	1.9688i	-21.663 +	1.1317i	-35.74 +	0.28675i
-8.0037 +	1.9035i	-21.633 +	0.9002i	-35.842 -	0.076141i
-7.8932 +	1.8255i	-21.64 +	0.63844i	-35.995 -	0.4786i
-7.8225 +	1.7577i	-21.673 +	0.41033i	-36.153 -	0.82684i
-7.7431 +	1.6534i	-21.765 +	0.041907i	-36.446 -	1.3898i
-7.7053 +	1.5854i	-21.853 -	0.21856i	-36.678 -	1.7909i
-7.6658 +	1.4902i	-22.021 -	0.62786i	-37.075 -	2.4294i
-7.6468 +	1.4329i	-22.154 -	0.91153i	-37.372 -	2.8794i
-7.6225 +	1.3566i	-22.38 -	1.3515i	-37.862 -	3.5908i

Columns 4 through 6

-42.339 +	0i	-44.842 +	0i	-101.48 +	0i
-42.339 +	0.0054487i	-44.842 -	0.0010161i	-101.48 +	0.010326i
-42.339 +	0.010897i	-44.842 -	0.0020329i	-101.48 +	0.020651i
-42.338 +	0.021788i	-44.842 -	0.0040718i	-101.48 +	0.041294i
-42.336 +	0.054382i	-44.839 -	0.010284i	-101.47 +	0.10312i
-42.33 +	0.10814i	-44.831 -	0.021299i	-101.46 +	0.20542i
-42.318 +	0.16071i	-44.816 -	0.033683i	-101.44 +	0.3062i
-42.303 +	0.21164i	-44.797 -	0.047936i	-101.42 +	0.40495i
-42.284 +	0.26057i	-44.773 -	0.064415i	-101.39 +	0.5013i
-42.262 +	0.307726i	-44.744 -	0.083374i	-101.35 +	0.59502i
-42.208 +	0.39314i	-44.675 -	0.12948i	-101.26 +	0.77376i
-42.144 +	0.46805i	-44.592 -	0.18755i	-101.16 +	0.93975i
-41.991 +	0.58101i	-44.391 -	0.34402i	-100.89 +	1.2287i
-41.727 +	0.64844i	-44.033 -	0.6892i	-100.4 +	1.5469i
-41.546 +	0.62281i	-43.777 -	0.99415i	-100.04 +	1.6831i
-41.291 +	0.48119i	-43.392 -	1.5588i	-99.458 +	1.7803i
-41.143 +	0.32156i	-43.148 -	2.0021i	-99.065 +	1.7791i
-40.97 -	0.00767i	-42.818 -	2.7573i	-98.48 +	1.6899i
-40.892 -	0.28169i	-42.628 -	3.3142i	-98.1 +	1.58i
-40.842 -	0.76499i	-42.397 -	4.2178i	-97.556 +	1.3529i
-40.854 -	1.13i	-42.283 -	4.8587i	-97.211 +	1.1675i
-40.942 -	1.7334i	-42.171 -	5.8672i	-96.725 +	0.84941i
-41.08 -	2.2813i	-42.132 -	6.743i	-96.346 +	0.55606i
-41.274 -	2.8653i	-42.142 -	7.6445i	-95.992 +	0.24365i
-41.468 -	3.3562i	-42.183 -	8.3811i	-95.723 -	0.016563i
-41.822 -	4.1291i	-42.295 -	9.5073i	-95.345 -	0.41908i
-42.096 -	4.66791i	-42.402 -	10.27i	-95.107 -	0.69315i
-42.562 -	5.51104i	-42.607 -	11.432i	-94.768 -	1.1097i
-42.906 -	6.0952i	-42.771 -	12.216i	-94.552 -	1.3897i
-43.472 -	7.0079i	-43.055 -	13.408i	-94.242 -	1.8114i

G(:,:,4) =

Columns 1 through 3

```
    -10.37 +         0i    -21.192 +         0i     -38.53 +         0i
    -10.37 +  0.013663i    -21.192 +  0.020087i     -38.53 +  0.024271i
    -10.37 +  0.027324i    -21.192 +  0.040173i     -38.53 +  0.048541i
   -10.369 +  0.054644i    -21.192 +  0.080341i    -38.529 +  0.097074i
   -10.367 +   0.13653i    -21.189 +   0.20075i    -38.527 +   0.24257i
   -10.358 +   0.27251i     -21.18 +   0.40082i    -38.517 +   0.48431i
   -10.344 +   0.40741i    -21.166 +   0.59954i    -38.502 +   0.72446i
   -10.325 +   0.54077i    -21.145 +   0.79638i    -38.481 +    0.9624i
   -10.301 +   0.67222i     -21.12 +   0.99087i    -38.455 +    1.1976i
   -10.272 +   0.80142i     -21.09 +    1.1827i    -38.424 +    1.4297i
   -10.201 +    1.0522i    -21.018 +     1.557i     -38.35 +    1.8835i
   -10.115 +    1.2916i     -20.93 +    1.9178i    -38.261 +    2.3221i
   -9.9053 +    1.7331i     -20.72 +     2.594i    -38.051 +    3.1479i
   -9.5247 +    2.2963i    -20.348 +    3.4851i    -37.685 +    4.2441i
   -9.2436 +    2.6054i    -20.084 +    3.9946i    -37.434 +    4.8758i
   -8.8036 +    2.9735i    -19.691 +    4.6361i     -37.08 +    5.6782i
    -8.508 +    3.1583i    -19.444 +    4.9858i    -36.873 +    6.1211i
   -8.0757 +    3.3509i    -19.112 +    5.4007i    -36.629 +    6.6566i
   -7.8023 +    3.4267i    -18.924 +    5.6087i    -36.517 +    6.9339i
   -7.4255 +    3.4698i    -18.704 +    5.8262i     -36.44 +    7.2414i
   -7.2019 +    3.4568i    -18.603 +    5.9137i    -36.452 +    7.3809i
   -6.9153 +    3.3855i    -18.526 +    5.9687i     -36.57 +    7.5033i
   -6.7253 +    3.2883i    -18.534 +    5.9535i     -36.76 +    7.5344i
   -6.5819 +    3.1666i    -18.608 +    5.8911i    -37.033 +    7.5086i
   -6.5006 +    3.0578i    -18.714 +    5.8121i     -37.31 +    7.4509i
   -6.4321 +    2.8862i     -18.95 +    5.6532i    -37.816 +    7.3078i
   -6.4193 +    2.7723i    -19.155 +    5.5245i    -38.211 +    7.1773i
   -6.4425 +     2.611i     -19.53 +    5.3038i    -38.886 +    6.9316i
   -6.4812 +     2.514i     -19.82 +     5.141i    -39.386 +    6.7351i
   -6.5643 +    2.3886i    -20.308 +    4.8767i    -40.205 +    6.3915i
```

Columns 4 through 6

```
   -66.608 +         0i    -106.65 +         0i    -125.23 +         0i
   -66.607 +  0.020623i    -106.65 +  0.0082443i   -125.23 +  0.0093637i
   -66.607 +  0.041245i    -106.65 +  0.016487i    -125.23 +  0.018726i
   -66.607 +  0.082483i    -106.65 +  0.032965i    -125.23 +  0.037442i
   -66.604 +   0.20607i    -106.65 +  0.082252i    -125.23 +  0.093425i
   -66.593 +   0.41116i    -106.63 +   0.16338i    -125.21 +    0.1856i
   -66.576 +   0.61442i    -106.61 +   0.24241i    -125.18 +   0.275548i
   -66.552 +   0.81514i    -106.58 +   0.31858i    -125.14 +   0.36229i
   -66.522 +    1.0128i    -106.54 +   0.39137i     -125.1 +   0.44552i
   -66.487 +     1.207i     -106.5 +   0.46041i    -125.04 +    0.5248i
   -66.403 +    1.5841i    -106.39 +   0.58615i     -124.9 +   0.67042i
   -66.302 +    1.9444i    -106.26 +    0.6938i    -124.73 +   0.79692i
    -66.06 +    2.6088i    -105.94 +   0.84756i    -124.32 +   0.98475i
   -65.639 +    3.4477i    -105.37 +   0.90802i    -123.54 +    1.0895i
   -65.346 +    3.8977i    -104.95 +   0.83201i    -122.96 +    1.0404i
   -64.928 +    4.4124i    -104.32 +   0.54851i    -122.03 +   0.79458i
   -64.681 +    4.6535i    -103.91 +   0.25214i    -121.39 +   0.52282i
   -64.381 +    4.8737i    -103.34 -   0.33934i    -120.44 -  0.029122i
    -64.24 +    4.9343i    -103.01 -   0.82139i    -119.82 -    0.4813i
   -64.127 +    4.9113i    -102.59 -    1.6574i    -118.92 -    1.2654i
   -64.123 +     4.829i    -102.36 -    2.2794i    -118.35 -    1.8476i
   -64.227 +    4.6189i    -102.12 -    3.2932i    -117.54 -    2.7932i
   -64.413 +     4.375i       -102 -    4.1998i     -116.9 -    3.6347i
   -64.687 +    4.0769i    -101.95 -    5.1527i     -116.3 -    4.5142i
```

```
      -64.968 +      3.8035i      -101.96 -      5.9432i      -115.85 -      5.2393i
      -65.485 +      3.3405i      -102.06 -      7.1689i      -115.21 -      6.3548i
       -65.89 +      2.9987i      -102.18 -      8.0094i      -114.81 -       7.113i
      -66.581 +      2.4385i      -102.42 -      9.3017i      -114.23 -       8.267i
      -67.094 +      2.0334i      -102.62 -      10.183i      -113.86 -      9.0455i
      -67.936 +      1.3775i      -102.98 -      11.534i      -113.33 -      10.225i
```

G(:,:,5) =

Columns 1 through 3

```
      -9.3535 +          0i      -15.411 +          0i      -37.344 +          0i
      -9.3534 +    0.020534i      -15.411 +    0.036198i      -37.344 +    0.048034i
      -9.3533 +    0.041068i      -15.411 +    0.072394i      -37.343 +    0.096067i
      -9.3528 +    0.082129i       -15.41 +     0.14478i      -37.343 +     0.19212i
       -9.349 +      0.2052i      -15.406 +      0.3618i      -37.338 +     0.48013i
      -9.3358 +     0.40954i      -15.392 +     0.72255i      -37.323 +     0.95902i
       -9.314 +     0.61223i      -15.368 +      1.0813i      -37.297 +      1.4355i
       -9.284 +     0.81254i      -15.336 +      1.4371i      -37.262 +      1.9087i
      -9.2462 +      1.0099i      -15.296 +      1.7894i      -37.218 +      2.3778i
      -9.2011 +      1.2038i      -15.247 +      2.1375i      -37.166 +      2.8422i
      -9.0908 +      1.5798i       -15.13 +      2.8199i       -37.04 +      3.7555i
      -8.9569 +      1.9383i      -14.989 +      3.4816i      -36.888 +      4.6459i
      -8.6306 +      2.5975i      -14.647 +      4.7361i       -36.52 +      6.3494i
      -8.0362 +      3.4321i      -14.033 +      6.4284i      -35.863 +      8.6876i
       -7.597 +      3.8845i      -13.591 +      7.4261i      -35.395 +      10.095i
      -6.9098 +      4.4131i      -12.926 +      8.7337i      -34.701 +      11.985i
      -6.4492 +      4.6706i        -12.5 +      9.4866i      -34.269 +      13.106i
      -5.7777 +      4.9246i      -11.914 +      10.451i      -33.696 +      14.594i
      -5.3548 +      5.0121i      -11.572 +      10.991i      -33.381 +      15.467i
      -4.7758 +      5.0334i      -11.148 +      11.662i      -33.025 +       16.62i
      -4.4356 +      4.9831i      -10.934 +      12.026i      -32.874 +      17.295i
      -4.0066 +      4.8279i      -10.725 +      12.465i      -32.783 +      18.188i
      -3.7308 +      4.6415i       -10.66 +      12.747i      -32.833 +      18.839i
      -3.5334 +      4.4193i      -10.695 +      12.966i      -32.996 +      19.418i
      -3.4323 +       4.226i      -10.795 +      13.106i      -33.205 +      19.836i
      -3.3727 +      3.9279i      -11.061 +      13.268i      -33.641 +        20.4i
      -3.3904 +      3.7338i      -11.312 +      13.353i      -34.009 +      20.737i
      -3.4934 +      3.4654i       -11.79 +      13.454i      -34.668 +       21.19i
      -3.6055 +       3.309i      -12.171 +       13.51i      -35.172 +      21.459i
      -3.8241 +      3.1176i      -12.824 +      13.583i      -36.017 +      21.812i
```

Columns 4 through 6

```
      -99.888 +          0i      -198.76 +          0i      -161.56 +          0i
      -99.888 +    0.046856i      -198.76 +    0.032699i      -161.56 +    0.019391i
      -99.887 +    0.093711i      -198.76 +    0.065396i      -161.56 +     0.03878i
      -99.887 +     0.18741i      -198.76 +     0.13078i      -161.56 +    0.077544i
      -99.881 +     0.46832i      -198.75 +     0.32671i      -161.55 +     0.19359i
      -99.863 +     0.93519i      -198.73 +     0.65175i      -161.52 +     0.38535i
      -99.833 +      1.3993i      -198.69 +     0.97367i      -161.48 +     0.57373i
      -99.792 +      1.8597i      -198.64 +      1.2914i      -161.42 +     0.75759i
      -99.741 +      2.3155i      -198.58 +      1.6041i      -161.34 +     0.93618i
      -99.681 +      2.7662i       -198.5 +      1.9112i      -161.25 +       1.109i
      -99.534 +      3.6507i      -198.32 +      2.5072i      -161.04 +      1.4354i
      -99.356 +      4.5103i       -198.1 +      3.0762i      -160.77 +      1.7331i
      -98.922 +      6.1447i      -197.55 +      4.1209i       -160.1 +      2.2297i
      -98.135 +      8.3571i      -196.53 +      5.4272i      -158.85 +       2.701i
      -97.565 +      9.6652i      -195.77 +      6.1177i      -157.89 +       2.827i
      -96.699 +      11.382i      -194.57 +      6.8859i      -156.33 +      2.7361i
      -96.143 +       12.37i      -193.76 +      7.2263i      -155.26 +      2.4969i
```

```
    -95.379 +      13.635i       -192.58 +      7.4992i       -153.63 +      1.8951i
    -94.934 +      14.346i       -191.84 +      7.5383i       -152.56 +      1.3495i
    -94.388 +       15.24i       -190.81 +      7.4113i       -150.98 +     0.34453i
    -94.111 +      15.735i        -190.2 +      7.2191i       -149.97 -     0.43379i
    -93.832 +      16.353i       -189.38 +      6.7946i       -148.51 -      1.7388i
    -93.725 +      16.769i        -188.8 +      6.3348i       -147.35 -      2.9333i
     -93.73 +       17.11i       -188.31 +      5.7941i       -146.25 -      4.2085i
     -93.81 +      17.336i       -187.98 +      5.3114i        -145.4 -      5.2779i
    -94.049 +      17.607i       -187.58 +      4.5153i       -144.19 -      6.9515i
    -94.283 +      17.747i       -187.37 +      3.9418i       -143.42 -      8.1073i
    -94.734 +      17.899i       -187.15 +      3.0238i       -142.31 -      9.8937i
    -95.097 +      17.964i       -187.04 +      2.3752i        -141.6 -      11.117i
    -95.725 +      18.004i       -186.96 +      1.3484i       -140.56 -      12.997i
```

G(:,:,6) =

Columns 1 through 3

```
    -29.344 +          0i       -63.031 +          0i       -101.84 +          0i
    -29.344 +     0.02574i       -63.031 +    0.026149i       -101.84 +    0.037193i
    -29.344 +    0.051479i        -63.03 +    0.052297i       -101.84 +    0.074383i
    -29.343 +     0.10295i       -63.029 +     0.10458i       -101.83 +     0.14875i
    -29.336 +     0.25717i       -63.021 +     0.26121i       -101.83 +     0.37158i
    -29.313 +     0.51294i       -62.992 +     0.52069i       -101.79 +      0.7411i
    -29.275 +     0.76601i       -62.944 +     0.77687i       -101.73 +      1.1067i
    -29.223 +      1.0152i       -62.878 +      1.0284i       -101.66 +      1.4667i
    -29.156 +      1.2596i       -62.794 +       1.274i       -101.56 +        1.82i
    -29.077 +      1.4982i       -62.694 +      1.5129i       -101.44 +      2.1655i
    -28.883 +      1.9562i        -62.45 +      1.9675i       -101.16 +      2.8302i
    -28.647 +      2.3851i        -62.15 +       2.388i       -100.81 +      3.4562i
    -28.067 +      3.1469i        -61.41 +      3.1145i       -99.944 +      4.5777i
    -27.002 +      4.0337i       -60.033 +      3.8947i       -98.322 +      5.9002i
    -26.209 +      4.4533i       -58.996 +      4.2057i       -97.095 +      6.5354i
    -24.954 +      4.8361i       -57.338 +      4.3738i       -95.127 +      7.1339i
    -24.105 +      4.9365i       -56.199 +      4.3003i       -93.772 +       7.313i
    -22.856 +      4.8752i       -54.491 +      3.9377i       -91.734 +      7.2856i
    -22.062 +      4.7078i       -53.376 +       3.545i       -90.398 +      7.0922i
    -20.964 +      4.2977i       -51.775 +      2.7635i       -88.459 +      6.5804i
    -20.307 +       3.939i       -50.769 +      2.1362i       -87.222 +      6.1154i
    -19.452 +      3.3115i       -49.371 +      1.0743i       -85.464 +      5.2712i
    -18.866 +      2.7409i       -48.314 +     0.11159i       -84.089 +      4.4633i
    -18.394 +      2.1601i       -47.357 -     0.89171i       -82.794 +       3.586i
    -18.095 +      1.7077i        -46.66 -      1.7062i       -81.812 +      2.8473i
    -17.757 +       1.081i       -45.718 -      2.9205i       -80.419 +      1.6965i
    -17.593 +      0.7141i       -45.152 -      3.7107i       -79.538 +     0.90943i
    -17.409 +     0.26097i       -44.378 -      4.8473i       -78.274 -     0.29084i
    -17.308 +    0.030637i         -43.9 -      5.5639i       -77.465 -      1.1006i
    -17.152 -      0.2049i       -43.221 -       6.568i       -76.294 -      2.3269i
```

Columns 4 through 6

```
    -127.21 +          0i       -162.81 +          0i       -377.84 +          0i
    -127.21 +    0.071998i       -162.81 +     0.11321i       -377.84 +    0.0757331i
    -127.21 +     0.14399i       -162.81 +     0.22642i       -377.84 +     0.15146i
    -127.21 +     0.28797i       -162.81 +     0.45282i       -377.84 +     0.30289i
     -127.2 +     0.71957i        -162.8 +      1.1316i       -377.83 +     0.75677i
    -127.16 +      1.4367i       -162.76 +      2.2604i       -377.78 +      1.5104i
     -127.1 +      2.1492i       -162.69 +      3.3839i       -377.71 +       2.258i
    -127.01 +      2.8553i        -162.6 +        4.5i        -377.6 +      2.9978i
     -126.9 +      3.5536i       -162.48 +      5.6073i       -377.48 +      3.7281i
    -126.78 +      4.2432i       -162.34 +      6.7049i       -377.32 +       4.448i
```

$-126.46 +$	$5.5929i$	$-162.01 +$	$8.8674i$	$-376.95 +$	$5.853i$
$-126.08 +$	$6.8993i$	$-161.59 +$	$10.982i$	$-376.49 +$	$7.2062i$
$-125.13 +$	$9.3646i$	$-160.56 +$	$15.044i$	$-375.34 +$	$9.7341i$
$-123.36 +$	$12.652i$	$-158.65 +$	$20.677i$	$-373.21 +$	$13.035i$
$-122.02 +$	$14.559i$	$-157.22 +$	$24.114i$	$-371.6 +$	$14.9i$
$-119.88 +$	$17.005i$	$-154.93 +$	$28.801i$	$-369 +$	$17.203i$
$-118.42 +$	$18.376i$	$-153.38 +$	$31.629i$	$-367.22 +$	$18.423i$
$-116.24 +$	$20.08i$	$-151.08 +$	$35.465i$	$-364.54 +$	$19.824i$
$-114.83 +$	$21.008i$	$-149.6 +$	$37.779i$	$-362.78 +$	$20.499i$
$-112.8 +$	$22.136i$	$-147.5 +$	$40.936i$	$-360.23 +$	$21.172i$
$-111.53 +$	$22.74i$	$-146.19 +$	$42.86i$	$-358.6 +$	$21.422i$
$-109.74 +$	$23.466i$	$-144.39 +$	$45.517i$	$-356.29 +$	$21.539i$
$-108.35 +$	$23.937i$	$-143.03 +$	$47.554i$	$-354.47 +$	$21.429i$
$-107.06 +$	$24.309i$	$-141.79 +$	$49.454i$	$-352.76 +$	$21.157i$
$-106.09 +$	$24.549i$	$-140.88 +$	$50.889i$	$-351.46 +$	$20.833i$
$-104.72 +$	$24.825i$	$-139.66 +$	$52.914i$	$-349.63 +$	$20.186i$
$-103.86 +$	$24.959i$	$-138.92 +$	$54.186i$	$-348.48 +$	$19.654i$
$-102.64 +$	$25.091i$	$-137.93 +$	$55.983i$	$-346.87 +$	$18.714i$
$-101.86 +$	$25.133i$	$-137.33 +$	$57.109i$	$-345.85 +$	$17.996i$
$-100.75 +$	$25.123i$	$-136.54 +$	$58.689i$	$-344.43 +$	$16.781i$

D(:,:,7) =

Columns 1 through 3

$-1.1879 +$	$0i$	$-1.5712 +$	$0i$	$-2.9117 +$	$0i$
$-1.1879 +$	$0.0064373i$	$-1.5711 +$	$0.00739111i$	$-2.9117 +$	$0.0096322i$
$-1.1878 +$	$0.012874i$	$-1.5711 +$	$0.014782i$	$-2.9116 +$	$0.019264i$
$-1.1876 +$	$0.025744i$	$-1.5708 +$	$0.029559i$	$-2.9113 +$	$0.038522i$
$-1.1859 +$	$0.06429i$	$-1.569 +$	$0.073815i$	$-2.9093 +$	$0.096209i$
$-1.1799 +$	$0.12809i$	$-1.5625 +$	$0.14707i$	$-2.9021 +$	$0.19176i$
$-1.1701 +$	$0.19097i$	$-1.5519 +$	$0.21928i$	$-2.8905 +$	$0.28612i$
$-1.1567 +$	$0.25263i$	$-1.5377 +$	$0.29011i$	$-2.8748 +$	$0.37892i$
$-1.1402 +$	$0.31285i$	$-1.5201 +$	$0.35938i$	$-2.8556 +$	$0.47001i$
$-1.1208 +$	$0.37152i$	$-1.4994 +$	$0.42701i$	$-2.833 +$	$0.55932i$
$-1.0739 +$	$0.48407i$	$-1.4498 +$	$0.55721i$	$-2.7787 +$	$0.73257i$
$-1.0173 +$	$0.59003i$	$-1.3898 +$	$0.68048i$	$-2.7131 +$	$0.89834i$
$-0.87844 +$	$0.7805i$	$-1.2425 +$	$0.90378i$	$-2.5517 +$	$1.2041i$
$-0.61981 +$	$1.0064i$	$-0.97075 +$	$1.1729i$	$-2.2565 +$	$1.5908i$
$-0.42451 +$	$1.1169i$	$-0.76964 +$	$1.3095i$	$-2.0395 +$	$1.8029i$
$-0.10588 +$	$1.2272i$	$-0.45028 +$	$1.4589i$	$-1.6979 +$	$2.0616i$
$0.1232 +$	$1.2634i$	$-0.22793 +$	$1.5246i$	$-1.4648 +$	$2.198i$
$0.4908 +$	$1.253i$	$0.11691 +$	$1.5729i$	$-1.1161 +$	$2.3544i$
$0.74815 +$	$1.1947i$	$0.35215 +$	$1.5711i$	$-0.88874 +$	$2.4316i$
$1.1386 +$	$1.0149i$	$0.70883 +$	$1.5159i$	$-0.5594 +$	$2.5179i$
$1.3891 +$	$0.82511i$	$0.94689 +$	$1.4431i$	$-0.34709 +$	$2.5622i$
$1.7197 +$	$0.42624i$	$1.3003 +$	$1.2761i$	$-0.034026 +$	$2.6182i$
$1.9211 -$	$0.011595i$	$1.5866 +$	$1.0792i$	$0.22985 +$	$2.6604i$
$2.0186 -$	$0.53594i$	$1.8582 +$	$0.82367i$	$0.50617 +$	$2.7i$
$1.9977 -$	$1.0036i$	$2.0579 +$	$0.57299i$	$0.7429 +$	$2.7279i$
$1.7621 -$	$1.7426i$	$2.3103 +$	$0.11417i$	$1.1355 +$	$2.7548i$
$1.4504 -$	$2.2232i$	$2.4329 -$	$0.24816i$	$1.428 +$	$2.7542i$
$0.73823 -$	$2.8458i$	$2.5191 -$	$0.87154i$	$1.9182 +$	$2.7059i$
$0.10884 -$	$3.1421i$	$2.4925 -$	$1.3321i$	$2.2786 +$	$2.628i$
$-1.0197 -$	$3.3157i$	$2.2946 -$	$2.0643i$	$2.8604 +$	$2.417i$

Columns 4 through 6

$-10.438 +$	$0i$	$-101.64 +$	$0i$	$-120.6 +$	$0i$
$-10.438 +$	$0.014994i$	$-101.64 +$	$0.01886i$	$-120.6 +$	$0.0086053i$
$-10.438 +$	$0.029987i$	$-101.64 +$	$0.037718i$	$-120.6 +$	$0.01721i$
$-10.438 +$	$0.059968i$	$-101.64 +$	$0.075429i$	$-120.6 +$	$0.034413i$
$-10.435 +$	$0.14981i$	$-101.64 +$	$0.18845i$	$-120.6 +$	$0.085926i$
$-10.428 +$	$0.29886i$	$-101.63 +$	$0.37604i$	$-120.59 +$	$0.17113i$

-10.415 +	0.44655i	-101.62 +	0.56215i	-120.59 +	0.2551i
-10.399 +	0.59254i	-101.6 +	0.74641i	-120.57 +	0.33757i
-10.378 +	0.73667i	-101.59 +	0.92873i	-120.56 +	0.4185i
-10.354 +	0.87892i	-101.56 +	1.1091i	-120.54 +	0.49788i
-10.297 +	1.1577i	-101.51 +	1.4639i	-120.5 +	0.65189i
-10.227 +	1.4285i	-101.45 +	1.8101i	-120.46 +	0.79879i
-10.057 +	1.942i	-101.31 +	2.4723i	-120.35 +	1.068i
-9.7494 +	2.6366i	-101.05 +	3.3882i	-120.17 +	1.4102i
-9.5249 +	3.0535i	-100.88 +	3.9523i	-120.04 +	1.6005i
-9.1733 +	3.6173i	-100.62 +	4.7343i	-119.86 +	1.831i
-8.9368 +	3.9555i	-100.46 +	5.2166i	-119.76 +	1.951i
-8.5927 +	4.4156i	-100.26 +	5.8943i	-119.63 +	2.0908i
-8.3764 +	4.6991i	-100.15 +	6.325i	-119.57 +	2.1642i
-8.0763 +	5.1056i	-100.04 +	6.9561i	-119.53 +	2.2551i
-7.8923 +	5.3736i	-100 +	7.3757i	-119.52 +	2.3084i
-7.6351 +	5.7861i	-99.98 +	8.0172i	-119.55 +	2.3861i
-7.4303 +	6.1503i	-99.998 +	8.5732i	-119.62 +	2.4551i
-7.2239 +	6.5415i	-100.04 +	9.1574i	-119.7 +	2.5334i
-7.0495 +	6.8772i	-100.09 +	9.6492i	-119.79 +	2.6052i
-6.7562 +	7.4213i	-100.17 +	10.432i	-119.94 +	2.731i
-6.5287 +	7.8095i	-100.22 +	10.986i	-120.05 +	2.8274i
-6.1214 +	8.4227i	-100.29 +	11.864i	-120.23 +	2.9895i
-5.7959 +	8.8445i	-100.31 +	12.478i	-120.35 +	3.1076i
-5.2131 +	9.4788i	-100.32 +	13.435i	-120.53 +	3.2957i

C.2 Pressure Distribution due to Oscillating Flap

The real and imaginary parts of the unsteady pressure distribution on the modified
DAST-ARW1 wing (Fig. C.1) at the reference flight condition Mach number 0.8072
and standard altitude 7.6 km and reduced frequency 0.8 are plotted in Figs. C.2–C.6.

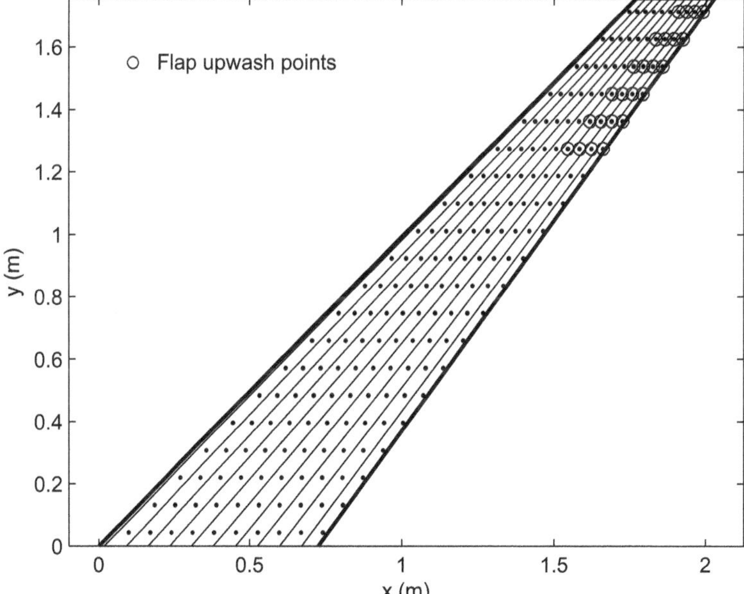

Fig. C.1 Modified DAST-ARW1 wing with flap

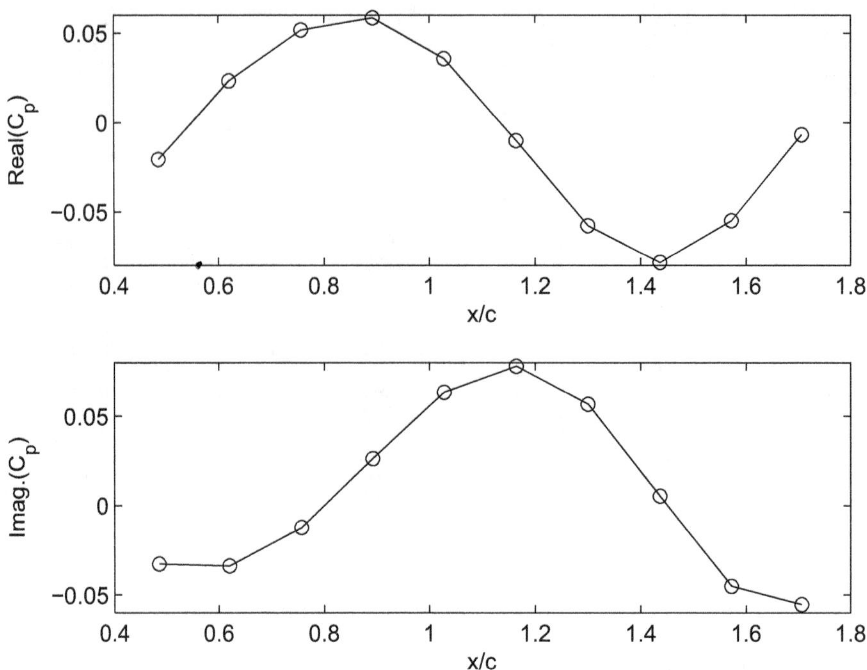

Fig. C.2 12.5 % semi-span, $M = 0.807$, $k = 0.8$

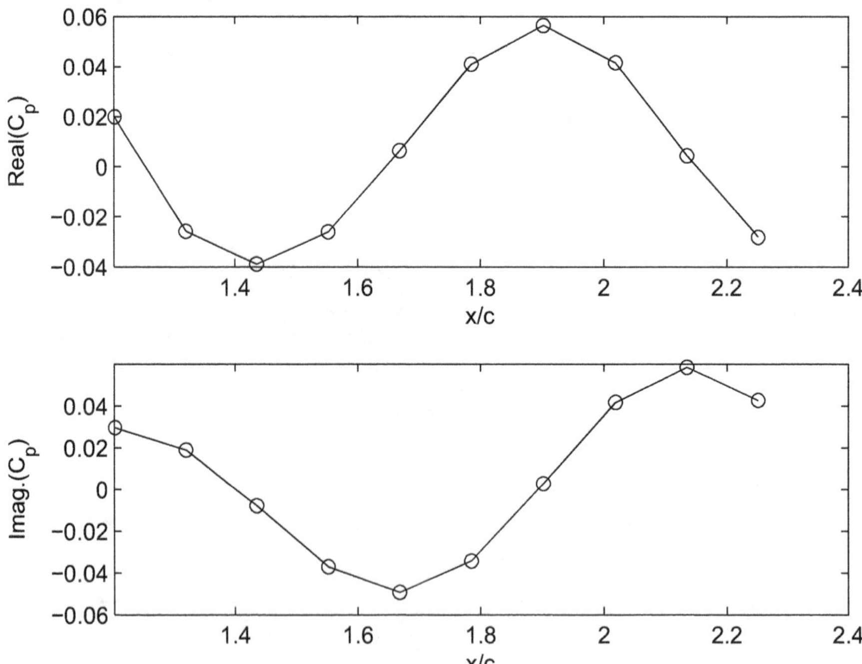

Fig. C.3 32.5 % semi-span, $M = 0.807$, $k = 0.8$

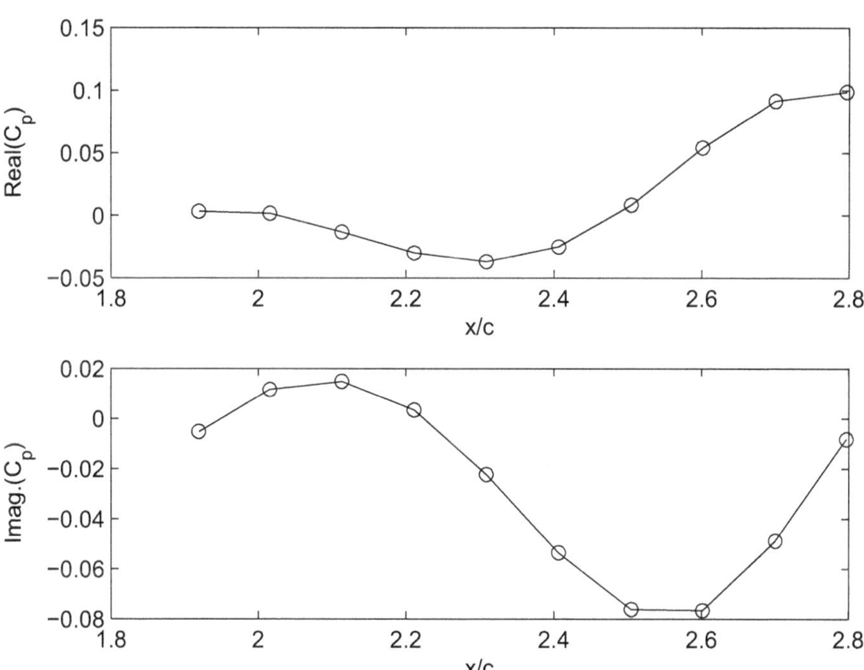

Fig. C.4 52.5 % semi-span, $M = 0.807, k = 0.8$

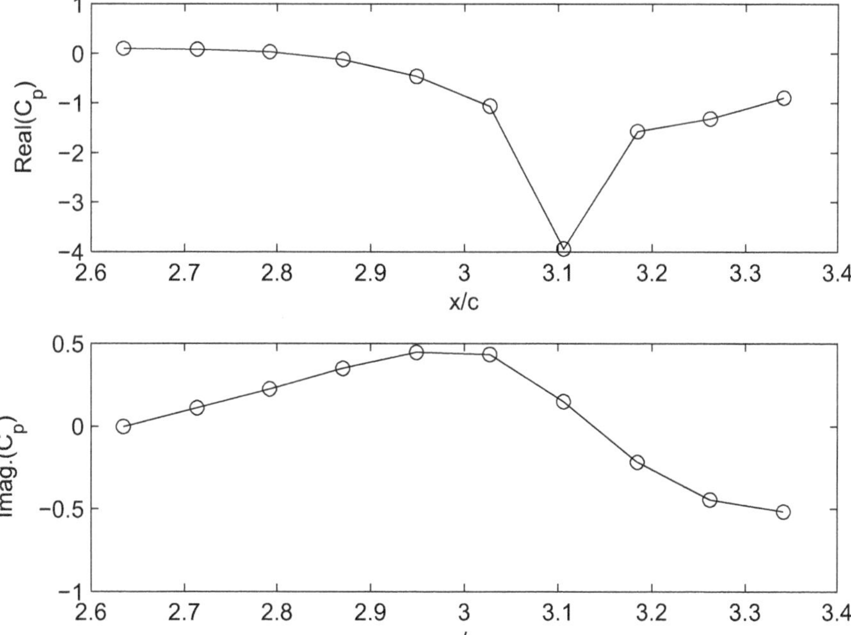

Fig. C.5 72.5 % semi-span, $M = 0.807, k = 0.8$

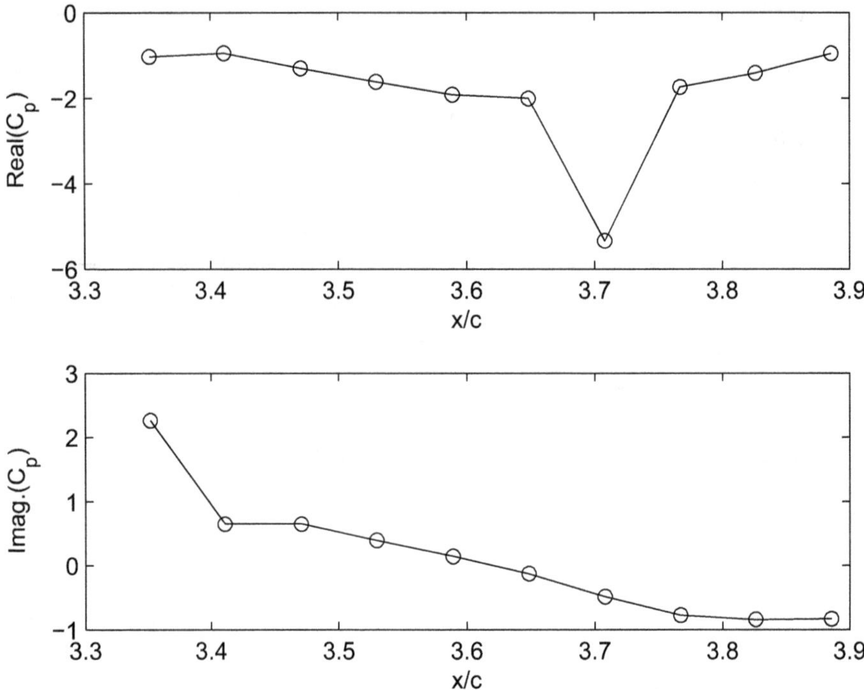

Fig. C.6 92.5 % semi-span, $M = 0.807$, $k = 0.8$

Appendix-D

D.1 Lyapunov Stability Theorem

For the homogeneous system described by

$$\dot{\mathbf{x}} = \mathbf{f}(\mathbf{x}, t); \qquad \mathbf{x}(t_0) = \mathbf{x}_0, \tag{D.1}$$

whose equilibrium point is $\mathbf{x}_e = \mathbf{0}$, the *Lyapunov stability theorem* can be stated as given below.

Definition D.1 A system described by Eq. (D.1) is said to be stable about the equilibrium point, $\mathbf{x}_e = \mathbf{0}$ in the sense of Lyapunov, if for each real and positive number, ϵ, however small, there exists another real and positive number, δ, such that

$$\mid \mathbf{x}(0) \mid < \delta \tag{D.2}$$

implies that

$$\mid \mathbf{x}(t) \mid < \epsilon; \qquad t \geq 0. \tag{D.3}$$

Definition D.2 A system described by Eq. (D.1) is said to be *asymptotically stable* about the origin, $\mathbf{x}_e = \mathbf{0}$, if it is stable in the sense of Lyapunov and if for each real and positive number, ϵ, however small, there exist real and positive numbers, δ and τ, such that

$$\mid \mathbf{x}(0) \mid < \delta \tag{D.4}$$

implies that

$$\mid \mathbf{x}(t) \mid < \epsilon; \qquad t > \tau. \tag{D.5}$$

Definition D.3 A system described by Eq. (D.1) is said to be *globally asymptotically stable* about the origin, $\mathbf{x}_e = \mathbf{0}$, if it is stable in the sense of Lyapunov and if for each real and positive pair, (δ, ϵ), there exists a real and positive number, τ, such that

© Springer Science+Business Media, LLC 2015
A. Tewari, *Aeroservoelasticity*, Control Engineering,
DOI 10.1007/978-1-4939-2368-7

$$| \mathbf{x}(0) | < \delta \tag{D.6}$$

implies that

$$| \mathbf{x}(t) | < \epsilon; \qquad t > \tau. \tag{D.7}$$

Theorem D.4 Let $V(\mathbf{x}, t)$ be a continuously differentiable, scalar function of the time as well as of state variables of a system described by Eq. (D.1), whose equilibrium point is $\mathbf{x}_e = \mathbf{0}$. If the following conditions are satisfied:

$$V(0, t) = 0, \qquad V(\mathbf{x}, t) > 0; \qquad \frac{dV}{dt}(\mathbf{x}) < 0; \qquad \text{for all } \mathbf{x} \neq 0 \tag{D.8}$$

$$\| \mathbf{x} \| \to \infty \text{ implies } V(\mathbf{x}, t) \to \infty, \tag{D.9}$$

then the origin, $\mathbf{x}_e = 0$, is globally asymptotically stable.

Proof of Lyapunov's theorem [158] is obtained from the unbounded, positive definite nature of $V(\mathbf{x}, t)$, and negative definite nature of $\dot{V}(\mathbf{x}, t)$, implying that for any initial perturbation from the origin, $\mathbf{x}(0) \neq \mathbf{0}$, the resulting solution satisfies $V(\mathbf{x}(t), t) \leq V(\mathbf{x}(0), t)\ t > 0$ (i.e., remains in a bounded neighborhood of the origin). Furthermore, the same also implies that $V(\mathbf{x}(t_2), t_2) \leq V(\mathbf{x}(t_1), t_1)\ t_2 > t_1$, which means a convergence of every solution to the origin.

References

1. Abbot, I., Doenhoff, A.: Theory of Wing Sections. Dover, New York (1959)
2. Abel, I., Perry, B., III, Murrow, H.N.: Two synthesis techniques applied to active flutter suppression on a flight research wing. J. Guid. Control **1**, 340–346 (1978)
3. Abel, I.: An analytical design technique for predicting the characteristics of a flexible wing equipped with an active flutter suppression system and comparison with wind-tunnel data. NASA TP-1367 (1979)
4. Abel, I., Perry, B., III, Newsom, J.R.: Comparison of analytical and wind-tunnel results for flutter and gust response of a transport wing with active controls. NASA TP-2010 (1982)
5. Abel, I., Noll, T.E.: Research and applications in aeroservoelasticity at NASA Langley Research Center. Proceedings 16th ICAS Congress. Tel Aviv (1988)
6. Abramowitz, M., Stegun, I.A.: Handbook of Mathematical Functions. Dover Publications, New York (1974)
7. Albano, E., Rodden, W.P.: A doublet-lattice method for calculating lift distributions on oscillating surfaces in subsonic flows. AIAA J. **7**, 279–285 (1969)
8. Ashley, H., Zartarian, G.: Piston theory—a new aerodynamic tool for aeroelastician. J. Aero. Sci. **23**, 1109–1118 (1956)
9. Appa, K.: Constant pressure panel method for supersonic unsteady airload analysis. J. Aircr. **24**, 696–702 (1987)
10. Astrom, K.J., Wittenmark, B.: Adaptive Control. Addison-Wesley, New York (1995)
11. Athans, M., Falb, P.L.: Optimal Control. Dover, New York (2007)
12. Ballhaus, W.F., Goorjian, P.M.: Computation of unsteady transonic flows by the indicial method. AIAA J. **16**, 117–124 (1978)
13. Ballhaus, W.F., Steger, J.L.: Implicit approximate-factorization schemes for the low-frequency transonic equation. NASA Technical Memorandum. TM X-73082 (1975)
14. Batina, J.T.: An efficient algorithm for solution of the unsteady transonic small-disturbance equation. AIAA Paper 87-0109 (1987)
15. Batina, J.T., Seidel, D.A., Bland, S.R., Bennett, R.M.: Unsteady transonic flow calculations for realistic aircraft configurations. J. Aircr. **26**, 131–139 (1989)
16. Beals, V., Targoff, W.P.: Control surface oscillatory coefficients measured on low-aspect ratio wings. U.S. Air Force, Wright Air Development Center. AFFDL-TR-53-64 (1953)
17. Bellman, R.: Dynamic Programming. Princeton University, Princeton (1957)
18. Bendiksen, O.O.: Transonic limit-cycle flutter/LCO. AIAA Paper 2004–2694 (2004)
19. Bennett, R.M., Abel, I.: Flight flutter test and data analysis techniques applied to a drone aircraft. J. Aircr. **19**, 589–595 (1982)
20. Birnbaum, W.: Das ebene problem des schlagenden flügels. Z. Angew. Math. Mech. **4**, 277–292 (1924)
21. Bisplinghoff, R.L., Ashley, H., Halfman, R.L.: Aeroelasticity. Addison-Wesley, Cambridge (1955)

© Springer Science+Business Media, LLC 2015

A. Tewari, *Aeroservoelasticity*, Control Engineering,

DOI 10.1007/978-1-4939-2368-7

22. Bisplinghoff, R.L., Ashley, H.: Principles of Aeroelasticity. Dover, New York (1962)
23. Blair, M.: A compilation of the mathematics leading to the doublet-lattice method. U.S. Air Force Wright Laboratory. WL-TR-92-3028 (1992)
24. Blair, M., Williams, M.H.: Time domain panel method for wings. J. Aircr. **30**, 439–445 (1993)
25. Bryson, A.E., Ho, Y.: Applied Optimal Control. Wiley, New York (1979)
26. Boely, N., Botez, R.M.: New methodologies for the identification and validation of a nonlinear F/A-18 model by use of neural networks. Proceedings AIAA Atmospheric Flight Mechanics Conference, Toronto (2010)
27. Burkhart, T.M.: Subsonic transient lifting surface aerodynamics. J. Aircr. **14**, 47–50 (1977)
28. Burkhart, T.M.: Numerical application of Evvard's supersonic wing theory to flutter analysis. Proceedings 21st. AIAA Structures, Structural Dynamics and Materials Conference, Seattle (1980)
29. Chen, P.C., Liu, D.D.: A harmonic gradient method for unsteady supersonic flow calculations. J. Aircr. **22**, 371–379 (1985)
30. Chiang, R.Y., Safonov, M.G.: Robust Control Toolbox. The Math Works Inc., Natick (2000)
31. Chipman, R.R.: An improved mach-box approach for supersonic oscillatory pressures. J. Aircr. **14**, 887–893 (1977)
32. Cho, J., Williams, M.H.: S-plane aerodynamics of nonplanar lifting surfaces. J. Aircr. **30**, 433–438 (1993)
33. Collar, A.R., Broadbent, E.G., Puttick, E.: An elaboration of the criterion for wing torsional stiffness. British A.R.C. **R&M 2154** (1946)
34. Cox, T.H., Gilyard, G.B.: Ground vibration test results for drones for aerodynamic and structural testing (DAST)/aeroelastic research wing (ARW-1R) aircraft. NASA Tech. Memo. **85906** (1986)
35. Cunningham, A.M., Jr.: Oscillatory supersonic kernel function method for interfering surfaces. J. Aircr. **11**, 664–669 (1974)
36. Cunningham, H.J., Desmarais, R.N.: Generalization of the subsonic kernel function in the s-plane, with applications to flutter analysis. NASA, TP-2292 (1984)
37. Denegri, C.M., Jr., Dubben, J.A.: F-16 limit-cycle oscillation analysis using transonic small-disturbance theory. AIAA Paper 2005–2296 (2005)
38. Desmarais, R.: A continued fraction representation for Theodorsen's circulation function. NASA, TM-81838 (1980)
39. Dietze, F.: Die luftkrafte des harmonisch schwingenden flügels in kompressibaren medium bei unterschallgeschwindigkeit (ebene problem) (The air forces of the harmonically vibrating wing in a compressible medium at subsonic velocity (plane problem). I. Method of computation, II. Numerical tables and curves. ZWB Forsch. Ber. **1733** (1943). (Translated by Air Materiel Command, U.S. Air Force, F-TS-506-RE and F-TS-948-RE (1947))
40. Dowell, E.H.: A simple approach of converting frequency domain aerodynamics to the time domain. NASA, TM-81844 (1980)
41. Dowell, E.H., I'lgamov, M.: Studies in Nonlinear Aeroelasticity. Springer, New York (1988)
42. Doyle, J.C.: Structured uncertainty in control system design. In: Proceedings 24th IEEE Conference on Decision and Control. Ft. Lauderdale, FL, 260–265 (1985)
43. Dunn, H.J.: An analytical technique for approximating unsteady aerodynamics in the time domain. NASA, TP-1738 (1980)
44. Edwards, J.W.: Applications of Laplace transform methods to airfoil motion and stability calculations. AIAA Paper 79-0772 (1979)
45. Edwards, J.W., Malone, J.B.: Current status of computational methods for transonic unsteady aerodynamic and aeroelastic applications. AGARD CP-507 (1992)
46. Edwards, J.W.: Transonic shock oscillations calculated with a new interactive boundary layer coupling method. AIAA Paper 93-0777 (1993)
47. Edwards, J.W.: Calculated viscous and scale effects on transonic aeroelasticity. J. Aircr. **45**, 1863–1871 (2008)
48. Etkin, B., Reid, L.D.: Dynamics of Flight: Stability and Control. Wiley, New York (1995)

49. Eversman, W., Tewari, A.: Consistent rational function approximations for unsteady aerodynamics. J. Aircr. **28**, 545–552 (1991)

50. Eversman, W., Tewari, A.: Modified exponential series approximation for the Theodorsen function. J. Aircr. **28**, 553–557 (1991)

51. Evvard, J.: Use of source distributions for evaluating theoretical aerodynamics of thin finite wings at supersonic speeds. NACA Rept. 951 (1950)

52. Fettis, H.E.: An approximate method for the calculation of nonstationary air forces at subsonic speeds. Wright Air Develop. Center, U.S. Air Force, Tech. Rept. 52–56 (1952)

53. Filippov, A.F.: Differential Equations with Discontinuous Right-hand Sides. Kluwer, Dordrecht (1988)

54. Fraeys de Veubeke, B.: Aerodynamique instationaire des profils minces deformables. Bulletin du Service Technique de l'Aeronautique, Brussels, 25 (1953)

55. Fung, Y.C.: An Introduction to the Theory of Aeroelasticity. Wiley, New York (1955)

56. Garrick, I.E.: On some reciprocal relations in the theory of non-stationary flow. NACA, Rept. 629 (1938)

57. Garrick, I.E., Rubinow, S.I.: Flutter and oscillating air force calculations for an airfoil in a two-dimensional supersonic flow. NACA, Rept. 846 (1946)

58. Garrick, I.E.: Nonsteady wing characteristics. Section F, Vol. VII of High Speed Aerodynamics and Jet Propulsion. Princeton University Press, Princeton (1957)

59. Geibler, W.: Ein numerisches verfahren zur berechnung der instationaren druckverteilung der harmonisch schwingenden tragflache mit ruder in unterschallströmung. Teil I: Theorie und ergebnisse für inkompressible strömung, Teil II: Theorie und ergebnisse für kompressible strömung. DLR DLR-FB-75-37 (1975) and DLR-FB-77-15 (1977)

60. Gibson, J.E.: Nonlinear Automatic Control. McGraw-Hill, New York (1963)

61. Giesing, J.P., Kalman, T.P., Rodden, W.P.: Subsonic unsteady aerodynamics for general configurations; Part I, Vol. I - Direct application of the nonplanar doublet-lattice method. U.S. Air Force, AFFDL-TR-71-5 (1971)

62. Glad, T., Ljung, L.: Control Theory–Multivariable and Nonlinear Methods. Taylor and Francis, New York (2000)

63. Glauert, H.: The force and moment on an oscillating airfoil. British A.R.C., R&M 1242 (1929)

64. Glover, K., Doyle, J.C.: State space formulae for all stabilizing controllers that satisfy an H_∞ norm bound and relations to risk sensitivity. Syst. Control Lett. **11**, 167–172 (1988)

65. Gupta, K.K., Brenner, M.J., Völker, L.S.: Integrated aeroservoelastic analysis capability with X-29A comparisons. J. Aircr. **26**, 84–90 (1989)

66. Hafez, M., South, J., Murman, E.: Artificial compressibility methods for numerical solutions of transonic full potential equation. AIAA J. **17**, 838–844 (1979)

67. Harder, R.L., Rodden, W.P.: Kernel function for nonplanar oscillating surfaces in supersonic flow. J. Aircr. **8**, 677–679 (1971)

68. Den Hartog, J.P.: Advanced Strength of Materials. McGraw-Hill, New York (1952)

69. Haskind, M.D.: Oscillations of a wing in a subsonic gas flow. Brown University Translation, A9-T-22 (1948) (Originally Prikl. Mat. i Mekh., Moscow, **XI**, No. 1, 1947.)

70. Hassig, H.J.: An integral transform for obtaining $A(p)$ from $A(ik)$. Aerospace Flutter and Dynamics Coucil Meeting, Williamsburg (1977)

71. Hayes, W.D., Probstein, R.F.: Hypersonic Flow Theory. Academic Press, New York (1959)

72. Hedman, S.G.: Vortex-lattice method for calculation of quasi steady state loadings on thin elastic wings. Aeronautical Research Institute of Sweden, Rept. 105 (1965)

73. Holst, T.L., Ballhaus, W.F.: Fast conservative schemes for the full potential equation applied to transonic flows. AIAA J. **17**, 145–152 (1979)

74. Hounjet, M.H.L.: Improved potential gradient method to calculate airloads on oscillating supersonic interfering surfaces. J. Aircr. **19**, 390–399 (1982)

75. Hughes, T.J.R.: The Finite Element Method. Prentice-Hall, Englewood-Cliffs (1987)

76. Huttsell, L., Shuster, D., Vol, J., Giesing, J., Mike Love, M.: Evaluation of computational codes for loads and flutter. AIAA Paper, 2001-569 (2001)

77. Ioannou, P.A., Sun, J.: Stable and Robust Adaptive Control. Prentice-Hall, Englewood-Cliffs (1995)

78. Isidori, A.: Nonlinear Control Systems. Springer, New York (1989)

79. Isogai, K., Suetsugu, K.: Numerical calculation of unsteady transonic potential flow over three-dimensional wings with oscillating control surfaces. AIAA J. **22**, 478–485 (1984)

80. Jones, R.T.: The unsteady lift of a wing of finite aspect ratio. NACA, Rept. 681 (1939)

81. Jones, W.P.: Aerodynamic forces on wings in non-uniform motion. British A.R.C., R&M 2117 (1945)

82. Jones, W.P., Appa, K.: Unsteady supersonic aerodynamic theory by the method of potential gradient. AIAA J. **15**, 59–65 (1977)

83. Jordan, P.F.: Remarks on applied subsonic lifting-surface theory. Martin Marietta Corp., Tech. Rept. 67-14 (1967)

84. Kailath, T.: Linear Systems. Prentice-Hall, Englewood Cliffs (1980)

85. Karpel, M.: Design of active and passive flutter suppression and gust alleviation. NASA, CR-3482 (1981)

86. Katz, J., Plotkin, A.: Low Speed Aerodynamics. McGraw-Hill, New York (1991)

87. Kautsky, J., Nichols, N.K., Van Dooren, P.: Robust Pole assignment in linear state feedback. Int. J. Control **41**, 1129–1155 (1985)

88. Kreyszig, E.: Advanced Engineering Mathematics. Wiley, New York (2001)

89. Krstić, M., Kanellakopoulos, I., Kokotović, P.V.: Nonlinear and Adaptive Control Design. Wiley, New York (1995)

90. Küssner, H.G.: Schwingungen von flugzeugflügeln. Luftfahrtforschung. **6**(6) (1929)

91. Küssner, H.G.: Zussamenfassender bericht über den instationären auftrieb von flügeln. Luftfahrtforschung. **13**(12) (1936)

92. Küssner, H.G., Schwarz, L.: Der schwingende flügel mit aerodynamisch ausgeglichenem ruder. Luftfahrtforschung. **17**, 377–384 (1940)

93. Küssner, H.G.: Allegemeine tragflächentheorie. Luftfahrtforschung. **17**(11, 12), 370–378 (1940). (Trans. NACA Tech. Memo. 979 (1941).)

94. Küssner, H.G.: A general method for solving problems of the unsteady lifting surface theory in the subsonic range. J. Aeronaut. Sci. **21**, 17–27 (1954)

95. Landahl, M.T.: Unsteady Transonic Flow. Pergamon, Oxford (1961)

96. Landahl, M.T.: Kernel function for nonplanar oscillating surfaces in subsonic flow. AIAA J. **5**, 1045–1046 (1967)

97. Landahl, M.T.: On the pressure loading functions for oscillating wings with control surfaces. AIAA J. **6**, 345–348 (1968)

98. Laschka, B.: Zur theorie der harmonisch schwingenden tragenden flache bei unterschallanströmung. Flugwissenschaften. **7**, 265–292 (1963)

99. Lee, B.H.K.: Oscillatory shock motion caused by transonic shock boundary-layer interaction. AIAA J. **28**, 942–944 (1990)

100. Leishman, J.G.: Unsteady lift of a flapped airfoil by indicial concepts. J. Aircr. **31**, 288–297 (1994)

101. Lighthill, M.J.: Oscillating airfoils at high Mach numbers. J. Aero. Sci. **20**, 402–406 (1953)

102. Livne, E.: Integrated aeroservoelastic optimization: Status and direction. J. Aircr. **36**, 122–145 (1999)

103. Lottati, I., Nissim, E.: Oscillatory subsonic piecewise continuous kernel function method. J. Aircr. **14**, 515–516 (1977)

104. Lottati, I., Nissim, E.: Nonplanar supersonic three-dimensional oscillatory piecewise continuous kernel function method. J. Aircr. **24**, 45–54 (1987)

105. Lu, S., Voss, R.: TDLM–A transonic doublet lattice method for 3D potential unsteady transonic flow calculation. DLR Forschungsbericht, DLR-FB 92-25 (1992)

106. Luke, Y., Dengler, M.A.: Tables of the Theodorsen circulation function for generalized motion. J. Aero. Sci. **18**(7), 478–483 (1951)

107. Mabey, D.G.: Unsteady transonic flows. Chap. 1 of Unsteady Transonic Aerodynamics. In: Nixon, D. (ed.) Progress in Aeronautics and Astronautics, 120, pp. 1–55. AIAA, Washington, D.C. (1989)
108. Maciejowski, J.M.: Multivariable Feedback Design. Addison-Wesley, Wokingham (1989)
109. Mangler, K.: Improper integrals in theoretical aerodynamics. R.A.E. Rept. Aero. **2424**(7) (1951)
110. MATLAB®Mathematics (Release 2014a). MathWorks Inc., Natick, M.A. (2014)
111. Mattheij, R.M.M., Molenaar, J.: Ordinary Differential Equations in Theory and Practice. SIAM Classics in Applied Mathematics 43. SIAM, Philadelphia (2002)
112. Mazelsky, B., Drischler, J.A.: Numerical determination of indicial lift and moment functions of a two-dimensional sinking and pitching airfoil at Mach numbers 0.5 and 0.6. NACA, TN 2739 (1952)
113. Miller, G.D.: Active flexible wing (AFW) technology. U.S. Air Force Wright Laboratory, AFWAL-TR-87-3096 (1987)
114. Morino, L., Kuo, C.-C.: Subsonic potential aerodynamics for complex configurations: a general theory. AIAA J. **12**, 191–197 (1974)
115. Morino, L., freedman, M.I., Deutsch, D.J., Sipcic, S.R.: An integral equation method for compressible potential flows in an arbitrary frame of reference. In: Morino, L. (ed.) Computational Methods in Potential Aerodynamics. Computational Mechanics Publications, Southampton (1985)
116. Mukhopadhyay, V., Newsom, J.R., Abel, I.: A method for obtaining reduced order control laws for high order systems using optimization techniques. NASA, TP-1876 (1981)
117. Mukhopadhyay, V.: Digital robust control law synthesis using constrained optimization. J. Guid. Control Dyn. **12**, 175–181 (1989)
118. Mukhopadhyay, V.: Transonic flutter suppression control law design and wind tunnel test results. J. Guid. Control Dyn. **23**, 930–937 (2000)
119. Mukhopadhyay, V.: Historical perspective on analysis and control of aeroelastic responses. J. Guid. Control Dyn. **26**, 673–684 (2003)
120. Murman, E.M.: Analysis of embedded shock waves calculated by relaxation methods. Proceedings AIAA CFD Conference, 27–40 (1973)
121. Nelder, J.A., Mead, R.: A Simplex method for function minimization. Comput. J. **7**, 308–313 (1965)
122. Newsom, J.R., Abel, I., Dunn, H.J.: Application of two design methods for active flutter suppression and wind-tunnel test results. NASA, TP-1653 (1980)
123. Obayashi, S.: Algorithm and code development for unsteady three-dimensional Navier-Stokes equations. NASA Contractor Report CR-192760 (1993)
124. Osher, S., Hafez, M., Whitlow, W., Jr.: Entropy conditions satisfying approximations for the full-potential equation of transonic flow. Math. Comput. **44**, 1–29 (1985)
125. Pan, Y., Dong, X., Zhu, Q., Yue, D.K.P.: Boundary-element method for the prediction of performance of flapping foils with leading-edge separation. J. Fluid Mech. **698**, 446–467 (2012)
126. Peloubet, R.P., Jr., Haller, R.L., Cunningham, A.M., Cwach, E.E., Watts, D.: Applications of three aeroservoelastic stability analysis techniques. U.S. Air Force, AFFDL-TR-76-89 (1976)
127. Perry, B., III, Cole, S., Miller, G.D.: A summary of the active flexible wing program. J. Aircr. **32**, 10–15 (1995)
128. Peterson, L.D., Crawley, E.F.: Improved exponential time series approximations for unsteady aerodynamic operators. J. Aircr. **25**, 121–127 (1988)
129. Pines, S., Dugundji, J., Neuringer, J.: Aerodynamic flutter derivatives for a flexible wing with supersonic and subsonic edges. J. Aeronaut. Sci. **22**, 693–709 (1955)
130. Pitt, D., Goodman, C.E.: Flutter calculations using doublet lattice aerodynamics modified by the full potential equations. AIAA Paper 87-0882 (1987)
131. Pontryagin, L.S., Boltyanskii, V., Gamkrelidze, R., Mishchenko, E.: The Mathematical Theory of Optimal Processes. Wiley, New York (1962)

132. Popov, V.M.: Hyperstability of Automatic Control Systems. Ed. Acad. Rep. Soc. Romania, Bucharest, Romania (1966)
133. Possio, C.: L' azione aerodinamica sul profilo oscillante alle velocita ultrasonore (Aerodynamic forces on an oscillating profile at supersonic speeds). Pontif. Acad. Sci. Acta I **11**, 93–106 (1937)
134. Possio, C.: L' azione aerodinamica sul profilo oscillante in un fluido compressibile a velocita ipsonora (Aerodynamic forces on an oscillating profile in a compressible fluid at subsonic speeds). Aerotecnica **18**, 441–458 (1938)
135. Prandtl, L.: Theorie desFlugzeugtragflügels im zusammenddrückbaren Medium. Luftfahrtforschung **13**, 313–319 (1936)
136. Raveh, D.E.: Identification of computational-fluid-dynamics based unsteady aerodynamic models for aeroelastic analysis. J. Aircr. **41**, 620–632 (2004)
137. Raveh, D.E.: CFD-based models of aerodynamic gust response. J. Aircr. **45**, 888–897 (2007)
138. Reddy, J.N.: Theory and Analysis of Elastic Plates. Taylor and Francis, Philadelphia (1999)
139. Reissner, E.: On the application of Mathieu functions in the theory of subsonic compressible flow past oscillating airfoils. NACA, Tech. Note 2363 (1951)
140. Richardson, J.R.: A more realistic method for routine flutter calculations. Proceedings AIAA Symposium on Structural Dynamics and Aeroelasticity, Boston (1965)
141. Rivera, J.A., Dansberry, B.E., Farmer, M.G., Eckstrom, C.V., Bennett, R.M., Seidel, D.A.: First benchmark model program test successfully completed. NASA Technical Memorandum, TM 102770, 52–53 (1991). (Wynne, E.C. (ed.): Structural Dynamics Division Research and Technology Accomplishments for F.Y. 1990 and Plans for F.Y. 1991)
142. Rodden, W.P., Stahl, B.: A strip method for prediction of damping in subsonic wind tunnel and flight flutter tests. J. Aircr. **6**, 9–17 (1969)
143. Roger, K.L., Hodges, G.E., Felt, L.: Active flutter suppression—A flight test demonstration. J. Aircr. **12**, 551–556 (1975)
144. Roger, K.L.: Airplane math modeling methods for active control design. AGARD, CP-228 (1977)
145. Rowe, W.S.: Collocation method for calculating the aerodynamic pressure distributions on a lifting surface oscillating in subsonic compressible flow. Proceedings AIAA symposium on Structural Dynamics and Aeroelasticity, Boston (1965)
146. Rowe, W.S., Winther, B.A., Redman, M.C.: Prediction of unsteady aerodynamic loading caused by control surface motion in subsonic compressible flow—Analysis and results. NASA, CR-2543 (1975)
147. Scanlan, R.H., Rosenbaum, R.: Introduction to the Theory of Aircraft Vibration and Flutter. Macmillan, New York (1951)
148. Schade, T.: Numerische losung des Poissiochen integralgliechung der schwingenden tragfläche in ebener unterschallströmung. ZWB, UM 3209 (1944)
149. Schade, T.: Beitrag zu zahlentafeln zur luftkraftberechnung der schwingenden tragfläche in ebener unterschallströmung. ZWB, UM 3211 (1944)
150. Schwarz, L.: Berechnung der druckenverteilung einer harmonisch sich verformenden tragfläche in ebener strömung. Luftfahrtforschung. **17**(11, 112) (1940)
151. Scott, R.C., Pado, L.E.: Active control of wind-tunnel model aeroelastic response using neural networks. J. Guid. Control Dyn. **23**, 1100–1108 (2000)
152. Sevart, F.D.: Development of active flutter suppression wind tunnel testing technology. U.S. Air Force, Wright Air Development Center, AFFDL-TR-74-126 (1975)
153. Shames, I.: Mechanics of Fluids. McGraw-Hill, New York (2002)
154. Shampine, L.F., Gladwell, I., Thompson, S.: Solving ODEs with MATLAB. Cambridge University Press, Cambridge (2003)
155. Shapiro, A.H.: The Dynamics and Thermodynamics of Compressible Fluid Flow, vol. I & II. Wiley, New York (1953)

156. Silva, R.G.A., Mello, O.A.F., Azevedo, J.L.F., Chen, P.C., Liu, D.D.: Investigation on transonic correction methods for unsteady aerodynamics and aeroelastic analyses. J. Aircr. **45**, 1890–1903 (2008)
157. Using Simulink®-Simulation and Model Based Design (Release 2014a). MathWorks Inc., Natick, MA (2014)
158. Slotine, J.E., Li, W.: Applied Nonlinear Control. Prentice-Hall, Englewood Cliffs (1991)
159. Smilg, B., Wasserman, L.S.: Application of three-dimensional flutter theory to aircraft structures. Air Materiel Command, U.S. Air Force, Tech. Rept. 4798 (1942)
160. Söhngen, H.: Die losungen der integralgleichung und deren anwendung in der tragflügeltheorie. Math. Z. **45**, 245–264 (1939)
161. Söhngen, H.: Bestimmung der auftriebsverteilung für beliebige instationäre bewegungen (ebenes problem). Luftfahrtforschung. **17**, 401–419 (1940)
162. Steger, J.L., Caradonna, F.X.: A conservative implicit finite difference algorithm for the unsteady transonic full potential equation. AIAA Paper 80-1368 (1980)
163. Stengel, R.F.: Optimal Control and Estimation. Dover, New York (1994)
164. Stewartson, K.: On the linearized potential theory of unsteady supersonic motion. Quart. J. Mech. Appl. Math. **III**, 182–199 (1950)
165. Stoer, J., Bulirsch, R.: Introduction to Numerical Analysis. Springer, New York (2002)
166. Tannehill, J.C., Anderson, D.A., Pletcher, R.H.: Computational Fluid Mechanics and Heat Transfer, 2nd ed. Taylor and Francis, Philadelphia (1997)
167. Tewari, A.: Doublet-point method for supersonic unsteady aerodynamics of nonplanar lifting surfaces. J. Aircr. **31**, 745–752 (1994)
168. Tewari, A.: Modern Control Design with MATLAB and Simulink. Wiley, Chichester (2002)
169. Tewari, A.: Atmospheric and Space Flight Dynamics–Modeling and Simulation with MATLAB and Simulink. Birkhäuser, Boston (2006)
170. Tewari, A.: Advanced Control of Aircraft, Spacecraft, and Rockets. Wiley, Chichester. (2011)
171. Tewari, A.: Automatic Control of Aerospace Vehicles. Birkhäuser, Boston (2011)
172. Tewari, A.: Output-rate weighted optimal control of aeroelastic systems. J. Guid. Control Dyn. **24**, 409–411 (2001)
173. Theodorsen, T.: General theory of aerodynamic instability and the mechanism of flutter. NACA, Rept. 496 (1935)
174. Theodorsen, T., Garrick, I.E.: Non-stationary flow about a wing-aileron-tab combination including aerodynamic balance. NACA, Rept. 736 (1942)
175. Tiffany, S.H., Adams, W.M., Jr.: Nonlinear programming extensions to rational function approximations of unsteady aerodynamics. Proceedings AIAA Symposium on Structural Dynamics and Aeroelasticity, Paper 87-0854, 406–420 (1987)
176. Tijdeman, H.: Investigations of the transonic flow around oscillating airfoils. National Aerospace Lab., NLR Technical Report, TR-77090U (1977)
177. Timman, R., Van de Vooren, A.I.: Theory of the oscillating wing with aerodynamically balanced control surface in a two-dimensional subsonic compressible flow. Natl. Aeronaut. Research Inst., Amsterdam, NLL Rept. F54 (1949)
178. Timman, R., Van de Vooren, A.I., Greidanus, J.H.: Aerodynamic coefficients of an oscillating airfoil in two-dimensional subsonic flow. Natl. J. Aero. Sci. **18**(12), 797–802 (1951)
179. Turner, M.J., Rabinowitz, S.: Aerodynamic coefficients for an oscillating airfoil with hinged flap, with tables for a Mach number of 0.7. NACA, Tech. Note 2213 (1950)
180. Ueda, T., Dowell, E.H.: A new solution method for lifting surfaces in subsonic flow. AIAA J. **20**, 348–355 (1982)
181. Ueda, T., Dowell, E.H.: Doublet-point method for supersonic unsteady lifting surfaces. AIAA J. **22**, 179–186 (1984)
182. Ueda, T., Dowell, E.H.: Flutter analysis using nonlinear aerodynamic forces. J. Aircr. **21**, 101–109 (1984)
183. Van Dyke, M.D.: Supersonic flow past oscillating airfoils including nonlinear thickness effects. NACA, Tech. Note 2982 (1953)

184. Vanderplaats, G.N.: CONMIN–A Fortran program for constrained function minimization-User manual. NASA Technical Memorandum, TM X-62282 (1973)
185. Vepa, R.: On the use of Padé approximants to represent unsteady aerodynamic loads for arbitrarily small motions of wings. AIAA Paper 76-17 (1976)
186. Vepa, R.: Finite state modeling of aeroelastic systems. NASA, CR-2779 (1979)
187. von Kármán, T., Sears, W.R.: Airfoil theory for non-uniform motion. J. Aeronaut. Sci. **5**, 379–390 (1938)
188. Voss, R.: Calculation of 3D unsteady transonic potential flows by a field panel method. Proceedings 2nd International Symposium on Aeroelasticity and Structural Dynamics. Aachen (1985)
189. Wagner, H.: Über die entstehung des dynamischen auftriebs von tragflügeln. Z. Angew. Math. Mech. **5**, 17–35 (1925)
190. Wang, Z.J.: Vortex shedding and frequency selection in flapping flight. J. Fluid Mech. **410**, 323–341 (2000)
191. Waszak, M.R.: Robust multivariable flutter suppression for benchmark active control technology wind-tunnel model. J. Guid. Control Dyn. **24**, 147–153 (2001)
192. Watkins, C.E., Runyan, H.L., Woolston, D.S.: On the kernel function of the integral equation relating the lift and downwash distributions of oscillating finite wings in subsonic flow. NACA, Rept. 1234 (1955)
193. Watkins, C.E., Berman, J.H.: On the kernel function of the integral equation relating the lift and downwash distributions of oscillating finite wings in supersonic flow. NACA, Rept. 1257 (1956)
194. Watkins, C.E., Woolston, D.S., and Cunningham, H.J.: A systematic kernel function procedure for determining aerodynamic forces on oscillating or steady finite wings at subsonic speeds. NASA, TR R-48 (1959)
195. Xiao, Q., Tsai, H.M., Liu, F.: Numerical study of transonic buffet on a supercritical airfoil. AIAA J. **44**, 620–628 (2006)
196. Yates, E.C., Jr.: Modified strip analysis method for predicting wing flutter at subsonic to hypersonic speeds. J. Aircr. **3**(1), 25–29 (1966)
197. Zaide, A., Raveh, D.E.: Numerical simulation and reduced-order modeling of airfoil gust response. J. Aircr. **44**, 1826–1834 (2006)
198. Zhu, Q., Peng, Z.: Mode coupling and flow energy harvesting by a flapping foil. Phys. Fluids **21**, 033601 (2009)
199. Zwaan, R.J.: Verification of calculation methods for unsteady airloads in the prediction of transonic flutter. J. Aircr. **22**, 833–839 (1985)

Index

© Springer Science+Business Media, LLC 2015
A. Tewari, *Aeroservoelasticity,* Control Engineering,
DOI 10.1007/978-1-4939-2368-7